Heat Exchanger Technologies for Sustainable Renewable Energy Systems

Heat Exchanger Technologies for Sustainable Renewable Energy Systems serves as a comprehensive resource on the cutting-edge advancements and applications of heat exchanger technologies in the realm of renewable energy. This book delves into the fundamental principles, design methodologies, and operational strategies for optimizing heat exchange processes in various sustainable energy systems.

Covering a wide range of topics, the book explores innovative heat exchanger designs, materials, and configurations that enhance thermal performance and efficiency. It examines the integration of heat exchangers in solar thermal systems, geothermal applications, and biomass energy systems, providing insights into their role in promoting energy conservation and sustainability.

The content encompasses both theoretical frameworks and practical applications, featuring case studies that illustrate successful implementations of heat exchanger technologies in real-world scenarios. Readers will gain a thorough understanding of performance evaluation metrics, modeling techniques, and experimental methodologies used to assess heat exchanger efficiency.

Key features of the book:

- Discusses the principles of heat transfer and fluid dynamics relevant to heat exchangers.
- Investigates emerging materials and design innovations for enhanced thermal performance.
- Explores the application of heat exchangers in various renewable energy systems, including solar, geothermal, and biomass.
- Provides in-depth analysis of modeling techniques and performance evaluation criteria.

- Highlights recent developments in heat exchanger technologies and their impact on sustainability.

Targeted at researchers, engineers, and students in the fields of renewable energy, mechanical engineering, and environmental science, this book is an essential guide for anyone seeking to advance their understanding of heat exchanger technologies and their vital role in sustainable energy systems.

Advances in Manufacturing, Design and Computational Intelligence Techniques

Series Editor: Ashwani Kumar, Professor and Head of Department, Mechanical Engineering at Technical Education Department Uttar Pradesh, Kanpur, India

The book series editor is inviting edited, reference and text book proposal submission in the book series. The main objective of this book series is to provide researchers a platform to present state of the art innovations, research related to advanced materials applications, cutting edge manufacturing techniques, innovative design and computational intelligence methods used for solving nonlinear problems of engineering. The series includes a comprehensive range of topics and its application in engineering areas such as additive manufacturing, nanomanufacturing, micromachining, biodegradable composites, material synthesis and processing, energy materials, polymers and soft matter, nonlinear dynamics, dynamics of complex systems, MEMS, green and sustainable technologies, vibration control, AI in power station, analog-digital hybrid modulation, advancement in inverter technology, adaptive piezoelectric energy harvesting circuit, contactless energy transfer system, energy efficient motors, bioinformatics, computer aided inspection planning, hybrid electrical vehicle, autonomous vehicle, object identification, machine intelligence, deep learning, control-robotics-automation, knowledge based simulation, biomedical imaging, image processing and visualization. This book series compiled all aspects of manufacturing, design and computational intelligence techniques from fundamental principles to current advanced concepts.

Hybrid Metal Additive Manufacturing: Technology and Applications
Edited by Parnika Shrivastava, Anil Dhanola, and Kishor Kumar Gajrani

Thermal Energy Systems: Design, Computational Techniques, and Applications
Edited by Ashwani Kumar, Varun Pratap Singh, Chandan Swaroop Meena, Nitesh Dutt

Heat Exchanger Technologies for Sustainable Renewable Energy Systems
Edited by Mukesh Kumar Awasthi, Ashwani Kumar, Nitesh Dutt, and Sivasakthivel Thangavel

https://www.routledge.com/Advances-in-Manufacturing-Design-and-Computational-Intelligence-Techniques/book-series/CRCAIMDCIT

Heat Exchanger Technologies for Sustainable Renewable Energy Systems

Edited by
Mukesh Kumar Awasthi, Ashwani Kumar,
Nitesh Dutt, and Sivasakthivel Thangavel

CRC Press
Taylor & Francis Group
Boca Raton London New York

CRC Press is an imprint of the
Taylor & Francis Group, an **informa** business

Designed cover image: Shutterstock

First edition published 2025
by CRC Press
2385 NW Executive Center Drive, Suite 320, Boca Raton FL 33431

and by CRC Press
4 Park Square, Milton Park, Abingdon, Oxon, OX14 4RN

CRC Press is an imprint of Taylor & Francis Group, LLC

ISBN: 978-1-032-87850-8 (hbk)
ISBN: 978-1-032-87851-5 (pbk)
ISBN: 978-1-003-53478-5 (ebk)

DOI: 10.1201/9781003534785

Typeset in Sabon
by KnowledgeWorks Global Ltd.

Contents

Aim and scope

The book *Heat Exchanger Technologies for Sustainable Renewable Energy Systems* aims to provide an in-depth exploration of cutting-edge heat exchanger technologies and their applications in the renewable energy sector. It focuses on the role of heat exchangers in enhancing the efficiency and sustainability of energy systems, highlighting both theoretical insights and practical applications. The book also addresses recent advancements, challenges, and optimization strategies in the design, development, and implementation of these technologies. By offering a comprehensive analysis, this resource is designed to cater to engineers, researchers, industry professionals, and students involved in renewable energy systems, with a specific focus on sustainability and energy management.

Chapter 1 introduces the fundamentals of heat exchanger technologies, laying the groundwork for understanding their role in renewable energy systems. This chapter offers a detailed explanation of the types, principles, and working mechanisms of heat exchangers, setting the context for the rest of the book. Chapter 2 delves into sustainable materials used in heat exchanger construction, emphasizing the importance of selecting eco-friendly materials that enhance thermal performance while minimizing environmental impact. The chapter discusses innovations in material science that contribute to the efficiency and durability of heat exchangers. Chapter 3 focuses on heat exchangers in solar thermal systems, examining how they are used to capture, store, and transfer solar energy for heating and power generation. This chapter highlights the design considerations specific to solar applications, including optimization techniques for maximizing energy capture. Chapter 4 explores the role of heat exchangers in wind energy systems. It discusses the integration of heat exchanger technologies in turbine cooling and energy storage systems, emphasizing how they contribute to improving the efficiency of wind power generation and overall system performance.

Chapter 5 provides insights into the application of heat exchangers in geothermal energy systems. The chapter explains the use of heat exchangers in extracting and utilizing geothermal heat, with an emphasis on the challenges and technological advancements specific to geothermal plants. Chapter 6 highlights bioenergy systems, discussing the use of heat exchangers in biomass conversion processes for energy production. It focuses on optimizing heat transfer in biorefineries and bioenergy plants, providing examples of how heat exchangers play a critical role in improving system efficiency. Chapter 7 presents an overview of heat exchangers in hydropower systems, focusing on their function in turbine cooling and waste heat recovery. The chapter explores various configurations that enhance the energy recovery process and reduce system losses. Chapter 8 discusses phase change materials (PCMs) and their integration with heat exchangers to enhance energy storage capabilities. This chapter examines how PCMs can be used in conjunction with heat exchangers to improve thermal management in renewable energy systems. Chapter 9 shifts the focus to nanofluids and their applications in heat exchangers for renewable energy systems. It discusses the role of nanotechnology in enhancing heat transfer performance, providing an overview of research advancements and practical applications. Chapter 10 deals with the optimization of heat exchanger designs using computational methods. This chapter covers numerical modeling, simulation techniques, and optimization algorithms used to improve the design and performance of heat exchangers in renewable energy applications.

Chapter 11 examines advanced manufacturing techniques for heat exchangers, including additive manufacturing and 3D printing. It discusses how these technologies contribute to the development of more efficient, customized, and sustainable heat exchangers. Chapter 12 focuses on economic and environmental aspects, offering an analysis of the cost-effectiveness and sustainability of heat exchanger technologies in renewable energy systems. The chapter also discusses policy and regulatory frameworks that influence the adoption of these technologies. Chapter 13 concludes the book with future trends and research directions, highlighting emerging technologies and areas for further exploration in the field of heat exchangers for renewable energy systems. This final chapter provides insights into ongoing research efforts and potential innovations that could shape the future of the industry.

Overall, this book serves as a valuable resource for anyone seeking to understand the current state and future potential of heat exchanger technologies in the context of renewable energy systems. It combines theoretical concepts, practical applications, and case studies to offer a holistic view of how these technologies are advancing sustainability in energy management.

Dr. Mukesh Kumar Awasthi
Dr. Ashwani Kumar
Dr. Nitesh Dutt
Dr. Sivasakthivel Thangavel

Preface

The global energy landscape is shifting towards sustainability and the role of heat exchanger technologies in renewable energy systems is more critical than ever. As renewable energy sources such as solar, wind, and geothermal continue to expand, efficient thermal management remains a cornerstone of system performance. This book, *Heat Exchanger Technologies for Sustainable Renewable Energy Systems*, offers a comprehensive exploration of the latest advancements, designs, and applications of heat exchangers in renewable energy systems. The 13 chapters of this book present cutting-edge research and industrial applications that aim to optimize energy efficiency, reduce environmental impacts, and contribute to a sustainable future.

Chapter 1 introduces the fundamental principles of heat exchangers, focusing on their role in renewable energy systems. It lays the groundwork for understanding the various types of heat exchangers and their operational mechanisms. Chapter 2 delves into advanced heat exchanger designs for solar energy applications. This chapter emphasizes how innovative configurations enhance heat transfer efficiency, making solar energy systems more viable. Chapter 3 explores heat exchanger technologies in wind energy systems, focusing on cooling techniques for wind turbine components to ensure their longevity and optimal performance. Chapter 4 covers the integration of heat exchangers in geothermal systems, discussing how these devices facilitate the extraction of thermal energy from the earth's crust with minimal environmental disruption.

Chapter 5 focuses on heat exchanger designs for biomass energy systems, with particular attention to improving combustion efficiency and reducing emissions through optimized thermal management. Chapter 6 addresses heat exchangers in hydrogen production systems, emphasizing their importance in managing the heat generated in processes like electrolysis, which is critical for producing clean hydrogen energy. Chapter 7 introduces nanofluids in heat exchangers, showcasing their superior thermal properties and potential to significantly enhance the performance of renewable energy systems. Chapter 8 discusses the use of PCMs in heat exchangers for energy storage applications. It highlights the role of PCMs in stabilizing energy supply by storing excess heat during peak production periods.

Chapter 9 presents innovations in heat exchanger technologies for waste heat recovery, focusing on their application in renewable energy plants to recycle and utilize excess thermal energy. Chapter 10 explores compact heat exchangers and their applications in distributed renewable energy systems, where space constraints and high efficiency are critical factors. Chapter 11 addresses the integration of artificial intelligence and machine learning in heat exchanger design and optimization. This chapter illustrates how AI-driven techniques can improve heat exchanger performance and predict operational failures. Chapter 12 discusses the environmental impact of heat exchangers in renewable energy systems, highlighting their role in reducing carbon footprints and ensuring sustainable operation. Finally, Chapter 13 provides case studies of industrial applications of heat exchangers in various renewable energy systems, demonstrating real-world solutions that have been successfully implemented across the globe.

This book is a valuable resource for engineers, researchers, and practitioners involved in the design, optimization, and implementation of heat exchanger technologies in renewable energy systems. By bridging the gap between theoretical principles and practical applications, we hope to contribute to the advancement of sustainable energy solutions that address the global energy challenge.

Dr. Mukesh Kumar Awasthi
Dr. Ashwani Kumar
Dr. Nitesh Dutt
Dr. Sivasakthivel Thangavel

About editors

Mukesh Kumar Awasthi has done his Ph.D. on the topic "Viscous Correction for the Potential Flow Analysis of Capillary and Kelvin-Helmholtz instability". He is working as an Assistant Professor in the Department of Mathematics at Babasaheb Bhimrao Ambedkar University, Lucknow. Dr. Awasthi is specialized in the mathematical modeling of flow problems. He has taught courses of Fluid Mechanics, Discrete Mathematics, Partial differential equations, Abstract Algebra, Mathematical Methods, and Measure theory to postgraduate students. He has acquired excellent knowledge in the mathematical modeling of flow problems, and he can solve these problems analytically as well as numerically. He has a good grasp of the subjects like viscous potential flow, electro-hydrodynamics, magneto-hydrodynamics, heat, and mass transfer. He has excellent communication skills and leadership qualities. He is self-motivated and responds to suggestions in a more convincing manner. Dr. Awasthi has qualified National Eligibility Test (NET) conducted on all India level in the year 2008 by the Council of Scientific and Industrial Research (CSIR) and got Junior Research Fellowship (JRF) and Senior Research Fellowship (SRF) for doing research. He has published 125 plus research publications (journal articles/books/book chapters/conference articles) in many national and international journals and conferences. Also, he has published 10 books. He is also a series editor of the series *Artificial Intelligence and Machine Learning for Intelligent Engineering Systems*. He has attended many symposia, workshops, and conferences in mathematics as well as fluid mechanics. He has got the "Research Awards" consecutively four times from 2013 to 2016 by the University of Petroleum and Energy Studies, Dehradun, India. He has also received the start-up research fund for his project "Nonlinear study of the interface in multilayer fluid system" from UGC, New Delhi. He is also listed among the top 2% of influential researchers in the world, as prepared by Stanford University based on Scopus data for the years 2022 and 2023. His ORCID is 0000-0002-6706-5226, Google Scholar web link is https://scholar.google.co.in/citations?user=Dj3ktGAAAAAJ and research gate web link is https://www.researchgate.net/profile/Mukesh-Awasthi-2.

Ashwani Kumar holds Ph.D. in Mechanical Engineering, specializing in Mechanical Vibration and Design with a diverse and impressive career spanning over 14 years. He currently holds the position of Head of Department Mechanical Engineering (Gazetted Officer Class 1) at the Technical Education Department (under Government of Uttar Pradesh) in Kanpur, Uttar Pradesh, India. Prior to his current role, Dr. Kumar served as an Assistant Professor in the Department of Mechanical Engineering at Graphic Era University, Dehradun (ranked 52nd by NIRF India) from July 2010 to November 2013. Beyond his academic contributions, Dr. Kumar possesses extensive leadership and administrative experience. He has held numerous key positions within the education sector, including Coordinator for AICTE-Extension of Approval, Nodal Officer for PMKVY-TI Scheme (Government of India), Internal Coordinator-CDTP Scheme (Government of Uttar Pradesh), Industry Academia Relation Officer, Assistant Centre Superintendent (ACS), Institute Examination Cell, Zonal Officer for Joint Entrance Examination (JEE-Diploma), and Sector Magistrate for State Assembly and General Elections. Dr. Kumar is a passionate researcher and a prominent figure in the field of academic publishing. His dedication to knowledge dissemination is evident in his role as Series Editor for seven distinguished book series published by international publishers of repute. These series encompass a wide range of topics and are titled *Advances in Manufacturing, Design and Computational Intelligence Techniques*, *Renewable and Sustainable Energy Developments*, *Smart Innovations and Technological Advancements in Mechanical and Materials Engineering*, *Solar Thermal Energy Systems: Advancements in Engineering, Ergonomics, and Sustainable Development*, *Artificial Intelligence and Machine Learning for Intelligent Engineering Systems*, *Computational Intelligence and Biomedical Engineering*, and *Renewable Green Technologies for Sustainable Energy Solutions*. Dr. Kumar's editorial expertise extends beyond book series. He serves as Editor-in-Chief for the *International Journal of Materials, Manufacturing and Sustainable Technologies* (IJMMST, ISSN: 2583-6625), and Editor of the *International Journal of Energy Resources Applications* (IJERA, ISSN: 2583-6617). He has also been a guest editor for a special issue of *Buildings* titled as *Sustainable Buildings*, *Resilient Cities*, and *Infrastructure Systems* (ISSN: 2075-5309, I.F. 3.1) and holds editorial board positions on eight international journals and acts as review board member of 20 prestigious (indexed in SCI/SCIE/Scopus) international journals with high impact factor. Dr. Kumar's dedication to scholarship is further demonstrated by his prolific authorship of over 165+ articles in prestigious journals, book chapters, and conference proceedings. Additionally, he has co-authored or edited over 40+ books in Mechanical Engineering, Materials Science, and Renewable Energy Engineering. Dr. Kumar's achievements have been recognized with prestigious awards, including a Best Teacher award for excellence in academics and research. He has also successfully guided

15 students through their B. Tech., M. Tech., and Ph.D. theses and currently serves as an external doctoral committee member at SRM University, New Delhi. Dr. Kumar is actively involved in cutting-edge research, focusing on areas such as Artificial Intelligence and Machine Learning in Mechanical Engineering, Smart Materials and Manufacturing Techniques, Thermal Energy Storage, Building Efficiency, Renewable Energy Harvesting, Sustainable Transportation, and Heavy Vehicle Dynamics. He holds four patents and frequently participates in international conferences as an invited speaker, session chair, and member of advisory boards, review boards, and program committees. Dr. Kumar's ORCID ID is 0000-0003-4099-935X and Google Scholar link: https://scholar.google.com/citations?hl=en&user=KOILpEkAAAAJ.

Nitesh Dutt is working as Assistant Professor in Department of Mechanical Engineering, College of Engineering, Roorkee, Uttarakhand, India. He has more than 7 years of teaching experience. He has done his bachelor's in mechanical engineering, master's and Ph.D. from IIT Roorkee. He has published more than 11 research articles in international journals and conferences. His main area of research is nuclear engineering, heat and mass transfer, thermodynamics, fluid mechanics, refrigeration and air conditioning, computational fluid dynamics (CFD).

Sivasakthivel Thangavel has been serving as an associate professor in the Department of Mechanical Engineering at the Global College of Engineering and Technology, Muscat, Sultanate of Oman since May 2018. He earned his Ph.D. in mechanical engineering from IIT Roorkee, India. Throughout his career, Dr. Siva worked as a JSPS Postdoctoral Fellow and researcher in China, Japan, and France, publishing more than 30 papers in refereed journals and conference proceedings. He is a certified higher education teacher and obtained a postgraduate certificate in academic practice in 2020 from the University of the West of England. Moreover, Dr. Siva boasts over 8 years of experience teaching undergraduate and postgraduate students at various higher educational institutes in India, China, Japan, and Oman. He has taught diverse modules across the field of mechanical engineering. His ORCID ID is 0000-0001-6802-4015 and Scopus ID is 55962921000.

Contributors

R.G. Abaszade is in Azerbaijan State Oil and Industry University, Baku, Azerbaijan and in Azerbaijan University of Architecture and Construction, Baku, Azerbaijan

R. Akshayaa is in Mathematics Division, School of Advanced Sciences, VIT University, Chennai Campus, Chennai, Tamil, Nadu, India

Shanmukh Sudhir Arasavelli is in Department of Mechanical Engineering, Gayatri Vidya Parishad College of Engineering (Autonomous), Visakhapatnam, Andhra Pradesh, India

Mukesh Kumar Awasthi is in Department of Mathematics, Babasaheb Bhimarao Ambedkar University, Lucknow, Uttar Pradesh, India

Govinda Rao Budda is in Department of Mechanical Engineering, Gayatri Vidya Parishad College of Engineering (Autonomous), Visakhapatnam, Andhra Pradesh, India

Vaibhav Chougule is in Department of Mechanical Engineering, Sharad Institute of Technology, College of Engineering, Yadrav, Ichalkaranji, Maharashtra, India

Kalidas Das is in Department of Mathematics, Krishnagar Government College, Nadia, West Bengal, India

Pinaki Ranjan Duari is in Department of Basic Science & Humanities (Mathematics Section), Asansol Engineering College, West Bengal, India

Nitesh Dutt is in Department of Mechanical Engineering, COER University Roorkee, Uttarakhand, India

K. Gnanaprasanna is in Mathematics Division, School of Advanced Sciences, VIT University, Chennai Campus, Chennai, Tamil Nadu, India

Vivek Goel is in F.E.T. Gurukula Kangri (DU), Haridwar, Uttarakhand, India

Yatika Gori is in Department of Mechanical Engineering, Graphic Era (Deemed to be University), Dehradun, Uttarakhand, India

Avesahemad S.N. Husainy is in Department of Mechanical Engineering, Sharad Institute of Technology, College of Engineering, Yadrav, Ichalkaranji, Maharashtra, India

D.B. Jani is in Government Engineering College Dahod-389151 under Gujarat Technological University Ahmedabad, Gujarat, India

Arpita Johri is in ABES Engineering College, Ghaziabad, Uttar Pradesh, India

Atharv Joshi is in Department of Mechanical Engineering, Sharad Institute of Technology, College of Engineering, Yadrav, Ichalkaranji, Maharashtra, India

Sonali Kale is in Department of AS & H, Pimpri Chinchwad College of Engineering, Pune, Maharashtra, India

E.A. Khanmamadova is in Azerbaijan State Oil and Industry University, Baku, Azerbaijan

Aman Kumar is in National Institute of Technology, Patna, Bihar, India

Ashwani Kumar is in Department of Mechanical Engineering, Technical Education Department, Kanpur, Uttar Pradesh, India

Mukesh Kumar is in Department of Mechanical Engineering, Indian Institute of Engineering Science and Technology (IIEST) Shibpur, Howrah, West Bengal, India

Sunil Kumar is in F.E.T. Gurukula Kangri (DU), Haridwar, Uttarakhand, India

A.G. Mammadov is in Azerbaijan State Oil and Industry University, Baku, Azerbaijan

Ambrish Maurya is in Department of Mechanical Engineering, National Institute of Technology, Patna, Bihar, India

Sunmoni Mudoi is in Department of Mathematics, Cotton University, Guwahati, Assam, India

C. Nirmalaa is in Mathematics Division, School of Advanced Sciences, VIT University, Chennai Campus, Chennai, Tamil Nadu, India

Kaustav Pradhan is in Department of Mechanical Engineering, Indian Institute of Science and Technology (IIEST) Shibpur, Howrah, West Bengal, India

Neha Singh Raghuvanshi is in ABES Engineering College, Ghaziabad, Uttar Pradesh, India

Mudit Saxena is in ABES Engineering College, Ghaziabad, Uttar Pradesh, India

Divyangana Sharma is in F.E.T. Gurukula Kangri (DU), Haridwar, Uttarakhand, India

Abhishek Kumar Singh is in Mathematics Division, School of Advanced Sciences, VIT University, Chennai Campus, Chennai, Tamil Nadu, India

Anand Kumar Solanki is in Department of Mechanical Engineering, Gayatri Vidya Parishad College of Engineering (Autonomous), Visakhapatnam, Andhra Pradesh, India

Sivasakthivel Thangavel is in Department of Mechanical Engineering at the Global College of Engineering and Technology, Muscat, Sultanate of Oman

Acknowledgments

The completion of this book, *Heat Exchanger Technologies for Sustainable Renewable Energy Systems*, would not have been possible without the invaluable contributions of many individuals and organizations.

First and foremost, we would like to express our heartfelt gratitude to the contributing authors. Their expertise, dedication, and meticulous research have enriched this book with cutting-edge knowledge and innovations in heat exchanger technologies. Without their commitment to advancing the field, this project would not have been successful.

We are deeply thankful to our academic and professional colleagues who have provided invaluable feedback and insights throughout the process. Their critical reviews and suggestions have significantly enhanced the quality of this work. Special thanks are also extended to the peer reviewers whose thorough assessments helped refine and shape each chapter.

We wish to extend our sincere appreciation to our institutions for their support, encouragement, and for providing the necessary resources that allowed us to undertake this ambitious project. The collaborative environment fostered by these institutions has been crucial to the completion of this book.

Our gratitude also goes to the publishers for their trust and for guiding us throughout the publication process. Their professional assistance, from editorial support to production management, has been instrumental in bringing this book to fruition.

Lastly, we want to express our deepest appreciation to our families and friends for their constant support and encouragement during the development of this book. Their patience and understanding, especially during the long hours of research and writing, have been a source of strength.

This book is a result of collective efforts, and we are truly grateful to everyone who played a part in its creation. We hope this work contributes meaningfully to the advancement of renewable energy technologies and fosters further innovations in the field.

Editors
Mukesh Kumar Awasthi, Ashwani Kumar,
Nitesh Dutt, and Sivasakthivel Thangavel

Chapter 1

Innovative technologies in heat exchangers for renewable energy systems

Neha Singh Raghuvanshi, Arpita Johri, Mudit Saxena, Ashwani Kumar, Nitesh Dutt, Mukesh Kumar Awasthi, and Sivasakthivel Thangavel

1.1 INTRODUCTION

1.1.1 Overview

Renewable energy sources are natural resources that are continually being exhausted and can be employed to generate power without exhausting their supplies. These sources include solar, wind, hydro, geothermal, and biomass energy. They are crucial as they offer a sustainable and eco-friendly alternative to fossil fuels, which are limited and significantly contribute to environmental pollution and climate change. By reducing dependence on non-renewable resources, renewable energy sources help in lowering down greenhouse gas emissions, decreasing air and water pollution, and enhancing energy security. Heat exchanger technologies are essential in renewable energy systems as they enable efficient transfer of heat between fluids, which is vital for optimizing energy production and improving system performance. Heat exchangers play a key role in various renewable energy applications by transferring heat from one medium to another, ensuring the systems operate effectively and reliably [1–4].

1.1.2 Types of renewable energy sources

Figure 1.1 demonstrates different types of renewable energy sources.

1.1.2.1 Solar energy

Solar energy harnesses the sun's radiation to generate power or provide heating, utilizing two primary technologies: photovoltaic (PV) cells and solar

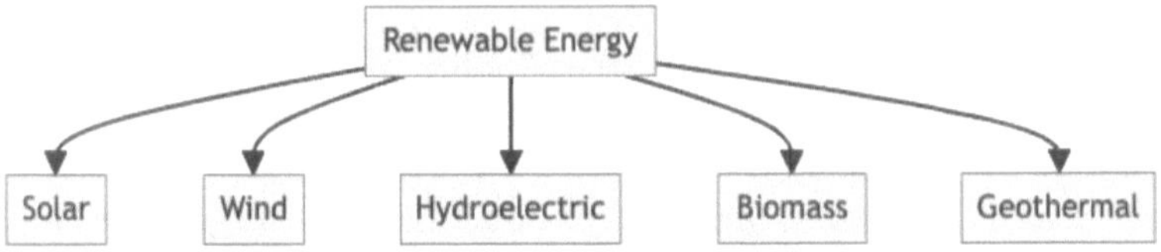

Figure 1.1 Types of renewable energy sources.

DOI: 10.1201/9781003534785-1

thermal systems. PV cells convert sunlight directly into electricity through the PV effect, a process that generates electric currents when light strikes semiconductor materials. Solar thermal systems, on the other hand, capture and convert solar energy into heat using solar collectors, which then transfer this heat to a fluid through heat exchangers. This fluid circulates to heat water or air in residential and commercial buildings. The abundance of solar energy, available across the globe, offers a sustainable alternative to fossil fuels, significantly reducing greenhouse gas emissions and contributing to environmental sustainability. Solar energy's versatility makes it an attractive option for both electricity generation and thermal applications, providing a crucial element in the transition toward cleaner energy sources [5–7].

1.1.2.2 Wind energy

Wind energy is produced by converting the kinetic energy of wind into electrical power using wind turbines. As wind flows over the blades of a turbine, it causes them to rotate, which drives a generator to produce electricity. Wind energy is a clean, renewable resource that helps mitigate climate change by reducing reliance on fossil fuels and lowering carbon emissions. Wind turbines often incorporate heat exchangers in their cooling systems to manage the operating temperatures of critical components, such as electrical generators and gearboxes. Wind energy's ability to provide a steady and reliable power source, coupled with its minimal environmental impact, makes it a pivotal technology in the global shift toward sustainable energy solutions [8, 9].

1.1.2.3 Hydro energy

Hydro energy is generated by capturing the kinetic energy of flowing water, typically through hydroelectric dams or run-of-river systems, to produce electricity. This method utilizes the movement of water to drive turbines connected to generators, converting the energy of flowing water into electrical power. Hydro energy is highly reliable and consistent, often used for baseload power generation, which helps stabilize the energy grid by providing a continuous supply of electricity. Heat exchangers in hydroelectric power plants play a crucial role in cooling systems for generators and other equipment, ensuring efficient operation and preventing overheating [10, 11].

1.1.2.4 Geothermal energy

Geothermal energy is derived from the Earth's internal heat, which can be utilized for electricity generation or direct heating applications. This form of energy leverages the heat stored beneath the Earth's surface accessed through geothermal wells. In geothermal power plants, heat exchangers are employed to transfer heat from geothermal fluids to a secondary fluid, which then drives a turbine to generate electricity. Alternatively, geothermal heat pumps use heat exchangers to transfer heat between the ground and buildings, providing efficient heating and cooling. Geothermal energy offers a stable and continuous power source with minimal environmental impact and low emissions, making it a highly sustainable option for energy production [12, 13].

1.1.2.5 Biomass energy

Biomass energy is produced from organic materials, including plant and animal waste, which are converted into biofuels, biogas, or heat through various processes. This renewable energy source helps mitigate waste and offers an alternative to fossil fuels in diverse applications. In biomass power plants, heat exchangers play a critical role in transferring heat generated from the combustion of biomass to a steam or fluid system, which then produces electricity or heat. The use of heat exchangers in biomass energy systems enhances efficiency by optimizing the transfer of thermal energy, thereby improving overall system performance [14–16].

1.1.3 Significance of conversion technologies

Conversion technologies are essential in the energy sector because they change different types of energy into usable power (Figure 1.2). They help make energy systems work better, especially when using renewable energy sources. They play an important role in following ways [17–19].

1.1.3.1 Efficient energy utilization

Efficient energy utilization is at the core of modern energy technologies, focusing on transforming raw energy sources into usable forms that can be

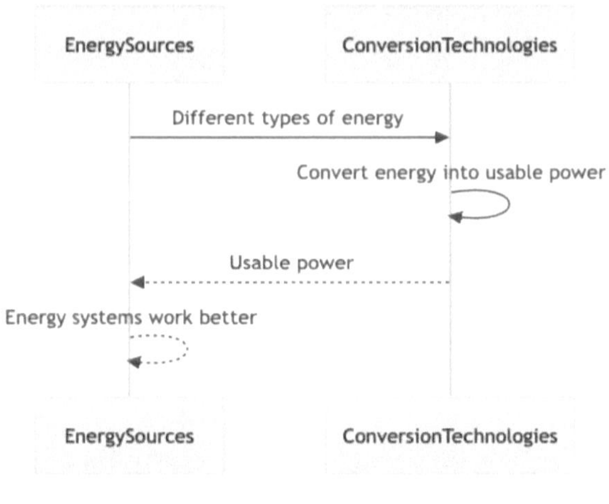

Figure 1.2 Role of renewable energy.

harnessed for practical applications. This process involves sophisticated technologies that convert various types of raw energy into forms that are easy to use and distribute. For example, solar panels, or PV cells, convert sunlight directly into electricity through the PV effect. This process involves semiconductor materials that generate an electric current when exposed to sunlight. Similarly, wind turbines convert the kinetic energy of wind into mechanical power through the rotation of turbine blades. This mechanical power is then converted into electricity by a generator. These technologies are crucial for optimizing energy use because they make it possible to capture and utilize the energy provided by natural sources in a way that can be integrated into electrical grids and used by consumers. By converting raw energy into electrical power, these systems enable the effective distribution and use of renewable energy sources, contributing to a more sustainable energy infrastructure. Efficient energy utilization not only maximizes the potential of available energy resources but also helps reduce reliance on fossil fuels, supporting the transition to a cleaner and more sustainable energy future [20–22].

1.1.3.2 Enhancing energy storage

Energy storage technologies play a pivotal role in managing the variability of energy supply and demand, particularly with renewable energy sources such as solar and wind power. These renewable sources can be intermittent, with energy production fluctuating based on weather conditions and time of day. By effectively storing energy, these technologies ensure that excess energy generated during peak production times can be retained and utilized during periods of low production or high demand. This capability is essential for maintaining grid stability and reliability, as it balances supply and demand and helps integrate renewable energy sources into the power grid. Enhancing energy storage not only improves the resilience of the energy system but also supports the broader adoption of renewable energy technologies by mitigating their inherent intermittency [23–25].

1.1.3.3 Improving system efficiency

Improving system efficiency is crucial for maximizing the performance and sustainability of energy systems. Conversion technologies that enhance system efficiency focus on optimizing the energy conversion process and minimizing energy waste. One notable example is combined heat and power (CHP) systems, which are designed to increase overall efficiency by capturing and utilizing the heat generated during electricity production. Additionally, advancements in energy conversion technologies, such as more efficient turbines, generators, and heat exchangers, contribute to reducing energy losses and enhancing performance. By focusing on improving system efficiency, these technologies help to reduce fuel consumption, lower greenhouse gas emissions, and minimize operational costs. Efficient systems are critical for achieving sustainability goals, as they ensure that the maximum amount of energy is extracted from available resources and utilized effectively, thereby supporting a more sustainable and economically viable energy infrastructure [26–28].

1.1.4 Integration with renewable sources

1.1.4.1 Adapting to intermittency

Renewable energy sources such as solar and wind are inherently intermittent due to their dependence on weather conditions and time of day. Solar power generation is contingent on sunlight, which varies throughout the day and is absent at night, while wind power relies on wind speed, which fluctuates. To address these challenges, conversion technologies are employed to manage the variability and ensure a consistent energy supply. One key solution is energy storage systems, such as batteries, which can store excess energy generated during periods of high production and release it during times when production is low. For example, a solar battery system can capture surplus electricity produced during sunny periods and provide power when the sun is not shining, effectively bridging the gap between supply and demand. Additionally, advanced grid management tools, such as demand response systems and smart grids, play a crucial role in balancing energy supply and demand. These technologies help to forecast energy production, adjust consumption patterns, and redistribute power efficiently across the grid. By integrating these systems, the energy grid can better accommodate the fluctuations of renewable sources, providing a steady and reliable energy supply despite their inherent intermittency [29–31].

1.1.4.2 Enhancing grid integration

Integration technologies are essential for connecting renewable energy sources to the existing power grid in a seamless and effective manner. For renewable energy sources to be successfully incorporated into the grid, they must be compatible with the grid's operational requirements and maintain system stability. Technologies such as inverters and controllers are crucial in this process. Inverters, used in solar panel systems, convert the direct current (DC) electricity generated by solar cells into alternating current (AC) electricity, which is compatible with the grid. Similarly, wind turbines use controllers to regulate the power output and ensure that it aligns with grid frequency and voltage standards. These technologies help to manage the variable output from renewable sources and synchronize it with the grid's requirements. Furthermore, grid operators use sophisticated software and hardware to monitor and control the flow of electricity, ensuring that renewable energy is integrated smoothly and does not disrupt the stability of the power grid. Enhancing grid integration facilitates the broader adoption of renewable energy sources by making it possible to incorporate their outputs reliably and efficiently [32, 33].

1.1.4.3 Supporting decentralized energy systems

Decentralized energy systems represent a shift from traditional large-scale power generation toward local energy production close to where it is consumed. This approach can significantly reduce energy losses associated with long-distance transmission and distribution, thereby increasing overall efficiency. Local generation systems, such as small-scale solar panels, wind turbines, and biomass digesters, provide energy at or near the point of use, which enhances energy security and reduces reliance on centralized power plants. By generating energy locally, communities can benefit from increased control over their energy sources, reduced transmission losses, and potentially lower energy costs [34].

1.1.4.4 Facilitating sustainable energy solutions

Conversion technologies are pivotal in advancing sustainable energy solutions by enhancing the efficiency and competitiveness of renewable energy systems. Innovations in these technologies drive improvements in performance, cost-effectiveness, and environmental impact. For instance, advancements in turbine design have led to more efficient wind turbines that can capture a greater amount of wind energy, while developments in solar cell technology have produced high-efficiency solar panels that generate more electricity from the same amount of sunlight. Similarly, waste-to-energy technologies have made it possible to convert various types of waste into usable energy, reducing the environmental impact of waste disposal and providing a renewable energy source [35].

1.2 HEAT EXCHANGER TECHNOLOGIES

Heat exchangers are key components in many industrial and energy systems, facilitating efficient heat transfer between fluids. Understanding their fundamental principles and their role in boosting energy systems is crucial for optimizing performance and efficiency [16]. Here's a brief overview.

1.2.1 Basic principles of heat exchangers

A heat exchanger is a device *designed to transfer heat from one fluid (liquid or gas) to another without allowing the fluids to mix*. Its main goal is to efficiently move thermal energy between the two fluids. Heat exchangers are used to heat or cool fluids, recover heat from exhaust gases, and enhance energy efficiency across various processes. They are commonly found in heating systems, refrigeration, air conditioning, and power plants [36, 37].

Heat exchangers are vital for capturing waste heat from processes and using it to preheat fluids or generate additional power (Table 1.1). For example, in a CHP plant, they recover heat from exhaust gases to enhance overall

Table 1.1 Comparison of various heat exchanging devices [20–22]

Type flow	*Arrangement*	*Heat transfer efficiency*	*Common applications*	*Advantages*	*Disadvantages*
Counterflow heat exchanger	Fluids flow in opposite directions	High, due to a large temperature gradient	Car radiators, industrial heat recovery systems	High efficiency, maximizes temperature difference	More complex design, can be larger and more expensive
Parallel flow heat exchanger	Fluids flow in the same direction	Lower compared to counterflow	Domestic water heaters, some heating, ventilation, and air conditioning (HVAC) systems	Simpler design, compact size	Lower heat transfer efficiency, less effective in temperature differential
Shell and tube heat exchanger	Fluids flow through separate tubes	Moderate to high, depending on design	Chemical processing, oil refining	Durable, handles high pressures and temperatures	Larger size, more complex maintenance, and higher cost
Plate heat exchanger	Fluids flow through alternating plates	High, due to large surface area	Dairy processing, food and beverage industries, HVAC systems	Compact size, high heat transfer efficiency	Potential for fouling, may require more frequent cleaning

efficiency. By optimizing heat transfer, heat exchangers reduce the amount of energy needed to heat or cool fluids, leading to lower energy consumption and cost savings in industrial processes and HVAC systems. In power plants, heat exchangers boost the efficiency of boilers and turbines by maintaining optimal temperatures and recovering heat from exhaust gases, leading to better performance and reduced fuel use. In HVAC systems, heat exchangers help maintain stable temperatures and improve comfort by effectively transferring heat between air or water streams. This result in better system stability and less wear on equipment [38–40].

1.3 HEAT EXCHANGERS IN RENEWABLE ENERGY SYSTEMS

Heat exchangers are pivotal in optimizing the performance and efficiency of renewable energy systems. They facilitate the transfer of thermal energy in various applications, from solar thermal collectors to geothermal systems. Here's a detailed look at their roles in different renewable energy technologies.

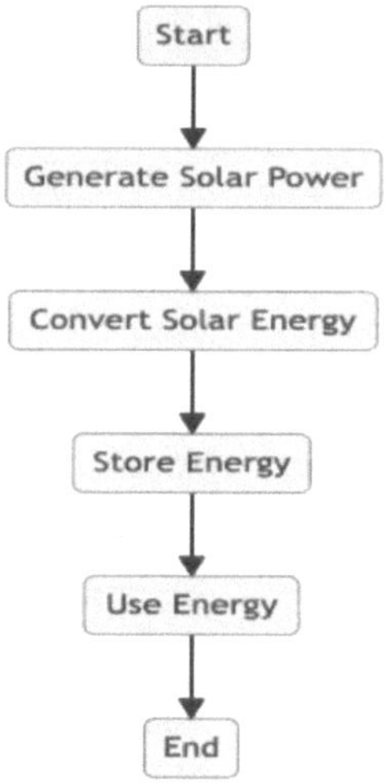

Figure 1.3 Mechanism of solar thermal systems.

1.3.1 Solar thermal systems

In solar thermal systems, heat exchangers play a critical role in transferring heat from the solar collector to the fluid, typically water or antifreeze, which is then used for space heating or hot water supply (Figure 1.3). The solar collector absorbs sunlight and converts it into thermal energy, heating the fluid that circulates through it. This heated fluid then flows through a heat exchanger, where it transfers its heat to another fluid medium, such as water in a storage tank or a hydronic heating system. This indirect heating process ensures that the water or medium used for domestic purposes is heated efficiently without being directly exposed to the solar collector. By using a heat exchanger, the system isolates the working fluid in the collector from the domestic water supply, preventing any potential contamination and maintaining a stable temperature control [41–43].

Heat exchangers also protect the system by isolating the collector fluid from the domestic water supply, preventing contamination and corrosion. Additionally, they allow for the use of antifreeze solutions in the collector loop, which is essential in colder climates to prevent the system from freezing. This versatility makes heat exchangers a crucial component in the effective and safe operation of solar thermal systems, enabling them to provide reliable and sustainable heating solutions [44–46].

1.3.2 Wind energy systems

Heat exchangers are integral to the cooling systems of wind turbines, ensuring the efficient and safe operation of the turbine's generator and other components. As wind turbines generate electricity, heat is produced, especially in

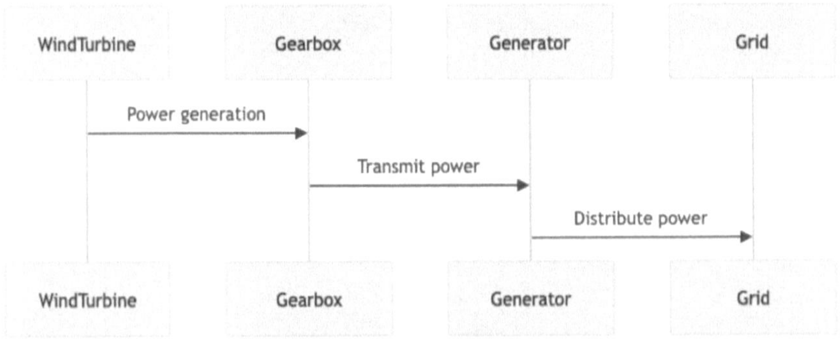

Figure 1.4 Wind energy systems mechanism.

the generator and gearbox. This heat needs to be managed effectively to prevent overheating, which could otherwise lead to reduced efficiency, increased wear and tear, and even potential failure of components. The primary role of the heat exchanger is to transfer this excess heat from the generator to a coolant system, which typically involves a fluid like oil or water-glycol mixture. The heated coolant then flows through the heat exchanger, where it releases the heat to the environment, either through air or water cooling mechanisms. This process keeps the generator and associated components at optimal temperatures, preventing them from reaching critical levels that could compromise performance or safety (Figure 1.4).

Mechanism: The thermal management system in a wind turbine relies on both conduction and convection processes within the heat exchanger. As the generator and gearbox operate, they generate significant heat due to mechanical and electrical losses. This heat is transferred by conduction to the cooling fluids circulating through the system. The heat exchanger then facilitates the transfer of this heat from the heated fluid to the external environment, usually by passing the fluid through a series of tubes or plates exposed to air or water. This mechanism ensures that the hydraulic fluids, lubricants, and other cooling mediums are maintained within their ideal operating temperature ranges, thereby safeguarding the turbine's components against thermal stress and degradation [47–49].

1.3.3 Hydropower systems

In hydropower plants, heat exchangers play a crucial role in managing the thermal loads of turbine and generator systems. These systems generate substantial heat during energy conversion, primarily due to friction and electrical resistance. Heat exchangers are essential for dissipating this excess heat to maintain optimal operating conditions and ensure the longevity of critical components such as generator bearings, transformers, and electrical circuits. This not only supports the continuous and efficient

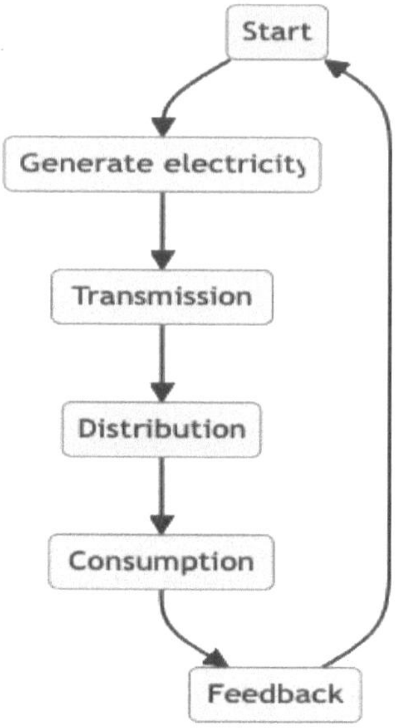

Figure 1.5 Hydropower systems mechanism.

operation of the plant but also contributes to its overall safety and reliability (Figure 1.5).

Mechanism: The mechanism of heat transfer in hydroelectric power plants typically involves conduction and convection processes facilitated by heat exchangers. As turbines and generators operate, friction and mechanical processes produce heat, particularly in areas like generator bearings and lubrication systems. This heat is transferred to a cooling medium, such as water or oil, which circulates through the heat exchanger. The heat exchanger then dissipates this heat to the environment or to a secondary cooling system, often using river water or a separate cooling loop. In this setup, the heat exchanger is crucial in maintaining the temperature of the turbine's lubrication system and generator bearings, ensuring that these components do not exceed their optimal operating temperature range. This thermal management is vital for preventing thermal stress, which can lead to material fatigue and potential system failures [50–52].

1.3.4 Geothermal energy systems

Heat exchangers are essential components in geothermal systems, facilitating the transfer of thermal energy between the ground (geothermal source) and the target application, such as a building or industrial process. In geothermal heat pumps, heat exchangers enable the efficient exchange of heat between the ground and the indoor environment. They extract heat from the ground during winter for heating and transfer excess indoor heat back to the ground in summer for cooling (Figure 1.6).

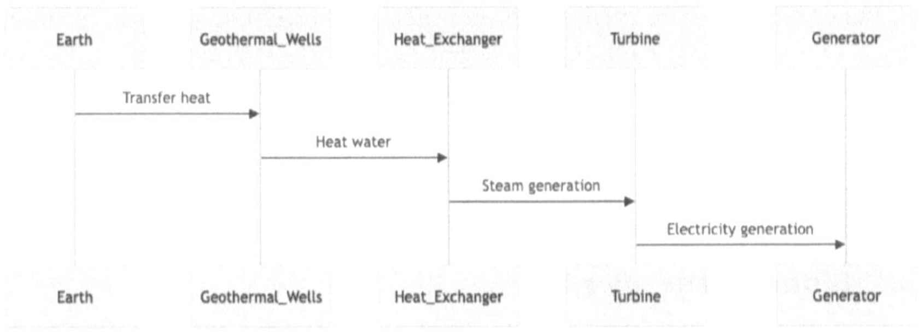

Figure 1.6 Geothermal energy systems mechanism.

Mechanism: The heat transfer mechanism in geothermal heat pumps involves a ground loop system and a heat exchanger to facilitate the movement of thermal energy. A closed-loop system typically consists of pipes buried underground, filled with a mixture of water and antifreeze. The heat exchanger within the heat pump extracts heat from the ground loop fluid in winter. The thermal energy is then transferred to a refrigerant within the heat pump, which is compressed to raise its temperature further before being used to heat the building. In summer, the process reverses: the indoor heat is absorbed by the refrigerant and transferred to the ground loop, where it dissipates into the cooler earth. This efficient exchange process allows the heat pump to maintain desired indoor temperatures with minimal energy use [10]. In geothermal power plants, heat exchangers play a critical role in harnessing geothermal energy for electricity generation. These plants typically use a binary cycle system, where heat from geothermal fluids is transferred to a secondary working fluid through a heat exchanger. The geothermal fluid, which can be steam or hot water, transfers its heat to the secondary fluid—usually a low-boiling-point organic compound. This fluid vaporizes and drives a turbine connected to a generator, producing electricity. After releasing its energy, the secondary fluid is condensed and recirculated through the heat exchanger, maintaining a continuous cycle. This process ensures efficient utilization of geothermal resources, maximizing energy extraction while minimizing the impact on the geothermal reservoir [53–55].

Heat exchangers in geothermal systems are designed to optimize the efficiency of heat extraction and utilization. By transferring heat from geothermal fluids to a secondary loop or working fluid, they ensure that the energy is effectively used for heating, cooling, or power generation. The choice of heat exchanger material and design is crucial for maximizing thermal conductivity and minimizing energy losses. For instance, materials like copper or specialized alloys are often used for their excellent thermal properties and resistance to corrosion. Additionally, the design must facilitate efficient fluid flow and heat transfer, often incorporating features like extended surfaces or turbulence enhancers to increase the contact area and improve heat exchange rates. The use of heat exchangers in geothermal systems not only improves energy efficiency but also contributes to the longevity and reliability of the system. By effectively managing the heat transfer process, these devices reduce the thermal stress on system components, preventing overheating and minimizing wear and tear [56–58].

1.3.5 Biomass energy systems

In biomass energy systems, heat exchangers are integral to enhancing the efficiency of the energy conversion process. They function by recovering and transferring heat generated during the combustion or gasification of biomass. This recovered heat is then utilized for various purposes, such as preheating incoming air or water, thereby boosting the overall efficiency of the system. By reusing the energy that would otherwise be wasted, heat exchangers reduce the fuel consumption required for the process, making the biomass energy system more sustainable and cost-effective. In biomass power plants, heat exchangers are typically employed to transfer heat from the flue gases produced during combustion to a water loop, which then supplies heat to the plant's steam generation system.

Mechanism: The mechanism of heat transfer in biomass energy systems primarily involves the movement of thermal energy from one medium to another through conduction, convection, or radiation. In a typical biomass power plant, the heat exchanger captures the high-temperature heat from flue gases, transferring it to a secondary medium such as water or air. For instance, a shell-and-tube heat exchanger might be used, where the flue gases pass through the tubes and transfer heat to water circulating in the shell. This heated water is then used to produce steam, which drives turbines to generate electricity. Alternatively, in a biomass gasification system, heat exchangers can capture heat from the hot syngas produced during gasification [11].

In biomass power plants, heat exchangers play a crucial role in optimizing the energy conversion process. They are used to transfer heat from flue gases to a water loop, which then supplies heat to the steam generation system. This is a key component of the Rankine cycle, where the generated steam drives a turbine connected to an electricity generator. Additionally, heat exchangers can be used to recover heat from the gasification process, utilizing it to generate steam for power production or for direct heating applications.

1.4 CONVERSION TECHNOLOGIES INVOLVING HEAT EXCHANGERS

Heat exchangers play a pivotal role in various conversion technologies, particularly those aimed at maximizing energy efficiency and recovering waste heat. This section explores three key areas where heat exchangers are crucial:

heat recovery systems, CHP systems, and thermoelectric generation. A comparative table will illustrate the different aspects and efficiencies of these technologies.

1.4.1 Heat recovery systems: Methods and technologies for recovering waste heat

1.4.1.1 Heat recovery steam generators (HRSGs)

Heat recovery steam generators (HRSGs) are specialized heat exchangers used to recover waste heat from industrial processes, particularly in gas turbine power plants, and convert it into steam. This steam can be used for generating electricity or for other heating applications, making the overall energy system more efficient. In a typical setup, exhaust gases from a gas turbine, which would otherwise be released into the atmosphere, are passed through the HRSG. The HRSG captures the thermal energy from these gases and uses it to produce high-pressure steam. This steam can then drive a steam turbine to generate additional electricity or be utilized for other industrial processes that require heat, such as district heating or desalination. The use of HRSGs significantly enhances the efficiency of combined-cycle power plants. In these plants, the electricity generation process consists of two stages: first, a gas turbine generates electricity by burning natural gas; then, the HRSG captures the high-temperature exhaust from the gas turbine to produce steam. This steam drives a second turbine, known as a steam turbine, to generate more electricity. This dual-stage process allows for greater energy extraction from the same amount of fuel, resulting in higher overall efficiency and reduced emissions. HRSGs are a critical component in industries where high-temperature processes are common, as they enable the capture and reuse of waste heat, thereby reducing fuel consumption and environmental impact [59–61].

1.4.1.2 Waste heat recovery units (WHRUs)

Waste heat recovery units (WHRUs) are designed to extract heat from various waste heat sources, such as flue gases, exhaust streams, or hot industrial fluids, and transfer it to another medium. This recovered heat can be used for a variety of applications, including preheating feedwater for boilers, district heating, or even for direct use in industrial processes. In a cement plant, for example, a WHRU can capture heat from the kiln's exhaust gases and use it to preheat the combustion air or feedwater for boilers. This not only improves the energy efficiency of the plant but also reduces fuel consumption and greenhouse gas emissions. WHRUs are versatile and can be integrated into many industrial systems to improve energy efficiency and sustainability. They are particularly effective in industries with high-temperature processes, such as steel production, chemical manufacturing, and petrochemical refining. By recovering heat that would otherwise be lost to the environment,

WHRUs help lower operational costs and reduce the environmental impact of these industries. Additionally, in some cases, the recovered heat can be converted into electricity through the use of Organic Rankine Cycle (ORC) systems or other power generation technologies, further enhancing the utility and efficiency of WHRUs in industrial applications.

1.4.1.3 Heat recovery from refrigeration systems

In large-scale refrigeration systems, such as those used in commercial buildings, supermarkets, and industrial facilities, significant amounts of waste heat are generated during the refrigeration cycle. This heat, often released from the compressor, can be captured and repurposed using heat exchangers to improve overall system efficiency. By recovering this waste heat, it is possible to provide heating for hot water generation, space heating, or other thermal applications, thus reducing the need for additional energy sources. For example, in a supermarket refrigeration system, the waste heat from the compressors, which would typically be expelled into the environment, can be captured and used to preheat water for cleaning purposes or to heat the store during colder months. This not only reduces the supermarket's heating costs but also decreases its overall energy consumption and carbon footprint. Similarly, in industrial refrigeration applications, waste heat recovery can be used to supply heat for processes such as pasteurization, drying, or even for maintaining comfortable working conditions within the facility. The integration of heat recovery systems in refrigeration not only improves energy efficiency but also contributes to sustainable energy management by reducing reliance on external heating sources. Moreover, by effectively utilizing the waste heat, businesses can lower their operational costs and contribute to reducing greenhouse gas emissions. This approach to energy recovery aligns with broader sustainability goals and helps industries and commercial operations meet environmental regulations and standards.

1.4.1.4 Heat recovery in industrial applications

Heat recovery technologies, including HRSGs, WHRUs, and systems integrated into refrigeration units, play a crucial role in various industrial applications. By capturing and reusing waste heat, these systems help improve the overall energy efficiency of industrial processes, reduce fuel consumption, and lower greenhouse gas emissions. In sectors like petrochemicals, cement production, and food processing, heat recovery is essential for optimizing energy use and reducing environmental impact. For example, in the petrochemical industry, heat exchangers are used to recover heat from distillation processes, using it to preheat incoming feedstock or to generate steam for power production. In cement plants, heat exchangers can recover heat from clinker coolers and kiln exhaust gases, using it to improve the efficiency of the calcinations process or to produce electricity through waste heat boilers. Similarly, in food processing, heat recovery can be used to capture heat from pasteurization or cooking processes, using it to preheat water or for other thermal needs within the facility. These applications demonstrate the versatility and importance of heat recovery systems in enhancing industrial efficiency and sustainability. By effectively utilizing waste heat, industries can achieve significant cost savings, reduce their environmental impact, and contribute to a more sustainable energy future.

1.4.2 Combined heat and power (CHP) systems

Figure 1.7 gives us the idea of CHP systems.

1.4.2.1 Integration of heat exchangers in CHP systems

CHP systems, also known as cogeneration systems, utilize heat exchangers to capture and repurpose waste heat generated during electricity production. This integration significantly enhances the efficiency of energy systems by simultaneously producing electricity and useful thermal energy from a single fuel soûrce. In a typical CHP setup, exhaust gases from a primary power generator, such as a gas turbine or reciprocating engine, are directed through a heat exchanger. This heat exchanger recovers thermal energy from the exhaust gases and converts it into steam or hot water. The recovered heat can then be utilized for various applications, including district heating, industrial processes, or domestic hot water supply. The effective use of heat exchangers in CHP systems ensures that the thermal energy, which would otherwise be wasted, is utilized productively. This dual generation of heat and power maximizes the energy output from the fuel, making CHP systems highly efficient. For instance, in a gas turbine-based CHP plant, the exhaust gases, which can reach temperatures as high as 500–600°C, are routed through a heat exchanger known as a heat recovery steam generator

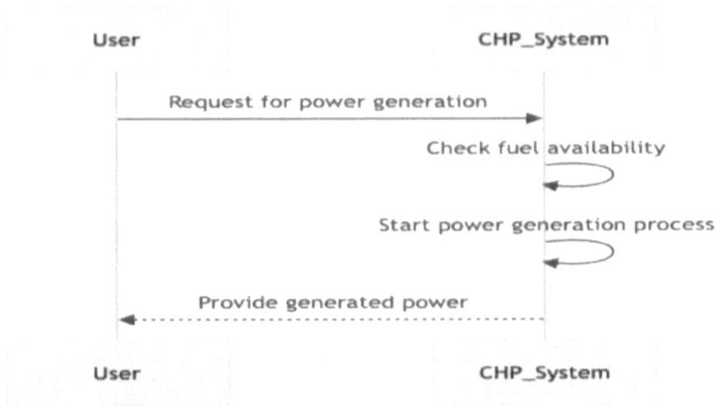

Figure 1.7 Combined heat and power (CHP) systems.

(HRSG). The HRSG captures this high-grade thermal energy to produce steam. This steam can be used to drive a secondary steam turbine for additional electricity generation or be supplied directly for heating purposes in industrial applications or residential buildings. The flexibility and adaptability of CHP systems make them suitable for a wide range of applications. They are commonly used in industrial facilities, hospitals, universities, and district heating networks, where both electricity and heat are required. By incorporating heat exchangers into CHP systems, facilities can achieve a seamless integration of power and thermal energy supply, optimizing the overall energy consumption and reducing dependency on separate heating systems [62–64].

1.4.2.2 Benefits and efficiency considerations

1.4.2.2.1 Increased efficiency

One of the primary benefits of CHP systems is their high efficiency. Traditional power generation systems typically operate at an efficiency of 30–50%, with a significant portion of the energy lost as waste heat. In contrast, CHP systems can achieve overall efficiencies of 70–90% by capturing and utilizing this waste heat. This is because the heat exchangers in CHP systems effectively recover thermal energy that would otherwise be dissipated into the environment. The integration of electricity generation and heat recovery allows CHP systems to extract more useful energy from the same amount of fuel, making them a more sustainable and economical choice for energy production. For example, in a large industrial facility, a CHP system can use the waste heat from its power generation processes to produce steam or hot water for use in various industrial processes, such as drying, pasteurization, or chemical reactions. This reduces the need for separate boilers or heating systems, thereby lowering fuel consumption and operational costs. The high efficiency of CHP systems not only contributes to cost savings but also enhances the overall reliability and resilience of the energy supply [65, 66].

1.4.2.2.2 Cost savings

CHP systems offer significant cost-saving potential by reducing the need for additional fuel for heating purposes. The ability to generate both electricity and thermal energy from the same fuel source means that facilities can reduce their fuel expenditures and lower their overall energy costs. For instance, a hospital using a CHP system can generate electricity for its power needs and simultaneously produce hot water for space heating and sterilization processes. This integrated approach reduces the hospital's reliance on external energy sources and provides substantial cost savings in the long term. In addition to direct fuel cost savings, CHP systems can also help reduce maintenance and operational costs. By using heat exchangers to efficiently manage waste heat, CHP systems minimize the wear and tear on auxiliary heating equipment, leading to longer equipment lifespans and reduced maintenance expenses. Furthermore, facilities that adopt CHP systems may benefit from lower utility bills and potential incentives or rebates offered by governments or utility companies for implementing energy-efficient technologies.

1.4.2.2.3 Environmental benefits

The high efficiency of CHP systems translates into lower greenhouse gas emissions compared to traditional power generation methods. By utilizing waste heat, CHP systems reduce the amount of fuel required to produce the same amount of energy, resulting in lower carbon dioxide (CO_2) and other pollutant emissions. Studies have shown that CHP systems can reduce CO_2 emissions by up to 30% compared to separate heat and power systems. This makes them a key technology for industries and institutions aiming to reduce their environmental footprint and comply with stringent emission regulations [13]. Additionally, the reduced fuel consumption associated with CHP systems leads to decreased demand for fossil fuels, contributing to lower overall emissions of pollutants such as sulfur dioxide (SO_2) and nitrogen oxides (NOx). These pollutants are major contributors to air quality issues and environmental problems like acid rain and smog. By implementing CHP systems with integrated heat exchangers, facilities can play a proactive role in promoting environmental sustainability and reducing their impact on the planet. In summary, the integration of heat exchangers in CHP systems provides a range of benefits, including enhanced efficiency, cost savings, and significant environmental advantages. These systems represent a smart and sustainable solution for industries and institutions looking to optimize their energy use and reduce their ecological impact.

1.4.3 Thermoelectric generation

1.4.3.1 Use of heat exchangers in thermoelectric generators

Thermoelectric generators (TEGs) utilize the Seebeck effect to convert temperature differences directly into electrical energy. Heat exchangers play a crucial role in these systems by efficiently transferring heat to and from the thermoelectric materials, which are often composed of semiconductors such as bismuth telluride. The primary function of heat exchangers in TEGs is to maintain a significant temperature gradient across the thermoelectric modules. This gradient is essential for generating electrical power, as the voltage produced by the Seebeck effect is directly proportional to the temperature difference across the thermoelectric material. In a typical TEG setup, a heat exchanger is placed on the hot side, where it captures heat from a high-temperature source, such as industrial waste heat, automotive exhaust gases, or geothermal energy. This heat is then transferred to the thermoelectric module. On the opposite side, another heat exchanger is used to dissipate heat, maintaining a low temperature. This configuration creates a substantial temperature difference across the thermoelectric material, enabling the efficient generation of electricity. For example, in a waste heat recovery system, heat exchangers capture heat from an industrial process and transfer it to the thermoelectric modules. The cold side heat exchanger, often coupled with a cooling system such as water or air cooling, ensures that the temperature differential is maximized.

1.4.3.2 Enhancing performance and efficiency

The efficiency and power output of TEGs heavily depend on the effectiveness of heat exchangers in maintaining the thermal gradient. High-performance heat exchangers are essential for minimizing thermal losses and ensuring that the maximum amount of heat is transferred to and from the thermoelectric modules. Enhancing heat exchanger performance involves optimizing their design and materials to achieve the best possible thermal conductivity and heat transfer characteristics. One of the primary challenges in thermoelectric generation is maximizing the temperature differential without causing excessive heat loss. Advanced heat exchanger designs, such as microchannel heat exchangers or those using high-conductivity materials like graphene or advanced ceramics, are employed to improve heat transfer rates. These materials have superior thermal conductivity and low thermal resistance, enabling them to maintain a large temperature gradient across the thermoelectric modules. Additionally, innovative geometries, such as finned or porous structures, increase the surface area available for heat transfer, further enhancing the efficiency of the heat exchangers.

In practical applications, the efficiency of TEGs can be significantly improved by integrating phase-change materials (PCMs) in the heat exchangers. PCMs absorb and release large amounts of heat during phase transitions, effectively stabilizing temperature fluctuations and maintaining a consistent thermal gradient across the thermoelectric modules. This approach is particularly beneficial in applications with variable heat sources, such as automotive exhaust systems or intermittent industrial processes. Another approach to enhancing TEG performance is the use of segmented thermoelectric modules, which combine different thermoelectric materials optimized for specific temperature ranges. Heat exchangers are designed to match these materials' thermal properties, ensuring that heat is transferred efficiently across the various segments. This configuration allows for a more efficient conversion of heat into electricity, as each segment operates at its optimal temperature. Moreover, advancements in additive manufacturing and computational modeling have enabled the development of custom-designed heat exchangers tailored to specific TEG applications. These technologies allow for precise control over the geometry and material composition of heat exchangers, optimizing them for maximum efficiency and performance (Table 1.2).

Table 1.2 Comparison of conversion technologies involving heat exchangers

Technology	*Heat exchanger role*	*Efficiency*	*Typical applications*	*Benefits*
Heat recovery systems	Captures and transfers waste heat	50–70% recovery of waste heat	Industrial processes, power plants, refrigeration	Reduces energy consumption, lowers operational costs, improves system efficiency
Combined heat and power (CHP) systems	Integrates heat recovery with power generation	70–90% overall efficiency	Industrial facilities, district heating, power plants	Reduces fuel use, cuts CO_2 emissions, lowers operational costs
Thermoelectric generation	Transfers heat to/from thermoelectric materials	5–10% electrical efficiency	Waste heat recovery, remote power generation	Converts waste heat to electricity, enhances efficiency of heat recovery systems

1.5 INNOVATIONS AND ADVANCES IN HEAT EXCHANGER TECHNOLOGIES

Heat exchangers are becoming more advanced with improvements in materials, design, and technology. These innovations are making heat exchangers more efficient and effective for a variety of uses. This section covers recent advancements in materials and designs, computational methods, and smart technology integration.

1.5.1 Emerging materials and designs

1.5.1.1 Advanced materials for better heat transfer

The heat exchanger industry faces an increasing demand for advanced materials to enhance efficiency and performance. Traditional metals like copper and aluminum have long been the standard for heat exchangers, but their thermal conductivity limits necessitate the exploration of new materials that can significantly improve heat transfer rates.

1.5.1.1.1 High thermal conductivity materials

Emerging materials such as graphene, carbon nanotubes, and advanced ceramics are now being integrated into heat exchanger designs. Graphene, in particular, stands out due to its remarkable thermal conductivity, which can reach up to 5000 W/mK. This capability allows graphene-based heat exchangers to transfer heat much more efficiently than conventional materials. For example, in applications where rapid heat dissipation is crucial—such as in high-performance electronics or advanced thermal management systems—graphene can offer significant advantages, leading to smaller and more efficient designs.

1.5.1.1.2 Phase-change materials (PCMs)

Another innovative class of materials, PCMs, is gaining traction in heat exchanger technology. PCMs can absorb and release heat during phase transitions, such as melting and solidifying. This unique property allows them to act as thermal buffers, managing temperature fluctuations effectively. In heat exchangers, PCMs can store excess heat when temperatures are high and release it when needed, enhancing overall energy efficiency. This capability is particularly beneficial in renewable energy applications, such as solar thermal systems, where heat management is crucial for optimizing energy storage and use.

1.5.2 Computational methods and simulation

Figure 1.8 demonstrates computational methods and simulation.

1.5.2.1 Role of simulations in optimizing performance

1.5.2.1.1 Computational fluid dynamics (CFD)

CFD uses numerical methods and algorithms to solve and analyze problems involving fluid flows. It allows engineers to visualize how fluids move and interact within a heat exchanger, predicting performance and identifying potential issues. CFD can simulate the flow patterns of fluids inside a heat

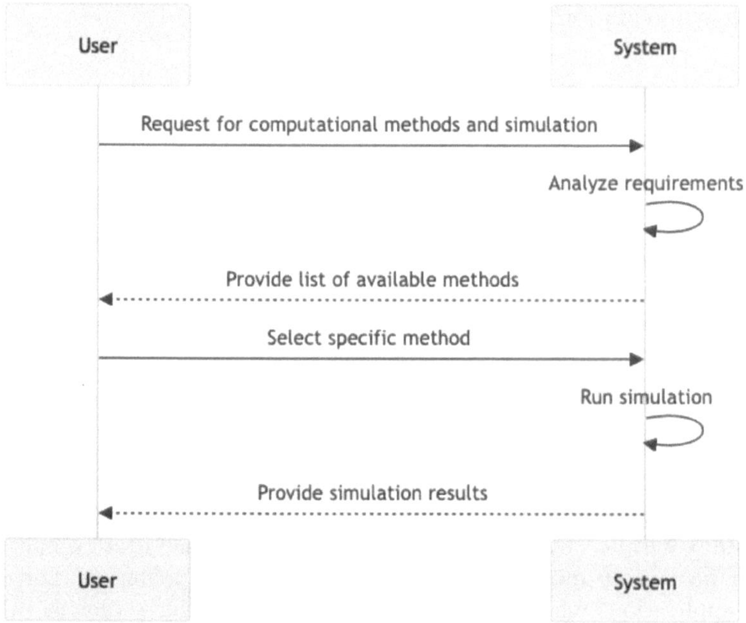

Figure 1.8 Computational methods and simulation.

exchanger, revealing areas of turbulence, dead zones, and non-uniform flow. For example, in a shell and tube heat exchanger, CFD can show how the shell-side fluid interacts with the tube bundle, helping optimize baffle designs to improve flow distribution and heat transfer. CFD models are employed to design heat exchangers for crude oil distillation and gas processing, ensuring they handle high temperatures and pressures effectively. Though CFD simulations require specialized software and expertise, which can be costly. The accuracy of CFD predictions depends on the quality of the input data and the assumptions made during modeling.

1.5.2.1.1.1 THERMAL-FLUID MODELING

Thermal-fluid modeling involves simulating both thermal and fluid flow behaviors to understand how heat is transferred in various conditions. These models integrate heat transfer equations with fluid dynamics to provide a comprehensive view of heat exchanger performance. These models consider variations in fluid properties with temperature and pressure, allowing for more precise predictions of heat exchanger performance under varying operational conditions. Advanced models can account for temperature-dependent viscosity and thermal conductivity, providing insights into how changes in operating conditions affect performance. Thermal-fluid modeling helps design heat exchangers for pasteurization processes, ensuring that all product is heated uniformly to meet safety standards. Models are used to design and optimize heat exchangers for heating and cooling systems, ensuring they perform efficiently across various environmental conditions. Creating accurate thermal-fluid models requires detailed data and can be complex, potentially leading to high computational demands. Models need to be validated against experimental data to ensure their accuracy, which can be time-consuming and costly.

1.5.2.2 Case studies of recent advancements

1.5.2.2.1 Microchannel heat exchangers

Microchannel heat exchangers are compact devices with small channels through which fluids flow. These designs have gained attention due to their ability to enhance heat transfer efficiency and manage high thermal loads in a compact space. Recent advancements reported are as follows:

i. *Improved cooling efficiency:* Recent studies have demonstrated that microchannel heat exchangers can improve cooling efficiency by up to 40% compared to traditional heat exchanger designs. This improvement is attributed to the high surface area-to-volume ratio of microchannels, which allows for more effective heat dissipation.
ii. *Applications in electronics:* Microchannel heat exchangers are increasingly used in cooling electronic devices, such as CPUs and GPUs, where high thermal loads need to be managed within tight spaces. The

enhanced cooling performance helps prevent overheating, extending the lifespan and reliability of electronic components. Example, Intel has incorporated microchannel heat exchangers in high-performance computing systems to manage the heat generated by advanced processors. The increased cooling efficiency has allowed for higher processor speeds and improved performance without compromising system stability.

1.5.2.2.2 Phase-change materials (PCMs)

PCMs are substances that absorb and release thermal energy during phase transitions (e.g., from solid to liquid). They are integrated into heat exchangers to enhance thermal energy storage and management. Recent advancements reported are as follows:

i. *Increased heat retention:* Recent studies have shown that integrating PCMs into heat exchangers for solar energy systems can increase heat retention by 25%. This enhancement improves the stability and efficiency of energy storage, making solar energy systems more reliable and effective.
ii. *Applications in solar energy systems:* PCMs are used to store excess solar energy, which can be released when sunlight is not available. This capability helps balance the supply and demand for solar energy, improving the overall efficiency and reliability of solar power systems. For instance, solar reserve has employed PCMs in their solar thermal energy storage systems. The integration of PCMs has enhanced the system's ability to store and release thermal energy, providing a more stable and continuous energy supply even when the sun is not shining.

1.5.3 Smart control systems and IoT in heat exchangers

1.5.3.1 Smart control systems

Contemporary heat exchangers are increasingly equipped with smart control systems that incorporate sensors and sophisticated algorithms. These systems enable real-time adjustments to operational parameters, such as flow rates and temperatures, based on current data. For example, if sensors detect deviations from optimal conditions, the system can automatically adjust settings to maintain efficiency. This capability allows the system to respond dynamically to changing conditions, improving overall performance and energy efficiency. In HVAC systems, smart controls adapt heat exchanger operations to varying indoor and outdoor temperatures, resulting in enhanced comfort and reduced energy consumption.

1.5.3.2 Internet of Things (IoT)

The Internet of Things (IoT) transforms heat exchanger management through remote monitoring and control capabilities. Sensors embedded in heat exchangers continuously gather performance data, including temperature, pressure, and flow rates. This information is sent to cloud-based platforms for real-time analysis. Facility managers can remotely access this data to monitor performance, optimize operations, and anticipate maintenance needs. For instance, IoT-enabled systems can provide early warnings of potential issues, enabling timely maintenance and reducing downtime. This proactive approach improves system efficiency and extends equipment lifespan [19].

1.5.4 Real-time monitoring and optimization

1.5.4.1 Real-time data analysis

The advancement of continuous monitoring systems has significantly transformed heat exchanger management by enabling real-time data analysis. These systems systematically collect and assess performance indicators, such as temperature, pressure, and flow rates. This ongoing analysis facilitates the prompt identification of anomalies or deviations in performance, allowing for swift corrective actions before problems escalate. For example, if a heat exchanger begins to overheat or shows signs of reduced efficiency, the system can immediately alert operators, enabling timely maintenance or adjustments. Such proactive measures help avert unexpected failures, minimize downtime, and ensure that the heat exchanger functions at optimal efficiency. Constant monitoring and analysis contribute to a more stable and reliable operation, thereby enhancing overall system performance and prolonging the equipment's lifespan.

1.5.4.2 Adaptive algorithms

Machine learning algorithms significantly boost heat exchanger performance through adaptive optimization. These algorithms process real-time data to uncover patterns and relationships that may not be visible through conventional methods. As they continuously learn from incoming data, they can automatically adjust operational parameters like flow rates and temperatures. For instance, in response to fluctuating environmental conditions or varying heat loads, these algorithms can fine-tune the heat exchanger's performance to maintain optimal efficiency and prevent overuse. This dynamic approach not only enhances operational efficiency but also extends the heat exchanger's lifespan by reducing wear and tear. Over time, the system becomes more proficient at managing performance, leading to lower energy consumption and reduced operational costs.

1.6 CHALLENGES AND FUTURE DIRECTIONS

Apart from the fact that there is immense growth in the industry, there are some considerable challenges too. These challenges include technicalities of design and material selection, compatibility, and maintenance. Economic consideration and environmental sustainability too need to be comprehended and dealt with caution.

1.6.1 Technical challenges

1.6.1.1 Design and material selection issues

Modern heat exchangers increasingly incorporate complex designs, such as microchannels and specialized surfaces, to enhance their thermal efficiency. While these innovations significantly improve performance, they introduce various challenges in manufacturing and maintenance. Microchannel heat exchangers, for instance, utilize tiny channels that maximize surface area while minimizing size.

1.6.1.2 Maintenance and durability

The maintenance requirements for advanced heat exchangers can be more demanding compared to traditional systems, primarily due to their complex structures and the materials used. For example, microchannel heat exchangers, despite their efficiency, are particularly prone to fouling. The small passageways can accumulate debris or sediment more easily than larger channels, leading to decreased performance and requiring more frequent and intensive cleaning. This not only raises operational costs but also necessitates specialized cleaning procedures that might not be readily available.

1.6.2 Economic and environmental considerations

1.6.2.1 Cost and lifecycle analysis

The adoption of advanced materials and designs in heat exchangers often comes with a higher initial cost. For instance, high-performance ceramics or microchannel heat exchangers, known for their superior efficiency, can be significantly more expensive to manufacture and purchase compared to traditional heat exchangers. While these advanced systems promise enhanced thermal performance and energy savings, a comprehensive cost and lifecycle analysis is essential to evaluate. The initial capital investment in advanced heat exchangers must be weighed against the total cost of ownership. This includes not only the purchase price but also installation costs, operational expenses, maintenance requirements, and eventual disposal costs.

1.6.2.2 Environmental impact and sustainability

The environmental implications of producing advanced materials, such as graphene or high-performance alloys, require careful consideration. The production processes for these materials can be resource-intensive and may generate considerable waste, raising questions about their overall sustainability. The manufacturing of advanced materials often involves significant energy consumption and the use of scarce resources, which can contribute to environmental degradation. Evaluating the environmental impact of these materials involves examining the entire supply chain, from raw material extraction to processing and fabrication. Sustainable sourcing practices and efficient manufacturing processes are essential to mitigate negative effects and enhance the overall sustainability of heat exchangers.

1.6.3 Future research and development

Research could focus on creating new materials with even better thermal conductivity, durability, and lower environmental impact. Advancements in nanomaterials or composite materials could lead to significant improvements in heat exchanger performance. Continued innovation in heat exchanger design, such as more efficient microchannel configurations or new surface treatments, could further enhance heat transfer rates and reduce system sizes. Developing technologies that self-clean or resist fouling could solve maintenance issues and extend the lifespan of heat exchangers. These innovations could reduce manual cleaning and improve efficiency. Finding better ways to integrate heat exchangers with renewable energy systems, like solar or geothermal, could improve the performance and efficiency of these systems. Advances in heat exchanger technology could lead to better energy capture and use in various renewable applications.

1.7 CONCLUSION

In summary, the progress in heat exchanger technology requires addressing technical challenges, weighing economic costs against long-term gains, and tackling environmental issues. Innovations such as microchannel designs, smart control systems, and PCMs are improving performance and efficiency. Nonetheless, continuous research and development are crucial to overcome problems like fouling, high pressure drops, and sustainability concerns. Future advancements in materials, artificial intelligence, and nanotechnology are expected to further enhance heat exchangers, making them more efficient, cost-effective, and eco-friendly. As technology evolves, it will remain vital across diverse industrial applications.

REFERENCES

1. Zhao, Y., & Wu, J. (2020). "Renewable energy sources and their integration into power systems: A review." Renewable and Sustainable Energy Reviews, 120, 109587. https://doi.org/10.1016/j.rser.2019.109587.
2. Bertsch, S. S., & Thome, J. R. (2021). "Heat exchanger technologies in renewable energy systems: Current advances and future directions." Energy, 223, 119979. https://doi.org/10.1016/j.energy.2021.119979.

3. Meena, C. S., Prajapati, A. N., Kumar, A., & Kumar, M. (2022). "Utilization of solar energy for water heating application to improve building energy efficiency: An experimental study." Buildings, 12, 2166. https://doi.org/10.3390/buildings12122166.
4. Meena, C. S., Kumar, A., Roy, S., Cannavale, A., & Ghosh, A. (2022). "Review on boiling heat transfer enhancement techniques." Energies, 15, 5759. https://doi.org/10.3390/en15155759.
5. Adams, T., & Brown, A. (2018). "Energy storage systems and their role in mitigating the intermittency of renewable energy sources." Journal of Renewable Energy, 135, 535–550. https://doi.org/10.1016/j.renene.2018.01.040.
6. Carrasco, J. M., Franquelo, L. G., Bialasiewicz, J. T., Galván, E., PortilloGuisado, R. C., Prats, M. M., ... & Moreno-Alfonso, N. (2006). Power-electronic systems for the grid integration of renewable energy sources: A survey. IEEE Transactions on industrial electronics, 53(4), 1002–1016, DOI: https://doi.org/10.1109/TIE.2006.878356.
7. Thangavel, S., Kumar, A., & Heidari, M. (2024). Strategic and Technical Analysis of Sustainable Smart Cities Development in Oman (Chapter 8, pp. 156–184). Boca Raton, FL: CRC Press. https://doi.org/10.1201/9781003496656-8.
8. Chitt, M., Thangavel, S., Kumar, A., Singh, V. P., & Swaroop, C. S. (2024). Building Performance Prediction and Optimization (Chapter 11, pp. 241–257). Boca Raton, FL: CRC Press. https://doi.org/10.1201/9781003496656-11.
9. García, I., & Gómez, A. (2019). "Decentralized energy systems: Benefits and challenges of local generation." Energy Policy, 129, 1123–1131. https://doi.org/10.1016/j.enpol.2019.03.005.
10. Kumar, R., & Patel, P. (2021). "Innovations in renewable energy technologies: Enhancing efficiency and sustainability." Renewable and Sustainable Energy Reviews, 143, 110875. https://doi.org/10.1016/j.rser.2021.110875.
11. Kalogirou, S. A. (2009). "Solar thermal collectors and applications." Progress in Energy and Combustion Science, 35(3), 231–295. https://doi.org/10.1016/j.pecs.2008.10.001.
12. Giebel, G., & Mann, J. (2012). "Wind turbine performance and cooling systems: A review of current technology." Renewable and Sustainable Energy Reviews, 16(8), 5850–5864. https://doi.org/10.1016/j.rser.2012.05.036.
13. Tiwari, S., Swaminathan, M., Harender, & Singh, D. B. (2022). "Performance enhancement of the photovoltaic system with different cooling methods." Environmental Science and Pollution Research, 29(30), 45107–45130. https://link.springer.com/article/10.1007/s11356-022-20330-x
14. Kumar, A., Thangavel, S., Awasthi, M. K., Singh, V. P., & Dutt, N. (2024). Safety and Environmental Considerations in Nanofluid-Based Systems (Chapter 6, pp. 99–122). Boca Raton, FL: CRC Press. https://doi.org/10.1201/9781003494454-6.
15. Manji, A. N. S. H. A., Thangavel, S., Kumar, A., & Heidari, M. (2024). A Study on Waste Management Practices in Oman and a Way Forward to a Circular Economy (Chapter 6, pp. 129–149). Boca Raton, FL: CRC Press. https://doi.org/10.1201/9781003502012-6.
16. Razaviyn, Z., & Heidari, M., Marine (2024). "Energy harvesting from fluid flow via vortex induced vibrations." International Journal of Thermofluids (In Press).
17. Lund, J. W., Freeston, D. H., & Boyd, T. L. (2010). "Direct utilization of geothermal energy 2010 worldwide review." Geothermics, 39(3), 159–180. https://doi.org/10.1016/j.geothermics.2010.07.001.
18. Hofbauer, H., & Schwarzkopf, R. (2013). "Biomass combustion and gasification: Technologies and applications." Biomass and Bioenergy, 51, 1–8. https://doi.org/10.1016/j.biombioe.2012.09.007.
19. Kaswan, M. S., Rathi, R., Cross, J., Garza-Reyes, J. A., Antony, J., & Yadav, V. (2022). "Integrating green lean six sigma and industry 4.0: A conceptual framework." Journal of Manufacturing Technology Management, 34(1), 87–121. http://dx.doi.org/10.1108/JMTM-03-2022-0115.

20. Sharma, S. K., Mallick, A., Singh, D. B., & Tiwari, G. N. (2022). "Experimental study of solar energy–based water purifier of single-slope type by incorporating a number of similar evacuated tubular collectors." Environmental Science and Pollution Research, 29(5), 6837–6856. http://dx.doi.org/10.1007/s11356-021-16123-3.
21. Razaviyn, Z., Heidari, M., Thangavel, S., Verma, V., Kumar, A., & Yadav, A. K. (2025). "Marine energy harvesting from fluid flow via vortex induced vibrations." International Journal of Thermofluids, 25. https://doi.org/10.1016/j.ijft.2024.101015.
22. Pimoli, D. S., Tomar, P., Kumar, A., Galindo, M. V., Valencia, M., Dwivedi, G., & Shukla, A. (2024). "Experimental investigation of friction factor and heat transfer enhancement using circular perforated twisted tape inserts in heat exchangers." International Journal of Energy for a Clean Environment. https://doi.org/10.1615/InterJEnerCleanEnv.2024053616.
23. Lundberg, J., & Wellinger, A. (2011). "Combined heat and power (CHP) systems: Efficiency improvements and technological advancements." Energy, 36(12), 7373–7384. https://doi.org/10.1016/j.energy.2011.06.016.
24. Neto, J. P., & Castro, M. A. (2020). "Economic and environmental impacts of combined heat and power (CHP) systems in industrial applications." Journal of Cleaner Production, 252, 119747. https://doi.org/10.1016/j.jclepro.2019.119747.
25. Zhang, X., & Wang, Z. (2018). "Advanced materials for heat exchangers: A review of recent developments." Renewable and Sustainable Energy Reviews, 82, 1598–1624. https://doi.org/10.1016/j.rser.2017.09.066.
26. Wang, H., & Zhang, X. (2019). "Phase change materials in heat exchangers: Applications and innovations." Applied Thermal Engineering, 147, 1034–1049. https://doi.org/10.1016/j.applthermaleng.2018.09.051.
27. Gharbi, M., & Aissani, K. (2021). "Microchannel heat exchangers: Recent advances in design and applications." International Journal of Heat and Mass Transfer, 165, 120817. https://doi.org/10.1016/j.ijheatmasstransfer.2020.120817.
28. Kang, H., & Chen, X. (2020). "Enhanced surfaces in heat exchangers: Design and performance." Energy Conversion and Management, 207, 112591. https://doi.org/10.1016/j.enconman.2020.112591.
29. Awasthi, M. K., Dutt, N., Kumar, A., & Kumar, S. (2024). "Electrohydrodynamic capillary instability of Rivlin–Ericksen viscoelastic fluid film with mass and heat transfer." Heat Transfer, 53(1), 115–133. https://doi.org/10.1002/htj.22944.
30. Verma, V. (2023). Ground and solar assisted heat pump systems for space heating and cooling applications in the northern region of India – A study on energy and CO2 saving potential, SETA, 59, 103405, ISSN 2213–1388, https://doi.org/10.1016/j.seta.2023.103405.
31. Bhutta, M. M. A., Hayat, N., Bashir, M. H., Khan, A. R., Ahmad, K. N., & Khan, S. (2012). CFD applications in various heat exchangers design: A review. Applied Thermal Engineering, 32, 1–12, https://doi.org/10.1016/j.applthermaleng.2011.09.001.
32. Cengel, Y. A., & Turner, R. H. (2020). Fundamentals of Thermal-Fluid Sciences. McGraw-Hill Education.
33. Khan, M. G., & Fartaj, A. (2011). A review on microchannel heat exchangers and potential applications. International journal of energy research, 35(7), 553–582, https://doi.org/10.1002/er.1720.
34. Sharma, K. K., Kumar, R., & Jha, P. K. (2024). Physical modeling and numerical investigation of fluid flow and solidification behavior in a slab caster mold using hexa-furcated nozzle. Heat Transfer, 53(3), 1050–1072. https://doi.org/10.1002/htj.22980.
35. Chakrabarty, D., Patil, P., & Sharma, K. (2017). Experimental study on fluid flow and heat transfer from a rectangular prism approaching the wall of a wind tunnel. Heat Transfer—Asian Research, 46(7), 793–814. https://doi.org/10.1002/htj.21244.

36. Lodhi, S. K., Hussain, H. K., & Hussain, I. (2024). Using ai to increase heat exchanger efficiency: an extensive analysis of innovations and uses. International Journal of Multidisciplinary Sciences and Arts, 3(4), 1–14, DOI: https://doi.org/10.47709/ijmdsa.v3i4.4617.
37. Dos-Reis-Delgado, A. A., Carmona-Dominguez, A., Sosa-Avalos, G., Jimenez-Saaib, I. H., Villegas-Cantu, K. E., Gallo-Villanueva, R. C., & Perez-Gonzalez, V. H. (2023). Recent advances and challenges in temperature monitoring and control in microfluidic devices. Electrophoresis, 44(1-2), 268–297, https://doi.org/10.1002/elps.202200162.
38. Kim, Y., & Lee, S. (2021). "Advances in real-time monitoring and data analysis for heat exchangers." International Journal of Heat and Mass Transfer, 172, 121160. https://doi.org/10.1016/j.ijheatmasstransfer.2021.121160.
39. Awasthi, M. K., Kumar, A., & Dutt, N. (2024). Modeling Rayleigh-Taylor Instability in Nanofluid Layers (Chapter 13, pp. 249–262). Boca Raton, FL: CRC Press. https://doi.org/10.1201/9781003465171-13.
40. Saxena, A. (2023). Water Consumption Optimization of Hybrid Heat Pump Water Heating System. Lecture Notes in Mechanical Engineering. Springer. https://doi.org/10.1007/978-981-99-1894-2_61.
41. Ahmed, G. B., Hossain, M. N., & Kundu, A. (2024). "Novel design and performance evaluation of a multi-stepped absorber plate fin in a flat plate solar collector." International Journal on Interactive Design and Manufacturing (IJIDeM) (In Press).
42. Li, H., & Wang, J. (2023). "Machine learning for adaptive control of heat exchangers: Techniques and applications." Chemical Engineering Science, 263, 118359. https://doi.org/10.1016/j.ces.2022.118359.
43. Singh, V. P., Jain, S., Karn, A., Dwivedi, G., Kumar, A., Mishra, S., ... & Kamel, S. (2022). "Heat transfer and friction factor correlations development for double pass solar air heater artificially roughened with perforated multi-V ribs." Case Studies in Thermal Engineering, 39, 102461. https://doi.org/10.1016/j.csite.2022.102461.
44. Patil, P. P., Gori, Y., Kumar, A., & Tyagi, M. R. (2021). "Experimental analysis of tribological properties of polyisobutylene thickened oil in lubricated contacts." Tribology International, 159, 106983. https://doi.org/10.1016/j.triboint.2021.106983.
45. Bhandari, P., Padalia, D., Ranakoti, L., Khargotra, R., András, K., & Singh, T. (2023). "Thermo-hydraulic investigation of open micro prism pin fin heat sink having varying prism sides." Alexandria Engineering Journal, 69, 457–468. https://doi.org/10.1016/j.aej.2023.02.016.
46. Sathish, T., Saravanan, R., Jeyanthi, C. E., Sabariraj, R. V., Hemachandu, P., Ağbulut, Ü., ... Soudagar, M. E. M. (2023). "PCM embedded square channel receiver for effective operation of net zero Emission/Energy PDC building heating system." Sustainable Energy Technologies and Assessments, 60, 103519, http://dx.doi.org/10.1016/j.seta.2023.103519.
47. Sharma, P., & Singh, R. (2022). "Emerging materials and design innovations for heat exchangers: A comprehensive review." Renewable and Sustainable Energy Reviews, 155, 111973. https://doi.org/10.1016/j.rser.2021.111973.
48. Zhao, L., & Wu, D. (2023). "Innovations in heat exchanger design: Advances in microchannel configurations and surface treatments." International Journal of Thermal Sciences, 175, 107585. https://doi.org/10.1016/j.ijthermalsci.2022.107585.
49. Awasthi, M. K., Kumar, A., Dutt, N., & Singh, S. (Editors). (2024). Computational Fluid Flow and Heat Transfer: Advances, Design, Control & Applications. Boca Raton, FL: CRC Press. https://doi.org/10.1201/9781003465171.
50. Kumar, A., Dutt, N., & Kumar Awasthi, M. (2024). Heat Transfer Enhancement Techniques: Thermal Performance, Optimization and Applications. Publisher Wiley Scrivener.
51. Kundu, A., Kumar, A., Kumar Awasthi, M., Dwivedi, G., & Kumar Shukla, A. (2024). "An explorative study of earth-air heat exchanger for sustainable building applications." International Journal of Energy for a Clean Environment. https://doi.org/10.1615/InterJEnerCleanEnv.2024052513.

52. Dutt, N., Hedau, A., Kumar, A., Awasthi, M. K., & Singh, V. P. (2024). Effect of duct length variation on solar air heater performance for smooth and D-shaped roughened absorber plate. Heat Transfer, 53(7), 3902–3930, https://doi.org/10.1002/htj.23115.
53. Dutt, N. (2024). "Effect of duct length variation on solar air heater performance for smooth and D-shaped roughened absorber plate." Heat Transfer. https://doi.org/10.1002/htj.23115.
54. Singh, K., Pandey, A. K., & Kumar, M. (2021). "Numerical solution of micropolar fluid flow via stretchable surface with chemical reaction and melting heat transfer using Keller-Box method." Propulsion and Power Research, 10(2), 194–207. http://dx.doi.org/10.1016/j.jppr.2020.11.006.
55. Kaur, H., Rai, K. N., & Upadhyay, S. (2022). "A numerical study of moving boundary problem involving dual phase lag model of heat mass transfer during immersion frying." Mathematics and Computers in Simulation, 202, 79–100. http://dx.doi.org/10.1016/j.matcom.2022.05.025.
56. Dutt, N. (2024). "Thermo-hydraulic performance investigation of solar air heater duct having staggered D-shaped ribs: Numerical approach." Heat Transfer, 53(3), 1501–1531. https://onlinelibrary.wiley.com/doi/10.1002/htj.22998.
57. Dutt, N. (2023). "Thermo-hydraulic performance of solar air heater having discrete D-shaped ribs as artificial roughness." Environmental Science and Pollution Research. https://doi.org/10.1007/s11356-023-28247-9.
58. Reddy, P. N., Verma, V., Kumar, A., & Awasthi, M. (2023). "CFD simulation and thermal performance optimization of flow in a channel with multiple baffles." Journal of Heat and Mass Transfer Research, 10(2), pp. 257–268. https://doi.org/10.22075/JHMTR.2023.31108.1458.
59. Awasthi, M. K., Dutt, N., Kumar, A., & Hedau, A. (2024). Introduction to Mathematical and Computational Methods (Chapter 1, pp. 01–20). Boca Raton, FL: CRC Press. https://doi.org/10.1201/9781003465171-1.
60. Mishra, A., Rawat, S. K., Yaseen, M., & Pant, M. (2023). Development of machine learning algorithm for assessment of heat transfer of ternary hybrid nanofluid flow towards three different geometries: Case of artificial neural network. Heliyon, 9(11). http://dx.doi.org/10.1016/j.heliyon.2023.e21453.
61. Saxena, A., Cuce, E., Singh, D. B., Sethi, M., Cuce, P. M., Sagade, A. A., & Kumar, A. (2023). Experimental studies of latent heat storage based solar air heater for space heating: A comparative analysis. Journal of Building Engineering, 69, 106282. http://dx.doi.org/10.1016/j.jobe.2023.106282.
62. Saxena, A., Joshi, S. K., Gupta, P., Tirth, V., Suryavanshi, A., Singh, D. B., & Sethi, M. (2023). An experimental comparative analysis of the appropriateness of different sensible heat storage materials for solar cooking. Journal of Energy Storage, 61, 106761. http://dx.doi.org/10.1016/j.est.2023.106761.
63. Pansare, R., Yadav, G., Garza-Reyes, J. A., & Raosaheb Nagare, M. (2023). Assessment of sustainable development goals through industry 4.0 and reconfigurable manufacturing system practices. Journal of Manufacturing Technology Management, 34(3), 383–413. http://dx.doi.org/10.1108/JMTM-05-2022-0206.
64. Kundu, A., Kumar, A., Dutt, N., Singh, V. P., & Meena, C. S. (2023). Modelling and Simulation of Thermal Energy System for Design Optimization (Chapter 7, pp. 103–140). Boca Raton, FL: CRC Press. http://dx.doi.org/10.1201/9781003395768-7.
65. Singh Pimoli, D., & Tomar, P. (2024). "Experimental investigation of friction factor and heat transfer enhancement using circular perforated twisted tape inserts in heat exchangers." International Journal of Energy for a Clean Environment (In Press).
66. Mokhtari, M., & Jamei, S. (2024). "Numerical and hydrodynamic analysis of a container vessel due to the change of/collision wave angle at different speeds in regular wave." Journal of Mechanical Engineering and Sciences, 18(4), pp. 10247–10260. https://doi.org/10.15282/jmes.18.4.2024.3.0809.

Chapter 2

Fundamentals of thermal management in renewable energy systems

R.G. Abaszade and E.A. Khanmamadova

2.1 INTRODUCTION

More than 50% of the total energy used in the world is for heating, and 25% of the produced energy comes from renewable sources (Sánchez-Pantoja et al., 2018). Solar energy has recently received attention as one of the feasible options to meet global demands without causing detrimental harm to the environment, such as global warming (Goswami, 2015). Practical application of solar energy requires reliable thermal storage systems, especially latent heat storage with phase-change materials due to the high energy density and uniform heat release (Molu et al., 2024; Vick et al., 1996). Its enhancement has been the focus of considerable research; among the latent heat storage systems, there have been extensive studies dealing with the performance of ice storage systems under different conditions. In the hybrid energy area, the solar-geothermal systems have exciting potential (Nelson et al., 1996). Solar energy preheats the fluids for the Rankine cycle or assists in water desalination systems like humidification-dehumidification (HDH), which is very helpful in small-scale water supply (Abaszade et al., 2020, 2021, 2022a,b, 2023, 2024a,b; Boychuk et al., 2022; Figarova et al., 2021, 2023; Guliyeva et al., 2023; Ivanichok et al., 2024; Khanmammadova et al., 2024; Mammadov et al., 2022, 2023; McTigue et al., 2020; Mohamed et al., 2020; Moradi et al., 2024; Sharma et al., 2024).

This thermal storage optimization cannot be accomplished without exergy analysis, which considers the loss of energy or heat temperature in order to provide a proper performance measure beyond what basic energy analysis could allow (Krane, 1987). In hydrogen energy, various nanomaterials, including carbon nanotubes and graphene, provide enhanced catalytic efficiency in fuel cells, and provide improvements in storage and transport systems (Kumar, 2024). The use of fossil fuels is underpinned by environmental issues such as acid rain and the increase in cancer rates, and echoes the climate objectives of net-zero emissions by 2050 (Luo et al., 2022; Meena et al., 2024; Nachtane et al., 2023). The energy crisis and resulting energy poverty remain critical in rural areas, given implications for social equity and economic access when there has been little access to more traditional forms of energy (Sharma et al., 2022).

DOI: 10.1201/9781003534785-2

2.2 RENEWABLE ENERGY SOURCES

2.2.1 Thermal power stations

Thermal energy storage (TES) in solar, geothermal, and hydrogen energy may provide a promising path to low-carbon, sustainable energy. This could add to reduce greenhouse gas emissions, improve energy security, and provide economic growth. Solar energy utilizes the energy of the Sun through versatile applications in thermal technology. Geothermal energy provides reliable power right from the Earth's core; further development of this resource is made possible through enhanced geothermal system (EGS). Hydrogen as an energy carrier has flexible storage options and various applications. In integrating these energies, intermittency and land and water uses become challenges besides being expensive. Research investment, innovation, and infrastructure building in TES continuously take place, while solar thermal plants serve industrial segments related to food, chemicals, and textiles. Flat plate collectors are efficient at low temperatures and, hence, are mainly used in various processes that require less heat (Nkwetta and Smythe, 2012) (Figure 2.1). Flat plate collectors have normally been recognized as being efficient at low temperatures and, hence, are popularly used for processes that require less heat.

Present industrial civilization is based upon conventional energy sources and involves large and relatively inefficient energy conversion systems (Abhat, 1983). Growing concerns over conventional fuel supplies and environmental consequences, notably greenhouse gas emissions with their impacts on climate change, have led to increased interest in renewable energy (RE) sources for electricity and heat production. It was a review that investigated the integration of renewable energy sources (RES) with TES systems, giving a focus on recent developments in both advanced and new approaches. Different forms

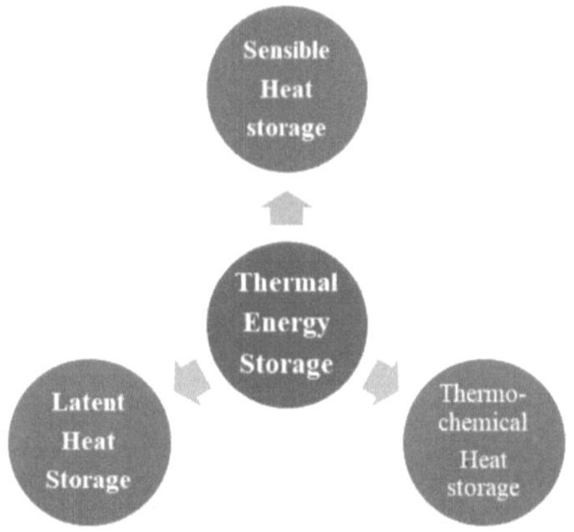

Figure 2.1 Schematic diagrams of thermal energy storage.

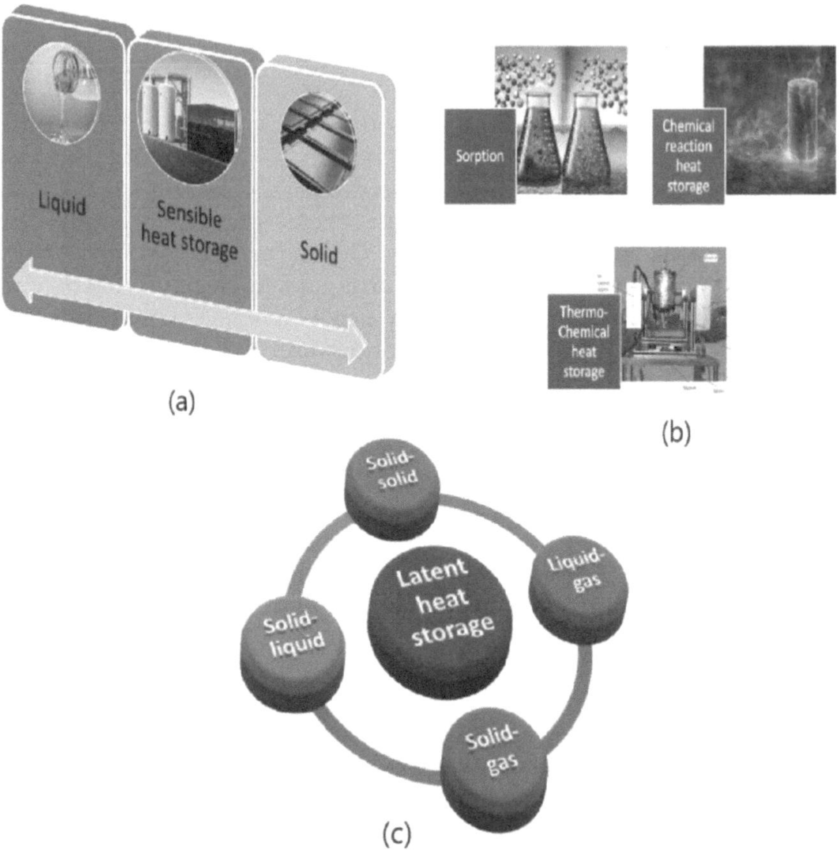

Figure 2.2 Different forms of renewable energy sources. (a) Forms of energy capture related to fluid, sensible heat storage, and solidity. (b) Sorption, heat storage through chemical reaction, forms of thermal-chemical heat storage. (c) Schematic representation of various basic energy sources.

of RES such as solar, wind, geothermal, and ocean are coupled to sensible and latent TES systems (Figure 2.2).

The present review outlines the advantages and issues related to the integration of the RES-TES system. It mentioned that further hybridization of more renewable power sources could make up for their deficiencies. The review presented discussions on different forms of useful outputs from integrated systems: hydrogen production, electrical power, and freshwater. Its general relevance to the global energy demand and resource problems was underlined, showing how these hybrid systems can replace sources based on fossil fuels. This review provides an in-depth overview of the current status regarding RES-TES integration and also gives several views on future development related to the optimum use of RES (Elkhatat and Al-Muhtaseb, 2023).

Present industrial civilization is based upon conventional energy sources and involves large and relatively inefficient energy conversion systems. Growing concerns over conventional fuel supplies and environmental consequences, notably greenhouse gas emissions with their impacts on climate change, have led to increased interest in RE sources for electricity and heat production. It was a review that investigated the integration of RES with TES systems, giving a focus on recent developments in both advanced and new approaches. Different forms of RES such as solar, wind, geothermal, and ocean are coupled to sensible and latent TES systems (Figure 2.2).

Industry accounts for 37% of global energy consumption, with over 80% of this coming from fossil fuels. Solar heat can replace fossil fuels in many industrial processes, but integrating TES systems is crucial for sustainability. Currently, only 741 solar heat industrial plants exist, covering a small portion of global capacity (Abhishek and Saini, 2020). However, significant potential remains untapped (Figure 2.3). The main challenge is developing cost-effective TES systems and materials. Sensible TES, the oldest and most developed method, is gaining interest due to growing demand for sustainability (Figure 2.4). This review explores the latest trends in sensible TES for solar industrial applications, focusing on sustainability and cost-effective solutions to meet the industrial sector's increasing energy needs (Burcu et al., 2020).

In the present work, energy and exergy analyses were conducted on an ice-on-coil TES system during its charging phase, using a thermal resistance network model (Kuravi et al., 2013). Key metrics, including stored energy, ice mass, and heat transfer fluid temperature, were validated against

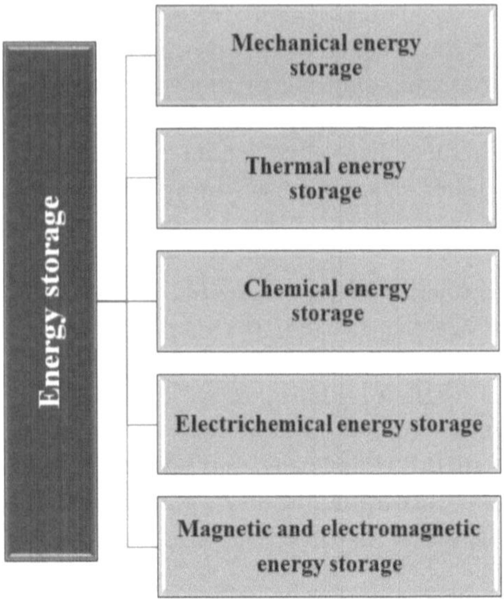

Figure 2.3 An overview of the different energy storage methods.

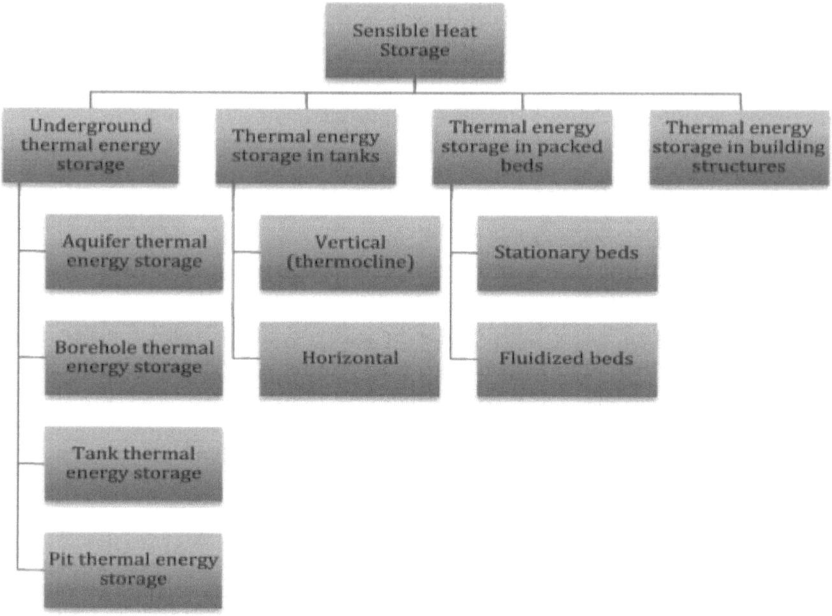

Figure 2.4 Classification of sensible heat storage technologies.

experimental data, showing that TES performance depends on specific working and design parameters. Results reveal that energy storage capacity and efficiency improve with lower inlet temperatures and longer tube lengths, while exergy efficiency benefits from higher inlet temperatures. The study also explores RE applications, proposing a novel trigeneration system combining geothermal and solar energy in a transcritical CO_2 (TRCC) cycle. This system efficiently produces power, hydrogen, and freshwater, with maximum outputs achieved in May. Under optimized conditions, the system demonstrates higher efficiency and lower costs compared to the Kalina cycle-based systems (Ezan et al., 2011; Li et al., 2022; Mossa et al., 2018; Verma et al., 2008).

A mathematical model has been presented by Dwarka and Kim in order to comparatively assess the thermal performance of randomly mixed and laminated phase change material (PCM) drywall system. The mathematical model was based on implicit enthalpy method, and the governing equations for both randomly mixed and laminated PCM systems have been solved using the finite-difference method. For randomly mixed PCM systems, the differential energy equation in rectangular coordinates (x, y, and z) is expressed as (Dwarka and Kim, 2005):

$$\frac{\partial}{\partial x}\left(K\frac{\partial T}{\partial x}\right)+\frac{\partial}{\partial y}+K\left(\frac{\partial T}{\partial y}\right)+\frac{\partial}{\partial z}\left(K\frac{\partial T}{\partial z}\right)=\rho\frac{\partial h}{\partial t}=\rho\frac{\partial}{\partial t}\int CdT \tag{2.1}$$

While Equation (2.1) is used for laminated PCM systems, the general energy equation reduces to one-dimensional due to small relative thickness. The surface temperature variations for the samples are obtained through a series of computer simulations with different sets of thermo-physical properties and environmental conditions. The major significant benefit that the laminated PCM wallboard systems have over those that were randomly mixed PCM types is the improved thermal performance and faster heat-transfer rates under narrow temperature swing. For example, the maximum instantaneous increase in heat flux was obtained between 20% and 50% higher during the phase-change process and about 18% more in their heat storage and release capabilities. In any case, experimental verification should be performed in order to validate and further develop the laminated system.

A design of thermodynamic analysis of TES systems using multiple PCMs is developed. Relative merits are investigated with respect to thermal exergy storage systems using two, three, as well as five PCMs. Energy efficiency is observed to improve much better using multiple PCMs compared to single PCM in a system (Dwarka and Kim, 2005):

$$\beta_c^1 = \frac{W_c^1}{W_T} = \frac{\left(1-e^{-N_c}\right)\left(T_{ci}-T_m\right)\left(1-\left(\frac{T_a}{T_m}\right)\right)}{T_{ci}-T_a-T_a ln\left(\frac{T_{ci}}{T_a}\right)} \tag{2.2}$$

The exergy efficiency of the discharge process, given in (2.2), defined by the ratio of the rate of exergy gain of the heat transfer fluid (HTF) and the rate of exergy release by the PCM through the discharge process for single PCM system is given by (Dwarka and Kim, 2005):

$$\theta_d^1 = \frac{\varpi_d^1}{\varpi_{Td}^1} = \frac{\theta_{d0}^1 - \theta_{di} ln\left(\theta_{d0}^1/\theta_{di}\right)}{\left(1-e^{-N_d}\right)\left(\theta_m-\theta_{di}\right)\left(1-\left(1/\theta_m\right)\right)} \tag{2.3}$$

This chapter examines the exergy efficiency of thermal storage systems, which is influenced by the number of PCMs, heat transfer units, and inlet temperatures during charge and discharge processes (Jegadheeswaran et al., 2009). Latent heat thermal energy storage (LHTES) systems store energy from solar or other sources, with a focus on PCMs. The chapter reviews mathematical modeling methods based on the first and second laws of thermodynamics. While the first law has been extensively studied, the second law (exergy analysis) offers deeper insights into thermodynamic efficiency, though further validation is needed (Devrim et al., 2015).

LHTES systems, especially those using PCMs, face limitations due to poor thermal conductivity, requiring enhancement techniques for performance optimization (Li, 2015). The chapter discusses the impact of these techniques and the importance of exergy-based performance evaluation. Exergy analysis is crucial for comparing LHTES systems and assessing the effects of operating and design parameters.

Thermoeconomic methodologies are also addressed, emphasizing the growing significance of heat storage systems in solar energy applications. Experimental studies reveal that solar-assisted latent heat storage is feasible during some months but less effective in winter. The study found energy efficiency of 70.4% and exergy efficiency of 2.5%.

Latent heat TES systems have potential for improving energy and exergy performance. Factors such as heat transfer fluid flow, PCM melting temperature, and heat exchanger design are critical for optimization. Further research is needed to enhance heat transfer mechanisms and PCM properties for better system efficiency (Guruprasad et al., 2017). Integrating TES with solar energy can address intermittency and improve energy reliability.

The chapter presented a review of research on closed thermodynamic cycles of ocean thermal energy conversion (OTEC) systems that included a description of both pure and mixture thermodynamic cycles and working fluids on their effects on cycle efficiency. In the case of the pure working fluid, in cycles, the efficiency changes owing to the change in the evaporation and condensation temperature caused by the difference in heat resources.

Methods for efficiency enhancement of cycles with mixture working fluids may include heat recovery of ammonia-depleted solution and its regeneration via a process of intermediate extraction. Moreover, the paper analyzes the performance of the ejector on cycle performance. Lastly, the techniques used for efficiency improvements is discussed and summarized. In a word, the thermodynamic efficiency can be improved by adopting proper working fluids and measures that could increase the utilization rate of ocean thermal energy. The related methods need to be compared and analyzed under the same working conditions to decide which is the best. In 2005, Wang et al. (2005) and Shuai (2006) proposed the closed solar energy reheating OTEC power generation system. Please refer to Figure 2.5 for its cycle schematic diagram. Solar energy was main heat source in this system, which reheat working fluid heated by warm water to superheated state (Shuai, 2006; Wang et al., 2005).

Similarly, Aydin et al. (2014) explored the use of solar energy as well for the enhancement of cycle efficiency. Based on the Rankine cycle, they made a comparative study of the following two schemes. Although in the first scheme, the solar energy had been used to heat the warm water, increasing the temperature of the heat source, in the second, it had been used to directly superheat the working fluid to superheated state at the inlet of the turbine. The second scheme is as shown in Figure 2.6. Comparing the results, it can be seen that the respective net thermal efficiencies are 1.9% and 3% for the first and second cycles under identical conditions. It is also by now established that the solar collector area in the second cycle was relatively small.

In 2017, Idrus et al. (2017) proposed the idea of geo-ocean thermal energy conversion as shown in Figure 2.7. The working fluid was first preheated and

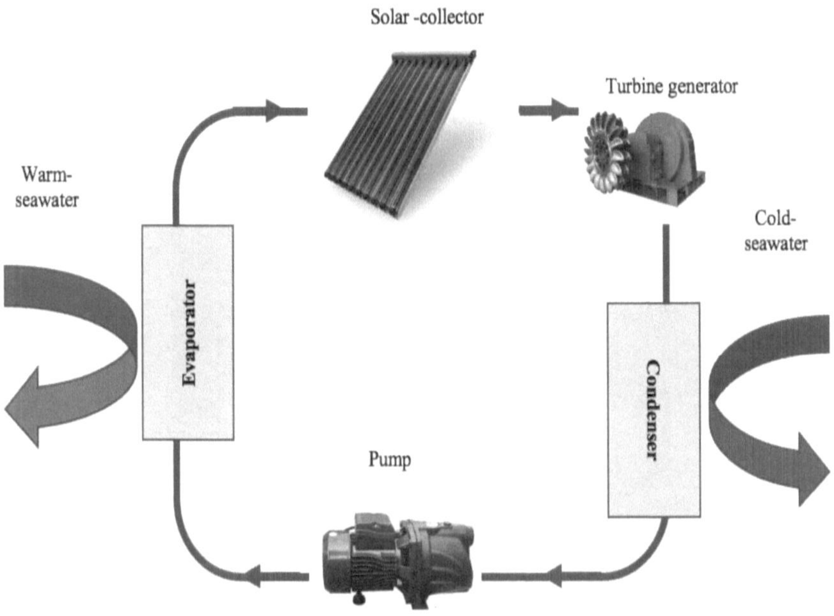

Figure 2.5 A solar energy-reheating ocean thermal energy conversion cycle.

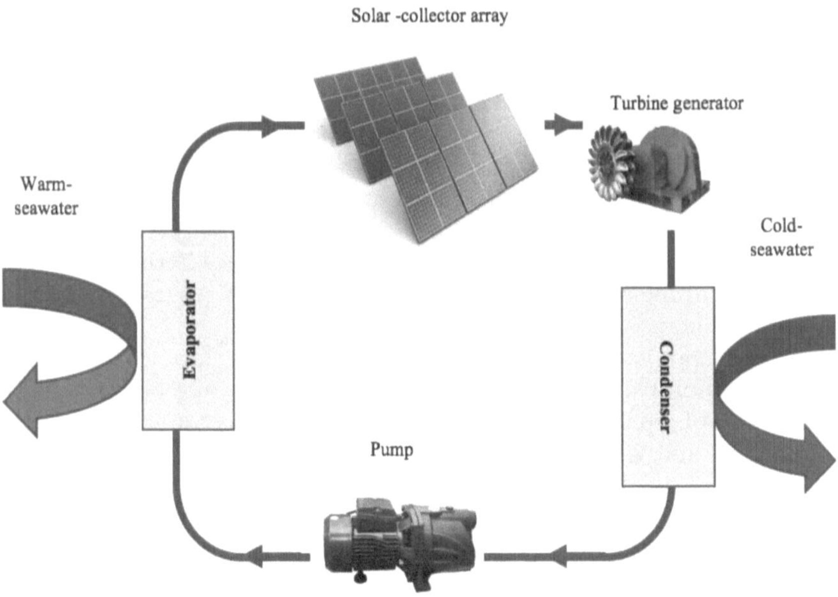

Figure 2.6 A diagram of solar-boosted ocean thermal energy conversion.

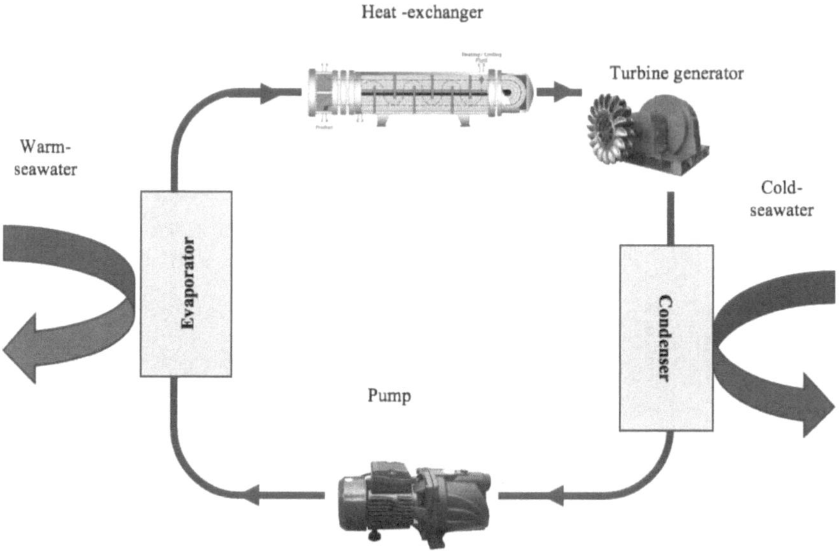

Figure 2.7 A diagram of geo-ocean thermal energy conversion.

vaporized by warm ocean water in the evaporator and then further superheated by geothermal energy.

Afterwards, the superheated steam drove the turbine to rotate and finally condensed in the condenser. The working fluid was pumped to the evaporator, and the cycle was finally completed. Their results indicated that the cycle thermal efficiency reached 4.61% when the ammonia temperature at the inlet of the turbine was 80°C. Furthermore, based on their estimation, the GeOTEC device cost was clearly lower than that of OTEC.

In 2014, Yoon et al. designed a new OTEC power cycle using ammonia as its working fluid; please refer to its flow diagram depicted in Figure 2.8. In this cycle, the superheated vapor at the outlet of evaporator passed through Turbine 1, which drove the generator to generate electricity. After that, the exhausted gas passed through the regenerator and cooler in turns to release the heat. Then, the cooled ammonia working fluid was separated into saturated gas and liquid in the separator.

Yuan et al. proposed in 2014 that the use of two ejectors would increase the condensation temperature required for the cycle so that both the power consumption of a pump for cold water and the pipeline cost of the water get reduced (Soto and Vergara, 2014).

2.2.2 Wind energy

Wind turbines operate by the power of the wind that drive an electrical generator. Wind over the blades forms areas of lower and higher air pressure, creating a turning force. The turning blades rotate a shaft inside the nacelle

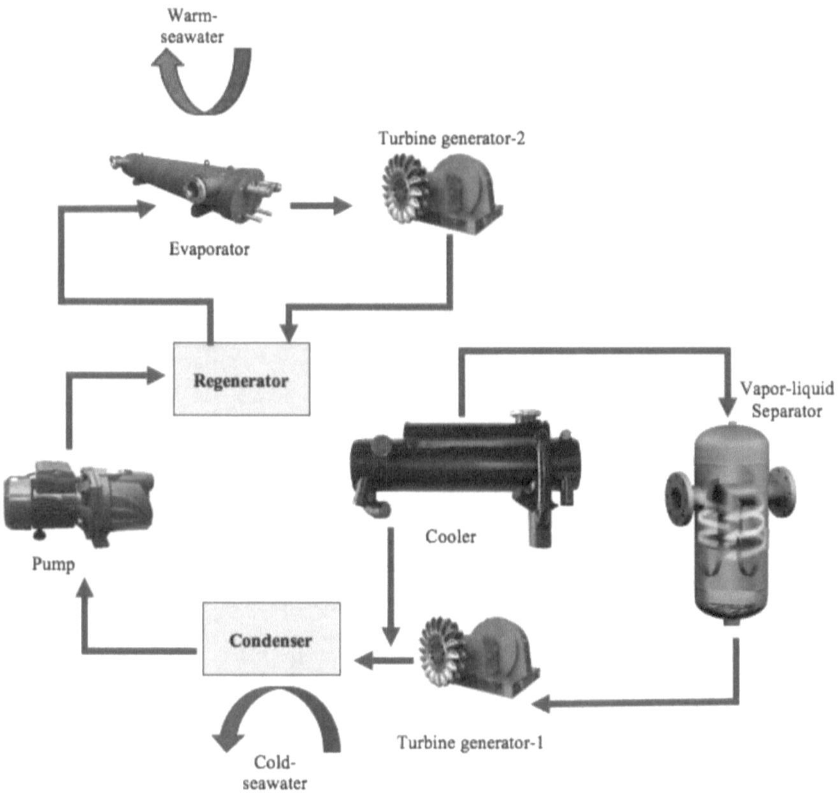

Figure 2.8 R717 OTEC power cycle.

that goes into a gearbox. The gearbox steps up the rotational speed to that appropriate for the generator. Using magnetic fields, the generator converts rotational energy to electrical energy. The power output feeds into a transformer, which steps up the electrical energy generated by the generator—typically generated at approximately 700 V—to the correct transmission voltage, usually 33 kV. A wind turbine draws kinetic energy from the swept area of the blades (Figure 2.9).

The power developed from the wind is given by

$$P_{air} = \frac{1}{2}\rho A \vartheta^3$$

where

ρ = air density (approximately 1.225 kgm^{-3})
A = swept area of rotor, m^2
ϑ = upwind free wind speed, ms^{-1}

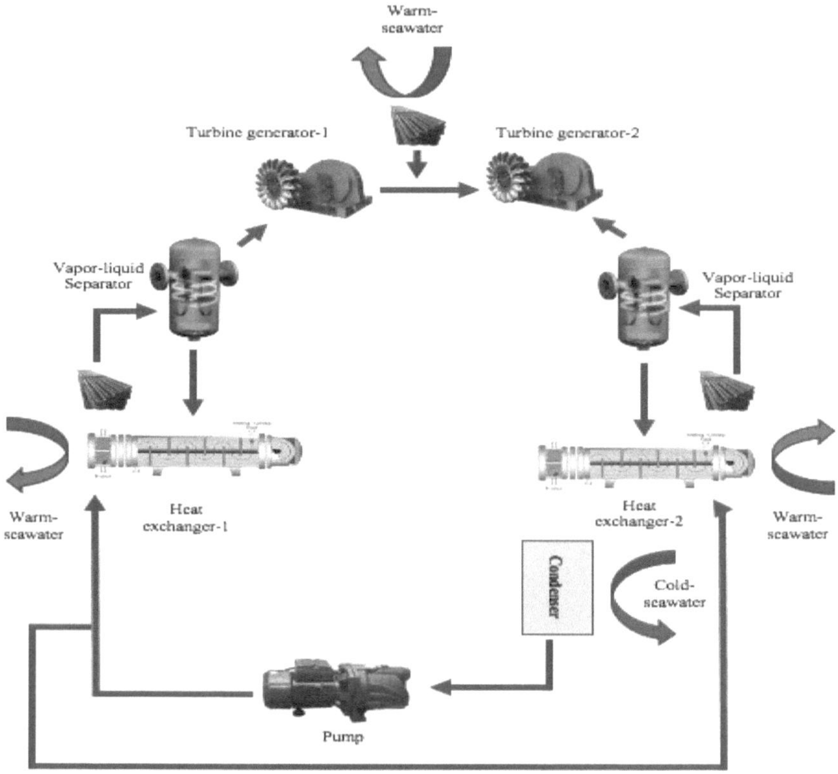

Figure 2.9 The absorption power cycle.

Although the above equation gives the power available in the wind, the power transferred to the wind turbine rotor is reduced by the power coefficient to the wind turbine rotor is reduced by the power coefficient, C_p:

$$C_p = \frac{P_{\text{wind turbine}}}{P_{\text{air}}}$$

$$P_{\text{wind turbine}} = C_p P_{air} = C_p \frac{1}{2} \rho A \vartheta^3$$

A maximum value of C_p is defined by the Betz limit, which states that a turbine can never extract more than 59.3% of the power from an air stream. In reality, wind turbine rotors have maximum C_p values in the range 25–45%. It is also conventional to define a tip-speed ratio, λ, as

$$\lambda = \frac{\varpi R}{\vartheta}$$

where ω = rotational speed of rotor; R = radius to tip of rotor; v = upwind free wind speed. The tip-speed ratio, λ, and the power coefficient, C_p, are dimensionless and so can be used to describe the performance of any size of wind turbine rotor (Dwarka and Kim, 2005).

2.2.3 Hydrogen energy

This section gives an overview of the intersection of nanomaterials and hydrogen energy, emphasizing a transformational capability. While studies on the use of metal hydrides for hydrogen storage began in the 1970s, recent works in production, storage, and applications are remarkable. Various hydrogen production routes, including water electrolysis and steam reforming, suffer from disadvantages such as waste heat and explosion hazards. Alcohol-based feedstock, such as methanol and ethanol, is quite promising due to its low feedstock cost and environmental friendliness.

Hydrogen storage methods under investigation generally include the storage of compressed gas, liquid hydrogen for aerospace applications, which is very costly, and solid-state storage by metal hydrides, the challenge of whose cost and processes of desorption remain. Fuel cells, although efficient and at zero emission, have gained significant momentum in automotive, marine, and aviation industries due to their high costs and concerns about durability. Hydrogen combustion is foreseen as an auspicious alternative, with its partial power efficiency improvement and better carbon footprint. However, one safety issue related to flammability needs consideration. Hydrogen-fueled internal combustion engines have the potential for more workable solutions regarding material durability and safety (Harish et al., 2022; Jena et al., 2023; Reilly et al., 1971; Xinyuan et al., 2024).

2.2.4 Nanomaterials used for energy storage

In this section, the basics concerning nanomaterials, hydrogen renewable energy systems, and hydrogen energy are studied (Harish et al., 2022). Nanomaterials are materials with dimensions between 1 to 100 nm and exhibit better properties compared to their bulk material (Jia and Lan, 2023; Osman, 2024; Ruska, 1980; Ruess and Vogt, 1948).

Although coloring of glass with metal nanoparticles had been done as early as the fourth century AD, the study of nanomaterials really started to take hold when the transmission electron microscope was invented in 1931.

Advanced inorganic nanomaterials in energy storage have great prospects to further improve Li-ion batteries and supercapacitors through size optimization, morphology, and composition. This review will detail just some of these accomplishments, and those yet to be achieved, so that scaling-up might become industrially possible.

The new integration of photovoltaic and wind systems in developing countries, offering increased reliability and cost benefits compared to single-generation systems, is also discussed in the chapter. While there is a clear set of advantages in this field, challenges still persist; hence, these technologies will require policy support and optimization. Successful deployment in this area is very much dependent on effective government support and subsidies.

The chapter considers how hydrogen energy is to be integrated into daily life, focusing on transport and industry as a cleaner and more efficient energy source. Hydrogen technology development will be an uphill task in such three critical areas: production, storage, and transportation. This will require heavy research and collaboration among governments, enterprises, and research institutions. With the advancement of technology, hydrogen energy may play an important role in the global energy system and contribute to environmental sustainability (Wu et al., 2024).

2.3 ECONOMIC EFFICIENCY OF RENEWABLE ENERGY SYSTEMS

The next section gives a feasibility study of a hybrid energy system for the south coast of KZN (where KZN stands for KwaZulu-Natal): it determines an optimal configuration based on Net Present Cost (NPC) criteria. The WE/Converter/Grid configuration would add 90 kW of wind energy to the site and completely offset the energy costs at the site, while offering off-grid capability. The next best system, which comes with higher energy costs, would be PV/Grid with an NPC of ZAR 235,846.80 and a renewable share of 74%. Through a year, the PV system can generate about 6.588 kWh, while grid purchases are very limited, standing approximately at about 2213 kWh and sales at about 2.154 kWh. This state of the art underpins pollution reduction and economic savings for the various stakeholders involved in renewable projects with significant environmental and economic benefits. It allows wind energy to be integrated in an optimal manner that is both technically and economically feasible. Therefore, this supports further sustainable electricity generation in South Africa and possibly beyond. Challenges for on-grid systems include limited storage and high energy waste, further suggesting that additional research needs to be done toward optimal storage sizing for enhancing stability and cost efficiency and hence managing excess energy production in hybrid systems (Abdulllahi et al., 2024; Aftab et al., 2024; Bedri, 2024; Burhan and Özlem, 2024; Chen, 2024; Mazibuko et al., 2024; Martinez, 2024; Mendoza-Rivera et al., 2023; Ozge et al., 2024; Rawat et al., 2021; Suhanyiova et al., 2023; Widyatmoko et al., 2023; Woo, 2024).

2.4 PROSPECTS OF RENEWABLE ENERGY SYSTEMS IN SOME COUNTRIES OF THE WORLD

2.4.1 United States

This chapter examines the impact of both non-renewable energy consumption, such as gasoline and gas, and renewable-energy consumption, such as solar and wind, along with CO_2 emissions on economic growth (EG) in North America during the period 1966–2020. In this respect, applications were carried out by using the VAR model and Granger causality tests, supported by impulse response functions. Current research shows that energy use is highly diversified among nations due to various levels of industrialization among Canada, the United States, and Mexico. The study brings special insights into North America and gives public policy recommendations toward the promotion of using RE for sustainable growth. (Mendoza-Rivera et al., 2023).

2.4.2 China

By 2040, China's transport system will still be expected to have heavy reliance on non-RE and limited RE. This is where, in the year 2020, the greenhouse gas (GHG) emissions from transport peaked and renewable adoption was required. This chapter further elaborates on the energy-related innovations within internal combustion engines, electric motors, hydrogen, and proposes biodiesel as an alternative fuel. This study now calls for policies that advance hydrogen fuel cell vehicles and clean energy at ports and freight (Bedri, 2024).

2.4.3 Canada

The aim presents the modeling, with the help of HOMER Pro, of the optimum grid-connected RE system for a community in southern Alberta. The "Tri-Brid" of wind, solar, and BESS system yields a competitive levelized cost of energy (LCOE) of 0.0705 CAD/kWh—a cost reduction of about 56%, with an RET fraction realized of 80%. The system modeled in this chapter responds to the global green initiatives that want a future with further integration of renewable sources and advanced battery technologies for greater efficiencies (Aftab et al., 2024).

2.4.4 Mexico

The research reveals the growing interest in MRE and SDG from Mexico, detailing the optimal sites that are considered appropriate for marine RE, taking into consideration environmental, socioeconomic, and social issues. Three regions are apt for MRE development but need biodiversity protection. This paper concludes resource management and community participation in MRE projects (Martinez et al., 2024).

2.4.5 Europe

This chapter presents the progress from non-renewable to RES in the EU-27 with emphasis on emission mitigation and energy poverty. Waste-to-energy is pioneered by Northern Europe, geothermal in Italy, and solar energy developing in colder countries. Collaboration between the various states in the EU is recommended for knowledge and expertise transfer and trade-off for renewable energies. The EU Strategy increased energy independence, protection of the environment, and R&D challenges now faced by the whole world include COVID-19 and the crisis situation in Ukraine (Suhanyiova et al., 2023).

2.4.6 Turkey

This section will depict an integrated multiple criteria decision making (MCDM) model by combining technique for order of preference by similarity to ideal solution (TAOV) and multi-attribute utility theory (MAUT) methods that rank the RE alternatives for Turkey (Burhan et al. 2024). The ranking was done in such a manner that hydroelectricity with low emission and high efficiency gained the highest ranking followed by wind, solar, biomass, and geothermal. Here, it can be mentioned that RE plays the main role in energy independence, environmental goals regarding targets, and job opportunities for Turkey (Durmus et al., 2023).

2.4.7 Southeast Asia

This section reviews the potentials of all forms of RES, including solar, wind, and hydroelectricity, in the Southeast Asia. For example, Thailand, Malaysia, and the Philippines have taken steps to promote solar energy. Hydropower also has vast potential; however, there are several environmental concerns about its development. Though the region has limited wind potential, countries such as Vietnam and the Philippines press ahead with wind energy projects. The paper discusses the role of policies and enforcement strategies in the adoption of RE (Ozge et al., 2024).

2.4.8 Indonesia

This research targets the application of IoT smart computing based DBMS for RE management data considered from 100 sources. The IoT systems were much more reliable and efficient compared to those obtained from conventional methods. This research also calls for adaptive mechanisms with respect to climate change and states that these emerging technologies are an enabler for key sustainable energy management (Shankar, 2023).

2.4.9 Nigeria

This chapter presents the techno-economic viability of wind turbine/battery, photovoltaic/battery, and hybrid systems of RE in Nigeria, conducting a comparison between six locations. The lowest LCOE value was from PV/battery systems, while the cost for wind turbine systems was high. This research provides the most critical output in terms of the economic viability of solar energy over wind, informing off-grid RE deployment.

2.4.10 Kenya

This part undertakes an analytical look at the contribution of RE to Kenya's economy using the Neo-Classical Solow—Swan Growth model. The authors carry out this study with the purpose of generating a two-way relationship between the use of RE and economic growth within an economy. It infers from these results that investments in RE are very crucial in realizing sustainable energy policies, together with policy incentives and demonopolization (Situma et al., 2024).

2.4.11 Brazil

Brazil has realized the increase in energy production, mainly because of its highly developed hydroelectric potential, which has grown by a great measure and contributed much toward the creation of clean energy. The weather ranges from equatorial to semi-arid, and this is influential in its energy outlook. All of the major electricity plants in Brazil comprise hydroelectric, thermoelectric, wind, and solar power plants; of these, 59% are hydroelectric plants. Hydroelectricity, as a source of renewable and clean energy, does present several concerns on its sustainability. The Brazilian river basins total 63.81% of the country's electrical potential. From 2016 to 2020, Brazil's energy matrix became more renewable, encouraging distributed generation—what might be understood as solar panels—alongside consumer autonomy since the energy would no longer be transmitted from centralized sources (Cruvinel da Sousa Mello et al., 2023).

2.4.12 India

Compared to past decades, the energy production in India has been on the increase; it is concentrated on hydroelectric power, which accounts for 59% of its electricity generation. The energy matrix is turning renewable; due to such changes, distributed generation and the use of solar panels have been developed. At the same time, hydroelectric power raises serious sustainability concerns (Singh et al., 2024).

2.4.13 Azerbaijan

Development of renewable sources of energy has also acquired a significant importance for Azerbaijan. As it is known, the demand for energy resources has increased day by day in recent years. In this respect, this is very important for determining the possibilities in using alternative energy sources. Selection of these sources is important for solving the energy demand. Currently, Azerbaijan is in the epicenter of world attention because of its rich hydrocarbon deposits. Its territory also has 9 of the 11 climate zones of the world, and hence, it is suitable for the use of RE. Due to the successful policy of the head, Ilham Aliyev, the possibility of using hydrocarbon deposits has become even more relevant at the same time for purchasing new materials. In this regard, different measures were taken and decarbonization policy is implemented in the regions. In this respect, for the first time ever in the world, the concept of remembering bright memory of villages and cities has found its application. After full restoration of sovereignty of Azerbaijan in 2020, this approach has been implemented in regions. For the first time, the concept of a Smart Village and a Smart City was introduced in Zangilan, and the continuous and fully RE of Agalı village, i.e., solar radiation, was used. Alternative energy sources are used here. In November 2024, Azerbaijan is also going to host COP29 (Conference of the Parties 29) in connection with the relevance of climate change and decarbonization in the world.

2.5 CONCLUSION

With the broad-ranging discussion on nanomaterials, in particular those applied within the automotive and transport industries, it would seem that nanomaterials are starting to represent a transformational capability to surmount the various limitations and further drive innovation. With graphene-based smart materials illustrating advantages associated with strength, conductivity, and flexibility, such benefits begin to position these materials quite aptly for automotive component applications. Application in vehicle weight reduction is integral to developing more sustainable options for transport and improving energy efficiency. It also identifies nanotechnology as one of the key emerging factors for cleaner energy technologies, including hydrogen energy. Nanomaterials are critical in improving hydrogen storage and fuel cell technologies, hence reducing environmental impacts in the transport sector and increasing energy security. With such benefits, the increasing penetration of nanotechnology into motor systems also introduces new challenges; an emerging concern within this frame will be that of cybersecurity. While modern transportation will increasingly use intelligent materials and digital systems, they will also be increasingly vulnerable to cyberattacks with potential impacts on safety and efficiency. Minimizing these risks is crucial for the full deployment of nanotechnology into transport. While the prospects of nanomaterials, particularly graphene-based materials, seem endless for automobile industries—from merely vehicle-component improvement to energy efficiency and sustainability—their roles are not dispensable in the way into the future of transportation. In this juncture, all efforts must be directed at mitigating emerging challenges such as cybersecurity and scalability of these technologies in order to get the full benefits of nanotechnology. Full exploitation of the powers offered by nanotechnology to the transport industry is possible only by encouraging interdisciplinary collaboration and continued research toward a safer, greener, more efficient future.

REFERENCES

Abaszade, R.G. (2022). Synthesis and analysis of flakes graphene oxide, Journal of Optoelectronic and Biomedical Materials, 14(3), 107–114.

Abaszade, R.G., Aliyev, E.M., Babanli, M.B., Kotsyubynsky, V.O., Zapukhlyak, R.I., Mamedov, A.G., Budak, H.F., Kasapoglu, A.E., Gur, E., Margitych, T.O., Stetsenko, M.O. (2023). Investigation of thermal properties of carbon nanotubes and carboxyl group—Functionalized carbon nanotubes, Physics and Chemistry of Solid State, 24(3), 530–535.

Abaszade, R.G., Aliyev, E.M., Mammadov, A.G., Khanmamadova, E.A., Guliyev, A.A., Aliyev, F.G., Zapukhlyak, R.I., Budak, H.F., Kasapoglu, A.E., Margitych, T.O., Singh, A., Arya, S., Gür, E., Stetsenko, M.O. (2024). Investigation of thermal properties of gadolinium doped carbon nanotubes, Physics and Chemistry of Solid State, 25(1), 142–147.

Abaszade, R.G., Kapush, O.A., Nabiyev, A.M. (2020). Properties of carbon nanotubes doped with gadolinium, Journal of Optoelectronic and Biomedical Materials, 12(3), 61–65.

Abaszade, R.G., Mammadov, A.G., Bayramov, I.Y., Khanmamadova, E.A., Kotsyubynsky, V.O., Kapush, O.A., Boychuk, V.M., Gur, E.Y. (2022a). Structural and electrical properties of sulfur-doped graphene oxide/graphite oxide composite, Physics and Chemistry of Solid State, 25(2), 256–260.

Abaszade, R.G., Mammadov, A.G., Bayramov, I.Y., Khanmamadova, E.A., Kotsyubynsky, V.O., Gur, E.Y., Kapush, O.A. (2022b). Modeling of voltage-ampere characteristic structures on the basis of graphene oxide/sulfur compounds, Technical and Physical Problems of Engineering, 14(2), 302–306.

Abaszade, R.G., Mammadov, A.G., Khanmamedova, E.A., Aliyev, F.G., Kotsyubynsky, V.O., Gür, E., Soltabayev, B.D., Margitych, T.O., Stetsenko, M.O., Singh, A., Arya, S. (2024). Photoconductivity of functionalized carbon nanotubes, Digest Journal of Nanomaterials and Biostructures, 19(2), 837–843.

Abaszade, R.G., Mammadov, A.G., Khanmammadova, E.A., Bayramov, I.Y., Namazov, R.A., Popal, K.M., Melikova, S.Z., Qasımov, R.C., Bayramov, M.A., Babayeva, N.I. (2023). Electron paramagnetic resonance study of gadoliniumum doped graphene oxide, Journal of Ovonic Research, 19(2), 259–263.

Abaszade, R.G., Mammadova, S.A., Kapush, O.A., Agayev, F.G., Nabiev, A.M., Mammadova, M.M., Budzulyak, S.I., Kotsyubynsky, V.O. (2021). Synthesis and characterization of graphene oxide flakes for transparent thin films, Physics and Chemistry of Solid State, 23(3), 595–601.

Abhat, A. (1983). Low temperature latent Heat thermal energy storage: Heat storage materials, Solar Energy, 30(4), 313–31.

Abhishek, G., Saini, R.P. (2020). A review on technical, applications and economic aspect of packed bed solar thermal energy storage system, Journal of Energy Storage, 27, 101046.

Aftab, M.A., Byrne, J., Hazendonk, P., Johnson, D., Spencer, L., Weis, T. (2024). Implementing tri-brid energy systems for renewable integration in Southern Alberta, Canada, Clean Technology, 6, 1038–1056.

Aydin, H., Lee, H.S., Kim, H.J., Shin, S.K., Park, K. (2014). Off-design performance analysis of a closed-cycle ocean thermal energy conversion system with solar thermal preheating and superheating, Renew Energy, 72(4), 154–63.

Bedri, S. (2024). Developments in the renewable energy system policies of china, Eurasian Research Journal, 6(1), 65–84.

Boychuk, V.M., Zapukhlyak, R.I., Abaszade, R.G., Kotsyubynsky, V.O., Hodlevsky, M.A., Rachiy, B.I., Turovska, L.V., Dmytriv, A.M., Fdorchenko, S.V. (2022). Solution combustion synthesized $NiFe_2O_4$/reduced graphene oxide composite

nanomaterials: Morphology and electrical conductivity, Physics and Chemistry of Solid State, 23(2), 815–824.
Burcu, K., Ana Ines, F., Halime, P. (2020). Review on sensible thermal energy storage for industrial solar applications and sustainability aspects, Solar Energy, 209, 135–169.
Burhan, A., Özlem, E.O. (2024). Determining the most useful renewable energy alternative for Turkey by combining MAUT and TAOV methods, International Journal of Energy Studies, 9(2), 291–308.
Chen, K., (2024). The application of renewable energy in transportation systems in China, Proceedings of the 4th International Conference on Materials Chemistry and Environmental Engineering, 28–33.
Cruvinel da Sousa Mello, T., Santos Silva, L.R., Sérgio da Luz, M. (2023). The current Brazilian energy model: 2016 to 2020, Brazilian journal of science, Technology and Innovation, 8, 96–105.
Devrim, A., Zafer, U., Olcay, K. (2015). Thermal performance analysis of a solar energy sourced latent heat storage, Renewable and Sustainable Energy Reviews, 50, 1213–1225.
Dwarka, K., Kim, J.S. (2005). Thermal analysis of composite phase change drywall systems, Journal of Solar Energy Engineering, 127, 352–6.
Elkhatat, A., Al-Muhtaseb, S.A. (2023). Combined, renewable energy–Thermal energy storage (RE–TES), systems: A review, Energies, 16, 4471.
Ezan, M.A., ErekA, D.I. (2011). Energy and exergy analyses of an ice-on-coil thermal energy storage system, Energy, 36, 6375–6386.
Figarova, S.R., Aliyev, E.M., Abaszade, R.G., Alekberov, R.I., Figarov, V.R. (2021). Negative differential resistance of grapheme oxide|sulphur compound, Journal of Nano Research, 67, 25–31.
Figarova, S.R., Aliyev, E.M., Abaszade, R.G., Figarov, V.R. (2023). Negative thermal expansion of sulphur-doped graphene oxide, Advanced Materials Research, 1175, 55–62.
Goswami, D.Y., (2022). Principle of solar engineering. CRC Press.
Guliyeva, N.A., Abaszade, R.G., Khanmammadova, E.A., Azizov, E.M. (2023). Synthesis and analysis of nanostructured graphene oxide, Journal of Optoelectronic and Biomedical Materials, 15(1), 23–30.
Guruprasad, A., Lingkun, L., Xiang, H., Guiyin, F. (2017). Thermal energy storage materials and systems for solar energy applications, Renewable and Sustainable Energy Reviews, 68, 693–706.
Harish, V., Ansari, M.M., Tewari, D., Gaur, M., Yadav, A.B., García-Betancourt, M.-L., Abdel-Haleem, F.M., Bechelany, M., Barhoum, A. (2022). Nanoparticle and nanostructure synthesis and controlled growth methods, Nanomaterials, 12, 3226.
Idrus, N.H.M., Musa, M.N., Yahya, W.J., Yahya, W.J. (2017). Geo-ocean thermal energy conversion (GeOTEC) power cycle/plant, Renew Energy, 111, 372–380.
Ivanichok, N., Kolkovskyi, P., Ivanichok, O., Kotsyubynsky, V., Boychuk, V., Rachiy, B., Bembenek, M., Warguła, Ł, Abaszade, R., Ropyak, L. (2024). Effect of thermal activation on the structure and electrochemical properties of carbon material obtained from walnut shells, Materials, 17(2514), 1–21.
Jena, N.K., Sahoo, S., Sahu, B.K., Naik, A.K., Bajaj, M., Misak, S., ... Prokop, L. (2023). Impact of a redox flow battery on the frequency stability of a five-area system integrated with renewable sources, Energies, 16(14), 5540. https://doi.org/10.3390/en16145540.
Jia, J., Lan, Y. (2023). Synthesis, characterization, and applications of nanomaterials for energy conversion and storage, Molecules, 28, 7383.
Khanmammadova, E., Abaszade, R., Muradi, R., (2024). Antiviral thin surface and the surface life, Miltifunctionla coatings and interfaces for tribological contacts (pp. 1–32).

Krane, R.J. (1987). A second law analysis of the optimum design and operation of thermal energy storage systems, International Journal of Heat and Mass Transfer, 30(l), 43–57.

Kumar, A. (2024). Solar-integrated hydrogen production and utilization in India. In Clean energy (pp. 243–258). CRC Press. https://doi.org/10.1201/9781003521341-15

Kuravi, S., Trahan, J., Goswami, D.Y., Rahman, M.M., Stefanakos, E.K. (2013). Thermal energy storage technologies and systems for concentrating solar power plants, Progress in Energy and Combustion Science, 39, 285–319.

Li, G. (2015). Energy and exergy performance assessments for latent heat thermal energy storage systems, Renewable and Sustainable Energy Reviews, 51, 926–954.

Luo, Y., Yinghong, W., Huang, C., Menon, C., Feng, S.-P., Chu, P.K. (2022). Plasma modifed and tailored defective electrocatalysts for water electrolysis and hydrogen fuel cells, EcoMat, 4(4), 25.

Mammadov, A.G., Abaszade, R.G., Babanli, M.B., Kotsyubynsky, V.O., Gur, E., Soltabayev, B.D., Margitych, T.O., Stetsenko, M.O. (2023). Photoconductivity of gadolinium-doped carbon nanotubes, International Journal on Technical and Physical Problems of Engineering, 15(3), 53–58.

Mammadov, A.G., Abaszade, R.G., Kotsyubynsky, V.O., Gur, E.Y., Bayramov, I.Y., Khanmamadova, E.A., Kapush, O.A. (2022). Photoconductivity of carbon nanotubes, Technical and Physical Problems of Engineering, 14(3), 155–160.

Mazibuko, T., Moloi, K., Akindeji, K. (2024). Techno-economic design and optimization of hybrid energy systems, Energies, 17, 4176.

McTigue, J.D., Wendt, D., Kitz, K., Gunderson, J., Kincaid, N., Zhu, G. (2020). Assessing geothermal/solar hybridization – Integrating a solar thermal topping cycle into a geothermal bottoming cycle with energy storage, Applied Thermal Engineering, 171, 115121.

Meena, Kumar, A., Singh, V. P., Ghosh, A. Sustainable technologies for energy efficient buildings. CRC Press. https://doi.org/10.1201/9781003496656

Mendoza-Rivera, R.J., García-Pérez, L.E., Martínez, F.V. (2023). Renewable and non-renewable energy consumption, CO_2 emissions, and responsible economic growth with environmental stability in North America, International Journal of Energy Economics and Policy, 13(4), 300–311.

Mohamed, A.S.A., Ahmed, M.S., Shahdy, A.G. (2020). Theoretical and experimental study of a seawater desalination system based on humidification dehumidification technique, Renewable Energy, 152(2020), 823–834.

Molu, R.J.J., Tripathi, B., Mbasso, W.F., Naoussi, S.R.D., Bajaj, M., Wira, P., Misak, S. (2024). Advancing short-term solar irradiance forecasting accuracy through a hybrid deep learning approach with Bayesian optimization, Results in Engineering, 23, 102461. https://doi.org/10.1016/j.rineng.2024.102461

Moradi, R., Pour Khalili, N., Khanmamadova, E., Abaszade, R. (2024). Functionalized Carbon Nanostructures for Flexible Electronics. In: Barhoum, A., Deshmukh, K. (eds.), Handbook of functionalized carbon nanostructures (pp. 2581–2614). Springer, Cham.

Mossa, R.W., Henshall, P., Arya, F., Shire, G.S.F., Hyde, T., Eames, P.C. (2018). Performance and operational effectiveness of evacuated flat plate solar collectors compared with conventional thermal, PVT and PV panels, Applied Energy, 216, 588–601.

Murshed, M., Ozturk, I. (2023). Rethinking energy poverty reduction through improving electricity accessibility: A regional analysis on selected African nations, Energy, 267, 126547.

Nachtane, M., Tarfaoui, M., Abichou, M.A., Vetcher, A., Rouway, M., Aâmir, A., Mouadili, H., Laaouidi, H., Naanani, H. (2023). An overview of the recent advances in composite materials and artifcial intelligence for hydrogen storage vessels design, Journal of Composites Science, 7, 119.

Nelson, D.J., Vick, B., Yu, X. (1996). Validation of the algorithm for ice-on-pipe brine thermal storage system, ASHRAE Transactions, 102, 55–62.
Nkwetta, D.N., Smythe, M. (2012). The potential applications and advantages of powering solar air-conditioning systems using concentrator augmented solar collectors, Applied Energy, 89, 380–6.
Osmani, P.S. (2024). Synthesis and characterization of advanced inorganic nanomaterials for energy storage devices, Journal for Research in Applied Sciences and Biotechnology, 3, 122–128.
Ozge, O., Francesco, F., Javanshir, F. (2024). Adoption of renewable heating systems and thermal energy communities in the Netherlands: An empirical study, Energy Reports, 11, 3815–3823.
Rawat,T.,Niazi,K.R.,Gupta,N.,&Sharma,S.(2021).Multi-objectivetechno-economic operation of smart distribution network integrated with reactive power support of battery storage systems, Sustainable Cities and Society, 75, 103359. http://dx.doi.org/10.1016/j.scs.2021.103359
Reilly, J.J., Holtz, A., Wiswall, R.H. (1971). A new laboratory gas circulation pump for intermediate pressures, Review of Scientific Instruments, 42(10), 1485–6.
Ruess, G., Vogt, F. (1948). HöchstlamellarerKohlenstoffausGraphitoxyhydroxyd. Monatshefte für Chemie und verwandte Teile anderer Wissenschaften, 78, 222–242.
Ruska, E. (1980). The early development of electron lenses and electron microscopy, Microscopica Acta, 59(Suppl. 5), 1–140.
Sánchez-Pantoja, N., Vidal, R., Pastor, M.C. (2018). Aesthetic impact of solar energy systems, Renewable and Sustainable Energy Reviews, 98, 227–238.
Shankar, A. (2023). Efficient data interpretation and artificial intelligence enabled IoT based smart sensing system, Artificial Intelligence Review, 56(12), 15053–15077. 10.1007/s10462-023-10519-y
Sharma, K. K., Kumar, R., & Jha, P. K. (2024). Physical modeling and numerical investigation of fluid flow and solidification behavior in a slab caster mold using hexa-furcated nozzle, Heat Transfer, 53(3), 1050–1072. https://doi.org/10.1002/htj.22980
Sharma, S. K., Mallick, A., Singh, D. B., & Tiwari, G. N. (2022). Experimental study of solar energy–based water purifier of single-slope type by incorporating a number of similar evacuated tubular collectors, Environmental Science and Pollution Research, 29(5), 6837–6856. http://dx.doi.org/10.1007/s11356-021-16123-3
Shuai, K. (2006). The solar energy-reheating ocean thermal energy conversion (pp. 17–18). Shanghai Jiao Tong University.
Singh, S., Pandey, V.K., Prabha, S., Kuldeep, S.K. (2024). The growth of renewable energy sources and review of microgrids in Indian perspective, E3S Web of Conferences, 540, 010.
Situma, M.P., Nelson, O., Nyongesa, Nyenyi, D. (2024). Effect of renewable energy consumption on economic growth in Kenya, International Journal of Research and Innovation in Social Science, 1848–1860.
Soto, R., Vergara, J. (2014). Thermal power plant efficiency enhancement with ocean thermal energy conversion, Applied Thermal Engineering, 62(1), 105–12.
Suhányiová, A., Korečko, J., Suhányi, L. (2023). Renewable energy sources management in the EU-27 countries, Acta MontanisticaSlovaca, 28(2), 479–495.
Verma, P., Varun, S.K., Singal (2008). Review of mathematical modeling on latent heat thermal energy storage systems using phase-change material, Renewable and Sustainable Energy Reviews, 12, 999–1031.
Vick, B., Nelson, D.J., Yu, X. (1996). Model of an ice-on-pipe brine thermal storage component, ASHRAE Transactions, 102, 45–54.
Wang, T., Gu, C.G., Yang, B., Shuai, K..(2005). OTEC-a solar energy-reheating cycle, CN 1673527 A.
West, J., Braun, J.E. (1999). Modeling partial charging and discharging of area-constrained ice storage tanks, HVAC&R Research, 5, 209–28.

Widyatmoko, S., Nanny, M., Hanifah, N.M. (2023). Implementation of edge computing IoT-based database management system for renewable energy management in Indonesia, West Science Information System and Technology, 01(02), 72–79.

Wu, Z., Yang, Y., Yin, C. (2024). Technology and Prospect of hydrogen energy, Highlights in Science, Engineering and Technology, 96, 144–151.

Xinyuan, C., Sh, J., Xu, J., Yu, K. (2024). Hydrogen energy: Production, storage and application, Highlights in Science, Engineering and Technology, 112, 137–145.

Yoon, J.I., Son, C.H., Baek, S.M., Ye, B.H., Kim, H.J., Lee, H.S. (2014). Performance characteristics of a high-efficiency R717 OTEC power cycle, Applied Thermal Engineering, 72(2), 304–8.

Yuan, H., Mei, N., Zhou, P. (2014). Performance analysis of an absorption power cycle for ocean thermal energy conversion, Energy Conversion and Management, 87, 199–207.

Zhang, L., Zhou, J., Li, H., Shen, L., Feng, Y.P. (2021). Recent progress and challenges in magnetic tunnel junctions with 2D materials for spintronic applications, Physical Review Applied, 8, 021308.

Chapter 3

Renewable energy sources and conversion technologies

R.G. Abaszade, A.G. Mammadov, and E.A. Khanmamadova

3.1 INTRODUCTION

The transition to renewable sources of energy assumes an imperative dimension in view of the increasingly daunting threats from climate change, a shrinking reserve of fossil fuels, and degradation of the environment. This source of energy—whose original substance is provided by natural processes continuously renovated—in its use gives a sustainable alternative compared to conventional methods of energy production, which are heavily reliant on limited resources such as coal, oil, and natural gas [Smil, 2017; Twidell and Weir, 2015].

This chapter covers the sources of various types of renewable energy: solar, wind, hydropower, biomass, geothermal, tidal, and wave energies; it also discusses the underlying physics with a thorough review of the energy conversion processes, efficiencies, and technological innovations that most of the different energy sources have undergone to provide an overview of how these technologies stand to help with the sustainability of energy production. It also covers the critical role of physics in advancing renewable energy technology and overcoming engineering challenges still existing in the wide diffusion of these resources.

Solar energy: It is the most available resource compared to all types of renewable resources. The sun's electromagnetic radiation can be converted into electrical or thermal energy with the use of photovoltaic (PV) cells and solar thermal systems, respectively [Kalogirou, 2009]. Both systems have their operation based on the interaction of photons with matter for power generation. Solar energy converters must develop their processes based on the PV effect and phenomena related to the absorption of heat and to thermodynamic processes so that improvements in efficiency and reductions in the costs of materials used can be achieved.

Wind energy: Turbines harness energy from the wind, converting the kinetic energy in the air movement into electrical energy. Wind turbines work on the principle of aerodynamics, fluid dynamics, and energy conservation; thus, their structuring and operation act as a means to maximize

DOI: 10.1201/9781003534785-3

energy capture while minimizing turbulence and losses [Manwell et al., 2010]. This chapter will delve into the efficiency of wind energy systems by harnessing wind predominantly through the Bernoulli principle and the Betz limit.

Hydropower is a very old resource developed from water potential. The act of releasing water from a height causes turbines to transfer this potential energy into kinetic energy and then to electrical energy. Hydropower systems are based on fluid dynamics, conversion of mechanical energy, and the use of Bernoulli's equation for an explanation of the mechanics involved.

Biomass energy basically involves organic material transformation into usable energy through such means as combustion, gasification, and anaerobic digestion. Thermodynamically, these processes and chemical reactions are bound by principles and chemical reactions that degrade biomass into released energy [Tester et al., 2005]. Biomass is a carbon-neutral source of energy when properly managed, and the physics supporting the different technologies of conversion will be discussed in this chapter.

Geothermal energy uses the internal heat of the Earth to produce electricity directly or heat. The production methodologies in this form of energy involve mechanisms of heat transfer: conduction, convection, and radiation [Fridleifsson, 2001]. It also requires basic knowledge of the Earth's geothermal gradients and thermodynamic cycles utilized in geothermal power plants.

Tidal and wave energy are two new fields of renewable energy produced by ocean movements. In tidal energy, the gravitational attraction of both the moon and the sun generates predictable cycles of rise and fall at sea level, while wave energy converts the kinetic energy of surface waves [Evans et al., 2010]. The chapter is involved in an in-depth look at the physics of renewable energies and how these naturally driven processes can be harnessed into meeting the rapidly increasing energy needs of humans without jeopardizing the future of the planet. Through ongoing research and development, renewable energy is set to play the starring role in the coming global transition in energies in the attempt to resolve the most pressing environmental concerns of this century [Tester et al., 2005].

3.2 CONVERSION TECHNOLOGIES

It is in this regard that the renewable sources of energy tap into natural processes with their underlying physics to generate energy. This chapter discusses the basic physical principles of different types of renewable sources, including PV systems, wind turbines, hydroelectric plants, bioenergy systems, and geothermal plants [Kumar et al., 2024]. Each of these technologies can help transform natural energy into electrical and heat forms that are quite useful, remarkably contributing to a global effort of mitigating climate change and reducing reliance on fossil fuels.

3.2.1 Photovoltaic systems (solar to electricity)

The PV effect is the direct conversion of sunlight into electricity, where it employs PV systems [Green, 2015]. When there is adequate incoming energy from photons in sunlight hitting and transferring energy to electrons in the semiconducting material—silicon being the most common one—the electrons will become excited by this incoming energy [Loferski, 1956]. The photographic description of this is given by the following equation:

$$E = h \cdot f$$

where E is the energy of the photon, h is Planck's constant ($6.626 \times 10^{-34}\,Js$), and f is the frequency of the incoming light [Kalogirou, 2014]. When these electron-hole pairs are separated by an internal electric field created at the p-n junction of the semiconductor, an electric current is generated.

The efficiency of PV cells is influenced by various factors, including:

Material properties: Different semiconductor materials have different band gaps, affecting their ability to absorb light. Silicon is the most common material used due to its optimal band gap and abundance.

Temperature: PV cell efficiency generally decreases with increasing temperature, requiring careful thermal management in system design.

Angle of incidence: The angle at which sunlight strikes the PV surface impacts energy capture; tracking systems can optimize this angle throughout the day. Advanced technologies, such as thin-film cells and multi-junction cells, are being developed to enhance efficiency. Thin-film technologies use less material and can be applied to flexible surfaces, while multi-junction cells stack multiple semiconductor layers to capture different light wavelengths more efficiently.

3.2.2 Wind turbines (wind to electricity)

Wind turbines convert the kinetic energy of moving air into mechanical energy and then into electricity [Manwell et al., 2009]. The physics of wind energy relies heavily on principles of aerodynamics and energy conversion.

3.2.2.1 Aerodynamics of wind turbines

Wind turbine blades are designed in such a way as to take advantage of the **Bernoulli principle**, which states that the increase in speed of a fluid—one may say, air—is simultaneously dealt with decrease in pressure [Burton et al., 2011]. Lift is created on the turbine blades due to the passing of the wind over the curved surface of the turbine blades, causing them to rotate. Rotational energy, in turn, is transferred into a generator, which ultimately converts the same into electrical energy [Sorensen, 2011].

Efficiency is a way of describing a wind turbine, and this may be quantified by something called the Betz limit [Betz, 1996]. The Betz limit states that no turbine is capable of capturing more than 59.3% of the kinetic energy in wind [Auerbach and Hamilton, 1958]. The relationship is given by:

$$Efficiency \leq \frac{16}{27} \approx 0.593$$

Optimizing turbine design, including blade shape, length, and material, is crucial to maximizing energy capture while minimizing turbulence and losses.

3.2.3 Hydroelectric plants (water to electricity)

Hydropower systems harness the potential energy of stored water, converting it into kinetic energy to drive turbines and generate electricity. The physics behind hydropower involves principles of fluid dynamics and mechanical energy conversion.

3.2.3.1 Potential and kinetic energy

Water stored at height in reservoirs has potential energy expressed by:

$$PE = m \cdot g \cdot h$$

where PE is potential energy, m is mass of the water, g is the acceleration due to gravity, and h is the height of the water column. When released, this potential energy converts to kinetic energy while water flows down through turbines:

$$KE = \frac{1}{2} m\vartheta^2$$

where KE is kinetic energy and ϑ is the velocity of the water.

3.2.3.2 Bernoulli's equation

The dynamics of water flow can be analyzed using Bernoulli's equation:

$$P + \frac{1}{2}\rho\vartheta^2 + \rho g h = \text{constant}$$

where P is pressure, ρ is the fluid density, ϑ is fluid velocity, and h is the height. This equation helps in designing efficient turbines that maximize energy extraction from flowing water.

Hydropower can be generated through storage systems (dams) that allow for energy regulation and peak power generation, and run-of-river systems that have less environmental impact but may provide lower energy output [Twidell and Weir, 2015].

3.2.4 Bioenergy-biomass to heat/electricity

Biomass energy is derived from organic materials, and through various processes can be converted into heat and electrical energy [Baxter, 2005; McKendry, 2002]. The fundamental physics involved includes concepts from thermodynamics and chemical reactions.

3.2.4.1 *Combustion process*

Burning biomass involves the breaking of chemical bonds holding energy, which is consequently released as heat. It may be modeled by the following equation:

$$Q = m\Delta H$$

Where Q is the heat energy produced, m is amount of mass of biomass being burnt, and ΔH is enthalpy of combustion. The efficiencies of combustion systems of biomass depend solely on temperature, oxygen supply, and moisture content [Al Asi, 2024].

Biomass gasification or anaerobic digestion, converting organic material into syngas or breaking down organic matter in the absence of oxygen to produce biogas, is another method through which biomass is transformed into biofuels [Ellabban et al., 2014]. Consequently, this process involves thermodynamic principles and chemical reactions in the conversion of biomass into usable energy forms.

3.2.5 Geothermal plants (Earth's heat to energy)

Geothermal energy simply consists of heat extracted from beneath the surface of the Earth. (Figure 3.1). The underlying physics in this source of geothermal energy is based on thermodynamics and mechanisms of the transfer of heat [Boyle, 2012; DiPippo, 2016].

3.2.5.1 *Mode of heat transmission*

Under the operating principles of geothermal systems, heat from underground reservoirs is moved onto the surface using conduction, convection and not forgetting radiant modes [Lund et al., 2010]. Using the hot water or the steam that has been tapped from the underground reservoirs, geothermal power plants may either use the steam for heating contact applications or convert it into helpful electricity. The efficiency of geothermal

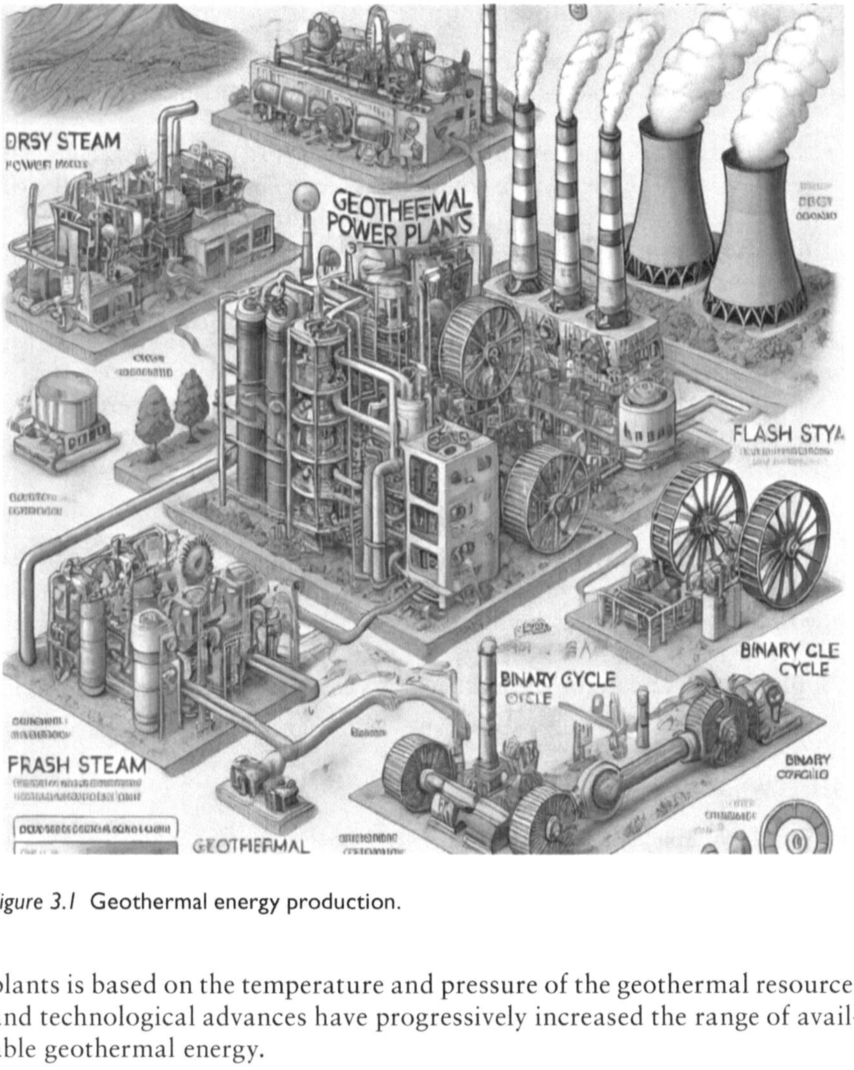

Figure 3.1 Geothermal energy production.

plants is based on the temperature and pressure of the geothermal resource, and technological advances have progressively increased the range of available geothermal energy.

3.3 STATISTICAL INSIGHTS ON RENEWABLE ENERGY: EFFICIENCY, SUSTAINABILITY, AND GLOBAL ADOPTION

There has been a dramatic increase of efficiency, cost efficiency, and global usage of renewable energy technology in the past 10 years. Since countries have a curious inclination toward the usage of less dirty energy, the mathematics of renewable energy describes the advancement in solar, wind, hydropower, bioenergy, and geothermal energy explorations [REN21, 2023].

3.3.1 Global renewable energy share and growth trends

Energy production from renewable sources has turned into an important factor in power generation and continued to increase in importance over the years:

- **Share of global generation:** Compared to 29% in global electricity generation in 2022, renewable sources of energy supplied 29% of all electricity generated, including solar, wind, hydro, and biomass [UNEP, 2023]. This trend has grown over a decade and is attributed to technologies and policies introduced to cut carbon emissions.

Hydropower's dominance contributes to the greatest fraction of the total global generation of renewable electricity and accounts for approximately 16% of the total generation of electricity, that is, 225 TWh. The infrastructure put up is also very good, besides there is good hydraulic efficiency of hydropower systems.

3.3.1.1 Rate of growth of solar and wind energy

Solar and wind energy are the renewables growing by the highest growth rates. In 2022, solar and wind generated 10% of the electricity produced in the world. These two once combined use wind energies [IPCC, 2023; SEIA, 2022]. The growth in Solar energy has been phenomenal, growing close to 22% per annum and highlighting that this energy source happens to be one of the extremes growing in the world.

3.3.1.2 Wind energy

The installed capacity of both offshore and onshore wind energy was up at 837 GW in 2022 and the onshore and offshore wind energy opting to continue to rise steadily [GWEC, 2023].

3.3.1.3 Types of geothermal power plants

There are no special designs for such plants; they are as follows general types (Table 3.1).

3.3.2 Solar energy efficiency and cost trends

Solar energy is one of the most promising renewable technologies due to its abundance and scalability [BNEF, 2022; IPCC, 2023]. Over the past decade, significant improvements in efficiency and cost reduction have driven its widespread adoption:

Photovoltaic efficiency: The efficiency of commercial silicon-based PV cells, the most common technology, typically ranges between 18% and 22%. However, research breakthroughs in multi-junction cells and

Table 3.1 Summarizing the different types of geothermal power plants along with their characteristics

Type of geothermal power plant	*Description*	*Operational principle*	*Advantages*	*Challenges*
Dry steam plants	Utilize steam directly from geothermal reservoirs	Steam from the reservoir drives the turbine directly, generating electricity	Simple design; high efficiency	Limited to high-temperature resources; fewer sites available
Flash steam plants	Bring hot water to the surface, causing it to flash into steam	Hot water from the reservoir is depressurized, allowing some of it to turn into steam, which drives the turbine	Can use moderate-temperature resources; versatile	Requires higher temperature reservoirs than binary systems
Binary cycle plants	Use a secondary fluid with a lower boiling point	Hot geothermal fluid transfers heat to a secondary fluid, vaporizing it to drive the turbine	Lower temperature resources can be utilized; closed-loop systems reduce environmental impact	Lower efficiency compared to dry and flash systems
Enhanced geothermal systems (EGS)	Created by injecting water into hot rock formations	Water is pumped into hot, dry rock formations, heated, and then extracted as steam to generate electricity	Can be developed in a wider range of locations; can utilize hot rocks that are not naturally permeable	High initial costs; requires significant drilling and reservoir creation

thin-film technologies have pushed laboratory efficiencies above 47% under concentrated sunlight. These multi-junction cells stack different semiconductor materials to capture a broader range of the solar spectrum, greatly enhancing energy conversion efficiency.

Thin-film solar cells: These cells offer lower material costs and greater flexibility. Though their efficiency (around 10–15%) is generally lower than that of silicon-based cells, they have gained popularity due to their lightweight and versatile applications, such as in building-integrated photovoltaics (BIPV).

Cost reduction: The cost of solar PV electricity has decreased by an astonishing 85% since 2010. This cost reduction has made solar one of the most affordable energy sources, with prices as low as $0.04 per kWh in sunny regions [WBG, 2022]. This dramatic decline is attributed to technological advancements, increased production capacity, and economies of scale.

3.3.3 Wind energy: Efficiency and growth

Wind energy, both onshore and offshore, has become a cornerstone of renewable energy strategies worldwide. It offers significant potential for large-scale power generation:

Capacity factor: The capacity factor, which measures the actual energy output compared to the maximum possible output, for modern onshore wind turbines is around 30–45%. Offshore wind turbines, benefiting from stronger and more consistent wind speeds, have higher capacity factors, often ranging between 50% and 60%.

Global capacity: The global installed capacity for wind power reached 837 GW by 2022, with onshore wind farms accounting for the majority. Offshore wind capacity is growing rapidly, especially in Europe and Asia, where deeper waters provide access to stronger winds. Offshore wind farms tend to be more expensive to build but offer higher energy yields due to their superior wind conditions.

Cost trends: The cost of onshore wind energy has fallen by approximately 50% since 2010, with the current cost averaging between $0.03 and 0.05 per kWh. Offshore wind energy is more expensive, with costs around $0.08–0.10 per kWh, but these costs are expected to decline as technology improves and larger turbines are deployed.

3.3.4 Hydropower

Hydropower remains one of the most efficient forms of renewable energy, with a long history of contributing to the global energy mix:

Efficiency: Hydropower plants, particularly large-scale dam-based systems, boast some of the highest conversion efficiencies of any renewable technology, typically ranging between 85% and 90% [IHA, 2023]. This high efficiency, combined with the ability to provide consistent baseload power, makes hydropower a reliable source of renewable energy [IEA, 2023].

Global capacity: Hydropower's global installed capacity reached approximately 1,360 GW by 2022, generating over 4,300 TWh of electricity annually. It continues to play a significant role in regions with abundant water resources and suitable geographical conditions.

Environmental considerations: While hydropower is efficient, large dam projects can have significant environmental impacts, including habitat disruption, changes in river ecosystems, and displacement of local communities [UNEP, 2023]. To address these concerns, run-of-river systems are being developed as a more sustainable alternative, offering smaller-scale power generation with minimal ecological disruption.

3.3.5 Bioenergy: efficiency and sustainability

Wood, agricultural crop residues, and other biomass is burned directly for heat (and sometimes power) or converted to fuel such as ethanol that can be used in transportation, etc.

Conversion efficiency: Traditional biomass combustion systems are based on steam cycle, the burn biomass pellet to provide heat or generate electricity types; thus, the overall energy efficiency for this kind of cycle is around 25–30%. This can be much higher for combined heat and power (CHP) systems that capture and utilize waste heat—up to 80–85% [IGA, 2022].

Bioenergy capacity: Bioenergy installations are often carbon-neutral provided they are well-regulated such that the amount of CO_2 released by combustion is offset by the CO_2 then absorbed during growth of plants. The use of waste material and organic residues also counts toward the circular economy by reducing landfill utilization, as well as bioenergy.

Global deployment: In 2022, bioenergy represented around 5% of global electricity generation [EIA, 2022]. The allure here is in its flexibility—bioenergy has the ability to provide power, heat, and transport fuels, all under one hat.

3.3.6 Geothermal energy: Efficiency and growth

Geothermal energy is generated through utilizing heat from the Earth's interior. This form of renewable energy is less common than solar and wind, but it provides dependable power in a consistent manner:

Efficiency: The efficiency of geothermal power plants depends on the temperature and pressure of the geothermal resource. On average, geothermal systems achieve an efficiency of 10–20%, with higher efficiencies possible in regions with higher-temperature resources.

Enhanced geothermal systems (EGS): Enhanced geothermal systems, which create artificial reservoirs by injecting water into hot rock formations, are expanding the potential for geothermal energy in areas without natural geothermal reservoirs [Bioenergy Europe, 2023; Lazard, 2022]. This technology could significantly increase geothermal capacity in regions where traditional geothermal systems are not feasible.

Global capacity: As of 2022, global geothermal capacity stood at 16 GW, providing consistent baseload electricity in regions such as Iceland, the Philippines, and the United States.

3.3.7 Carbon emissions and environmental benefits

The environmental benefits of renewable energy technologies extend beyond their ability to generate electricity without direct carbon emissions [The

Carbon Trust, 2022]. The broader adoption of renewables could have a profound impact on global emissions:

Carbon footprint: Solar and wind energy technologies have a much lower carbon footprint compared to fossil fuels. On average, solar energy generates between 20 and 60 g CO_2/kWh, while wind energy produces around 10 and 30 g CO_2/kWh, depending on the system and location. In contrast, coal-fired power plants emit 400–900 g CO_2/kWh, making renewables a crucial tool in reducing global greenhouse gas emissions [McKinsey, 2023].

Emissions reduction potential: According to the International Renewable Energy Agency (IRENA), transitioning to renewable energy could reduce global carbon dioxide (CO_2) emissions by approximately 70% by 2050. Achieving this would require significant investments in renewable technologies, energy storage, and grid modernization [EES, 2023; IRENA, 2023; REN21, 2023].

3.4 ENERGY STORAGE BATTERY STORAGE

Another challenge has also emerged from the rapid growth in the supplies of renewable energies, like solar and wind, because of their intermittent and variable nature. Solar energy is dependent on daylight hours, whereas wind energy comes when winds change. The key technologies that are discussed in this section include battery storage, smart grids, and balancing mechanisms, and how these enable the integration of renewable energy into the grid system [Dunn et al., 2011].

3.4.1 Battery storage systems

Battery storage has been very instrumental in taping the intermittency of renewable sources, storing extra amounts of electricity produced at each time of high production and releasing during the period of low production or high demand. The applications of battery storage systems have become imperative for ensuring flexibility and resilience in energy systems toward the better management of energy resources. Lithium-ion batteries are indeed the most popular form of energy storage over the last couple of decades and are critically responsible for changing the way we store and utilize electrical power [Liu et al., 2021]. This chapter outlines key characteristics, advantages, challenges, and prospects of lithium-ion batteries, pinpointing their critical role in relation to the integration of renewable energy sources and further grid stability. Among the advantages of Li-ion technology, high-energy density is one of the most important; this means a battery can store more energy in less space and with less weight compared with other technologies. That property has great relevance to applications in electric vehicles because of weight and space concerns. Lithium-ion batteries can achieve an energy density of around 150–250 Wh/kg, thereby making them perfect for use in portable and mobile applications [Rawat et al., 2021]. Generally, Li-ion batteries have comparatively low self-discharge rates, that is, if they are not used, they lose a charge over longer period. The monthly self-discharge rate of Li-ion batteries normally varies between 2% and 5% and is much lower when compared to nickel-cadmium or lead-acid batteries. That property ensures that energy stored in Li-ion batteries can be accessed when needed without substantial losses [Jena et al., 2023].

3.4.1.1 *Flow batteries*

Flow battery is a kind of energy storage technology that has been well known for their attractive features, especially when it comes to long duration energy storage applications [Soloveichik, 2011]. It consists of liquid electrolytes stored in external tanks, which is different to a typical Li-ion battery system (Figure 3.2) [Whittingham, 2004].

The flow battery systems are easily scalable for specific energy storage requirements. Since the design is modular, there is a provision to alter the size of the electrolyte tanks that enable operators to increase capacity without needing major changes in the core technology. This scalability helps in projects that might need storage capacity expansion over time (Figure 3.3).

Figure 3.2 Li-ion batteries.

Figure 3.3 Redox flow batteries (VRFB).

3.4.1.2 Solid-state batteries

Solid-state batteries represent a significant advancement in energy storage technology, offering several advantages over conventional lithium-ion (Li-ion) batteries. The defining characteristic of solid-state batteries is their use of a solid electrolyte instead of the liquid or gel electrolytes found in traditional Li-ion batteries. This solid electrolyte can be made from various materials, including ceramics, polymers, and glass, which enhances the battery's overall stability and performance [Jiang et al., 2020, Luo et al., 2015, Whittingham, 2004]. Solid-state batteries have the potential to achieve higher energy densities compared to Li-ion batteries. By allowing the use of more energy-dense materials in the anode and cathode, solid-state batteries can store more energy in the same physical volume. This characteristic is particularly advantageous for applications where space and weight are critical, such as electric vehicles and portable electronic devices (Figure 3.4).

The solid electrolyte allows for faster ion movement, enabling quicker charging times. This improvement could significantly reduce the time required to recharge batteries, addressing one of the major limitations of conventional batteries.

3.4.1.3 Pumped hydroelectric storage

Pumped hydroelectric storage at present represents the most established and widely used form of large-scale energy storage, with more than 95% of the

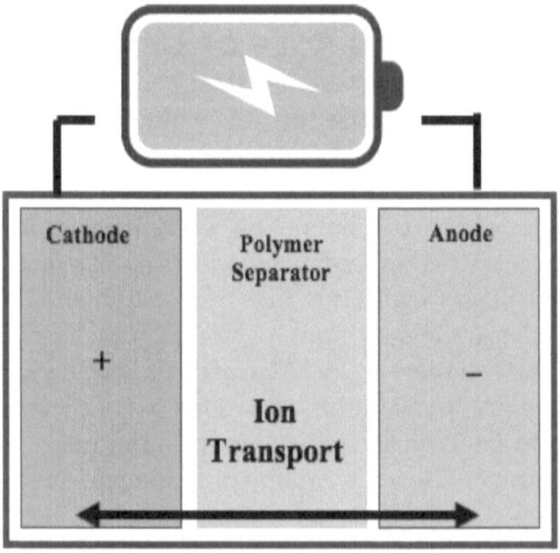

Figure 3.4 Principal work solid-state batteries.

total global energy storage capacity. The technology thus constitutes an important factor in the energy landscape, especially concerning the need to balance supply and demand reliably in view of increasing renewable energy generation. Pumped hydroelectric storage (PHS) technology is the most mature and widely deployed of all large-scale energy storage technologies, representing more than 95% of the current global installed capacity. It constitutes an important solution to balancing supply and demand with increasing renewable generation.

The working principle of PHS is simple yet effective, utilizing gravitational potential energy for storing and generating electrical energy. A PHS essentially consists of two water reservoirs at different elevations: electricity serves to pump water from the lower reservoir into the upper one in periods of low demands or excessive generation, usually supplied by renewable sources like solar or wind. This action essentially converts electrical energy into gravitational potential energy, an energy storage process for later use. The water is released back from the upper reservoir during peaks of electricity demand or when renewable generation is low. In that respect, the falling of water through the turbines turns the potential energy back into electrical energy. This can be moderated to meet peak demand and gives quite a rapid and reliable source of electricity. In normal operation, pumped hydroelectric storage systems exhibit an efficiency rate between 75% and 85%. That is to say, the greater portion of energy used for the process of pumping water back upwards is recoverable at the generation of electricity, making PHS a very efficient storage solution. PHS is better suited for long-duration applications in energy storage. The capabilities regarding storing huge amounts of water in uplifted reservoirs allow for sustained energy generation over long intervals, which assists in balancing fluctuating renewable energy sources [Luo et al., 2015].

3.5 BALANCING SUPPLY AND DEMAND IN RENEWABLE ENERGY SYSTEMS: THE ROLE OF GRAPHENE

Keeping the supply and demand of renewable energy sources in delicate balance, like solar and wind energies, calls for very nimble management and an extremely high level of technology [Abaszade et al., 2020a,b; Khanmammadova et al., 2024]. Solar and wind energies are vast resources, but their intermittency makes them generally ill-matched with electricity demand. There have been various methods of research and different technologies developed to ensure the maximum performance of the grid in relation to an uninterrupted and reliable supply [Abaszade et al., 2024a,b]. This includes the innovative use of graphene in this area. This part reflects on the balancing of energy supply and demand strategies, underlining improvements graphene brings to energy storage, efficiency, and grid management.

3.5.1 Energy storage as a buffer

Energy storage systems provide a critical buffer between generation from renewable sources and the actual demand for electricity. Graphene-based materials have started being integrated into energy storage technologies with increasing efficiencies [Figarova et al., 2021, Guliyeva et al., 2023]. This graphene can be used to manufacture high-capacity supercapacitors for storing energy that could release really fast. Such supercapacitors can provide complementary functionality to conventional battery systems through quick bursts of power at needed moments and are, therefore, becoming crucial in balancing short-term fluctuations in energy supply and demand [Abaszade et al., 2021]. Their capability for rapid charging and discharging also provides grid stability at times of peak demand [Abaszade et al., 2022]. Graphene increases the performance of lithium-ion batteries by enhancing conductivity and energy density. Graphene-composite batteries can offer faster charging times and more extended life cycle, which is currently a limitation in classical battery technologies [Abaszade et al., 2022a,b]. Integrating graphene into the respective energy storage solutions, we will be able to store excess energy generated at high renewable generation and discharge it when needed. Graphene-based materials can be used to achieve better performance of energy management systems [Abaszade et al., 2023a,b,c]. Such a system is able to monitor in real-time and regulate energy use based on changing conditions within renewable energy output. This, in turn, enables the shifting of loads by providing economic incentives for the consumers during periods of low demand when renewable generation has surplus capacity [Sharma et al., 2024].

3.5.2 Grid-scale flexibility

The variability in renewable energy sources demands the flexibility of modern electrical grids [Abaszade et al., 2023a,b,c; Boychuk et al., 2022; Ivanichok

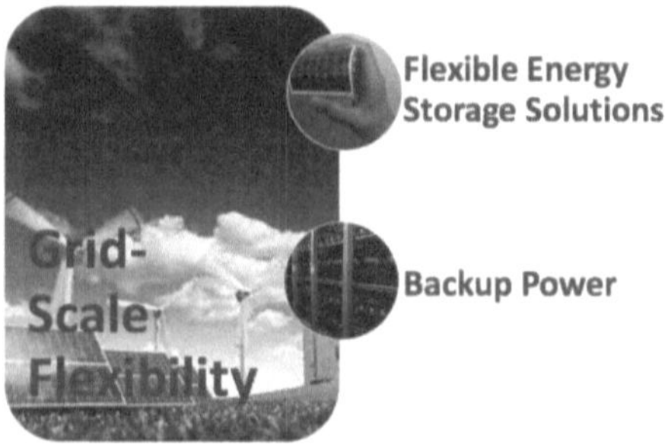

Figure 3.5 The structure of network-wide resilience.

et al., 2024] (Figure 3.5). Graphene technology could contribute to imbuing modern electrical grids with the flexibility they require in a variety of ways, which include the following:

Flexible energy storage solutions: Energy storage systems enhanced with graphene can deliver the fast-response capability called for in stabilizing the grid [Abaszade et al., 2023a,b,c; Boychuk et al., 2022; Figarova et al., 2023; Mammadov et al., 2023]. With graphene in their energy storage devices, utilities would have the capability to ramp up or ramp down electricity generation in response to changing demand or renewable output.

Backup power: Graphene-based materials will enable efficiency improvements in backup power systems that can provide energy when demand is at a peak or when renewable generation is low [Abaszade et al., 2024a,b]. These can work in concert with traditional natural gas plants or other flexible generation sources to maintain grid stability.

3.5.3 Virtual power plants—VPPs

Graphene is one single layer of carbon atoms arranged in a two-dimensional lattice and is considered for its unique properties, arousing tremendous interest in various fields: outstanding electrical conductivity, robust mechanical strength, and being lightweight [Zhou and Wang, 2021]. Virtual power plants (VPPs) are networks that allow a large number of dispersed energy resources, like solar panels, wind turbines, battery storage systems, and flexible loads to act together as a single generating station. VPPs utilize sophisticated software and communication technologies to monitor and manage output/uptake across the network, ensuring optimized performance and stability of the grid [Martinez and Kouramajian, 2020].

3.5.4 Data analytics and optimization

With the energy world increasingly going into a decentralized renewable energy environment, management becomes all the more important. VPPs do this by interconnecting different types of distributed energy resources—solar panels, wind turbines, and battery storage devices. Graphene's lightweight nature makes it ideal for developing portable and flexible computing devices [Khanmammadova et al., 2024]. These can be deployed at several locations, therefore, easing the collection of data from remote energy resources. This flexibility makes sure that VPPs can handle distributed resources with efficiency, no matter the geographical constraint.

3.6 CONCLUSION

This chapter has highlighted the fundamentally important role that conversion technologies play in the effective harnessing of renewable energy sources. Each one of these technologies is based on well-understood principles of physics and therefore provides scientific grounding for the design and understanding of efficiency within the technology. This enables innovation and optimization of the system by both researchers and engineers toward the goal of maximum energy capture with minimum environmental impact. Such technologies might, therefore, be considered, with regard to the fast-growing global demand for clean and sustainable energy, major drivers toward the future path of building a sustainable energy environment. Continuous investment in research and innovation will no doubt make renewable energy conversion technology more efficient and affordable to meet ever-growing global demand for energy, while at the same time meeting challenges related to climate change. The statistics presented here further drive home the pace with which renewable energy technologies are advancing, especially regarding gains in efficiency, cost, and the environment. In fact, all solar, wind, hydropower, bioenergy, and even geothermal energy systems have shown remarkable development, which again increases the share of renewable energy in the world's electricity mix. And insofar as these technologies continue to mature, they will lie at the very heart of reducing carbon emissions and mitigating the impacts of climate change that allow for a more sustainable energy future. Furthermore, energy storage and grid integration play a very important role in overcoming the issues related to intermittency in these renewable sources of energy. Technologies like battery storage, smart grids, and supply-demand balancing mechanisms do enable the utilization of renewable energy sources effectively by maintaining stability in the grid and reliability therein. Advanced energy storage and grid management technologies will play a very important role in the development and deployment process, particularly in a future where renewable energy makes up an increasingly larger portion of global electricity generation. Investing in such technologies, combined with fostering flexible and intelligent energy systems, will provide a globally transitioning energy sector with the ability to move toward a renewable future without sacrificing grid reliability and operational efficiency. Within this constantly shifting energy environment, graphene now clearly appears as the game-changing material that will make a difference by boundlessly upgrading VPP performance. By enabling an increase in data analysis capabilities, allowing for effective and accurate forecasting, graphene lets VPPs attain knowledgeable decision-making to balance out supply and demand in the increasingly renewable energy-dominated grid. In this way, the implementation of advanced materials like graphene into the energy management system will play a key role in accelerating renewable energy transition toward sustainability, efficiency, and resilience in the energy ecosystem. The super conductivity and lightweight properties of graphene open exciting opportunities for enhancements in data analytics and optimization at VPPs to eventually contribute to robust renewable energy infrastructure. With continuous development and discovery, graphene might make a vital change in energy storage and data analytics and hence will play a significant role in deciding the fate of renewable energy systems. Building from the very latest advance, we can ensure that not only do our energy systems meet the immediate high-pressing demands but also are in confluence with our long-term goals about sustainability and care for the environment toward a greener and robust energy future.

REFERENCES

Abaszade, R.G. (2022). Synthesis and Analysis of Flakes Graphene Oxide. Journal of Optoelectronic and Biomedical Materials, 14(3), 107–114.

Abaszade, R.G., Aliyev, E.M., Babanli, M.B., Kotsyubynsky, V.O., Zapukhlyak, R.I., Mammadov, A.G., Budak, H.F., Kasapoglu, A.E., Gur, E., Margitych, T.O., Stetsenko, M.O. (2023a). Investigation of Thermal Properties of Carbon Nanotubes and Carboxyl Group—Functionalized Carbon Nanotubes. Physics and Chemistry of Solid State, 24(3), 530–535.

Abaszade, R.G., Aliyev, E.M., Mammadov, A.G., Khanmamadova, E.A., Guliyev, A.A., Aliyev, F.G., Zapukhlyak, R.I., Budak, H.F., Kasapoglu, A.E., Margitych, T.O., Singh, A., Arya, S., Gür, E., Stetsenko, M.O. (2024a). Investigation of Thermal Properties of Gadolinium Doped Carbon Nanotubes. Physics and Chemistry of Solid State, 25(1), 142–147.

Abaszade, R.G., Babanli, M.B., Kotsyubynsky, V.O., Mammadov, A.G., Gür, E., Kapush, OA., Stetsenko, M.O., Zapukhlyak, R.I. (2023b). Influence of Gadolinium Doping on Structural Properties of Carbon Nanotube. Physics and Chemistry of Solid State, 24(1), 153–158.

Abaszade, R.G., Kapush, O.A., Mamedova, S.A., Nabiyev, A.M., Melikova, S.Z., Budzulyak, S.I. (2020a). Gadolinium Doping Influence on the Properties of Carbon Nanotubes. Physics and Chemistry of Solid State, 21(3), 404–408.

Abaszade, R.G., Kapush, O.A., Nabiyev, A.M. (2020b). Properties of Carbon Nanotubes Doped with Gadolinium. Journal of Optoelectronic and Biomedical Materials, 12(3), 61–65.

Abaszade, R.G., Mammadov, A.G., Bayramov, I.Y., Khanmamadova, E.A., Kotsyubynsky, V.O., Kapush, O.A., Boychuk, V.M., Gur, E.Y. (2022a). Structural and Electrical Properties of Sulfur-Doped Graphene Oxide/Graphite Oxide Composite. Physics and Chemistry of Solid State, 25(2), 256–260.

Abaszade, R.G., Mammadov, A.G., Bayramov, I.Y., Khanmamadova, E.A., Kotsyubynsky, V.O., Gur, E.Y., Kapush, O.A. (2022b). Modeling of Voltage-Ampere

Characteristic Structures on the Basis of Graphene Oxide/Sulfur Compounds. Technical and Physical Problems of Engineering, 14(2), 302–306.
Abaszade, R.G., Mammadov, A.G., Khanmamedova, E.A., Aliyev, F.G., Kotsyubynsky, V.O., Gür, E., Soltabayev, B.D., Margitych, T.O., Stetsenko, M.O., Singh, A., Arya, S. (2024b). Photoconductivity of Functionalized Carbon Nanotubes. Digest Journal of Nanomaterials and Biostructures, 19(2), 837–843.
Abaszade, R.G., Mammadova, S.A., Kapush, O.A., Agaev, F.G., Nabiev, A.M., Mammadova, M.M., Budzulyak, S.I., Kotsyubynsky, V.O. (2021). Synthesis and Characterization of Graphene Oxide Flakes for Transparent Thin Films. Physics and Chemistry of Solid State, 23(3), 595–601.
Abaszade, R.G., Mammadova, A.G., Khanmammadova, E.A., Bayramov, I.Y., Namazov, R.A., Popal, K.M., Melikova, S.Z., Qasımov, R.C., Bayramov, M.A., Babayeva, N.I. (2023c). Electron Paramagnetic Resonance Study of Gadoliniumum Doped Graphene Oxide. Journal of Ovonic Research, 19(2), 259–263.
Al Asi, A., Khashehchi, M., Thangavel, S., Kumar, A., Prasad, A. (2024). Strategic Approach to Biomass Energy Production: Feasibility, Challenges, and Opportunities (pp. 87–112). CRC Press. http://dx.doi.org/10.1201/9781003502012-4
Auerbach, S.I., Hamilton, W.J. (1958). The Betz Limit and the Wind Energy Conversion. The Physics Teacher, 36(7), 405–411.
Baxter, L. (2005). Biomass to Renewable Energy Processes. CRC Press.
Betz, A. (1966). Introduction to the Theory of Flow Machines. Pergamon Press.
Bioenergy Europe (2023). Bioenergy Statistical Report 2023. Bioenergy Europe, Brussels.
Bloomberg New Energy Finance (BNEF) (2022). Global Renewable Energy Market Outlook 2022. Bloomberg LP.
Boychuk, V.M., Zapukhlyak, R.I., Abaszade, R.G., Kotsyubynsky, V.O., Hodlevsky, M.A., Rachiy, B.I., Turovska, L.V., Dmytriv, A.M., Fdorchenko, S.V. (2022). Solution Combustion Synthesized NiFe2O4/reduced Graphene Oxide Composite Nanomaterials: Morphology and Electrical Conductivity. Physics and Chemistry of Solid State, 23(2), 815–824.
Boyle, G. (2012). Renewable Energy: Power for a Sustainable Future. Oxford University Press.
Boyle, G., Everett, B., Ramage, J. (2012). Energy Systems and Sustainability: Power for a Sustainable Future. Oxford University Press.
Burton, T., Sharpe, D., Jenkins, N., Bossanyi, E. (2011). Wind Energy Handbook. John Wiley & Sons.
DiPippo, R. (2016). Geothermal Power Plants: Principles, Applications, Case Studies and Environmental Impact. Butterworth-Heinemann.
Dunn, B., Kamath, H., Tarascon, J.M. (2011). Electrical Energy Storage for the Grid: A Battery of Choices. Science, 334(6058), 928–935.
Ellabban, O., Abu-Rub, H., Blaabjerg, F. (2014). Renewable Energy Resources: Current Status, Future Prospects and Their Enabling Technology. Renewable and Sustainable Energy Reviews, 39, 748–764.
Energy & Environmental Science (2023). Advances in Photovoltaics and Solar Cell Efficiency. The Royal Society of Chemistry.
Evans, A., Strezov, V., Evans, T.J. (2010). Assessment of Sustainability Indicators for Renewable Energy Technologies. Renewable and Sustainable Energy Reviews, 13(5), 1082–1088.
Figarova, S.R., Aliyev, E.M., Abaszade, R.G., Alekberov, R.I., Figarov, V.R. (2021). Negative Differential Resistance of Grapheme Oxide|Sulphur Compound. Journal of Nano Research, 67, 25–31.
Figarova, S.R., Aliyev, E.M., Abaszade, R.G., Figarov, V.R. (2023). Negative Thermal Expansion of Sulphur-Doped Graphene Oxide. Advanced Materials Research, 1175, 55–62.
Fridleifsson, I.B. (2001). Geothermal Energy for the Benefit of the People. Renewable and Sustainable Energy Reviews, 5(3), 299–312.

Global Wind Energy Council (GWEC). (2023). Global Wind Report 2023. Retrieved from https://gwec.net/
Green, M.A. (2015). Photovoltaic Technology and Its Evolution. Wiley Interdisciplinary Reviews: Energy and Environment, 4(2), 183–205.
Guliyeva, N.A., Abaszade, R.G., Khanmammadova, E.A., Azizov, E.M. (2023). Synthesis and Analysis of Nanostructured Graphene Oxide. Journal of Optoelectronic and Biomedical Materials, 15(1), 23–30.
Intergovernmental Panel on Climate Change (IPCC). (2023). Climate Change 2023: Mitigation of Climate Change. Cambridge University Press.
International Energy Agency (IEA). (2023). Renewable Energy Market Update 2023. IEA Publications, Paris.
International Geothermal Association (IGA). (2022). Global Geothermal Energy Market Outlook. Retrieved from https://www.geothermal-energy.org/
International Hydropower Association (IHA). (2023). Hydropower Status Report 2023. Retrieved from https://www.hydropower.org/publications
International Renewable Energy Agency (IRENA). (2023). Renewable Power Generation Costs in 2022. Retrieved from https://www.irena.org/Publications
Ivanichok, N., Kolkovskyi, P., Ivanichok, O., Kotsyubynsky, V., Boychuk, V., Rachiy, B., Bembenek, M., Warguła, Ł, Abaszade, R., Ropyak, L. (2024). Effect of Thermal Activation on the Structure and Electrochemical Properties of Carbon Material Obtained from Walnut Shells. Materials, 17, 2514–21.
Jiang, X., Dou, H., Pan, H., Chen, Z. (2020). Recent Advances in Solid-State Batteries: Material Design and Application. Energy Storage Materials, 29, 161–189.
Jena, N.K., Sahoo, S., Sahu, B.K., Naik, A.K., Bajaj, M., Misak, S., ... Prokop, L. (2023). Impact of a Redox Flow Battery on the Frequency Stability of a Five-Area System Integrated with Renewable Sources. Energies, 16(14), 5540. http://dx.doi.org/10.3390/en16145540.
Kalogirou, S.A. (2009). Solar Energy Engineering: Processes and Systems. Academic Press.
Kalogirou, S.A. (2014). Solar Energy Engineering: Processes and Systems. Academic Press.
Khanmammadova, E., Abaszade, R., Muradi, R. (2024). Antiviral Thin Surface and the Surface Life. In Miltifunctionla Coatings and İnterfaces for Tribological Contacts (pp. 124–134). CRC Press.
Kumar, A., Gangwar, A.K.S., Patra, B.B., Dwivedi, G. (2024). Optimizing Solar Air Heater Performance: A Comparative Analysis of Artificial Roughness Geometries. In Book Clean Energy: Technology, Advances, and Applications. http://dx.doi.org/10.1201/9781003521341-12
Lazard. (2022). Levelized Cost of Energy Analysis—Version 16.0. Retrieved from https://www.lazard.com/
Liu, K., Wu, X., Yuan, S. (2021). Lithium-Ion Battery Technologies for Grid-Scale Energy Storage. Renewable and Sustainable Energy Reviews, 146, 111133.
Loferski, J.J. (1956). Theoretical Considerations Governing the Choice of the Optimum Semiconductor for Photovoltaic Solar Energy Conversion. Journal of Applied Physics, 27(7), 777–784.
Lund, J.W., Freeston, D.H., Boyd, T.L. (2010). Direct Utilization of Geothermal Energy 2010 Worldwide Review. Geothermics, 39(3), 189–214.
Luo, X., Wang, J., Dooner, M., Clarke, J. (2015). Overview of Current Development in Electrical Energy Storage Technologies and the Application Potential in Power System Operation. Applied Energy, 137, 511–536.
Mammadov, A.G., Abaszade, R.G., Babanli, M.B., Kotsyubynsky, V.O., Gur, E., Soltabayev, B.D., Margitych, T.O., Stetsenko, M.O. (2023). Photoconductivity of Gadolinium-Doped Carbon Nanotubes. International Journal on Technical and Physical Problems of Engineering, 15(3), 53–58.
Mammadov, A.G., Abaszade, R.G., Kotsyubynsky, V.O., Gur, E.Y., Bayramov, I.Y., Khanmamadova, E.A., Kapush, O.A. (2022). Photoconductivity of Carbon Nanotubes. Technical and Physical Problems of Engineering, 14(3), 155–160.

Manwell, J.F., McGowan, J.G., Rogers, A.L. (2009). Wind Energy Explained: Theory, Design and Application. Wiley.

Manwell, J.F., McGowan, J.G., Rogers, A.L. (2010). Wind Energy Explained: Theory, Design and Application. Wiley.

Martinez, C., Kouramajian, V. (2020). Advanced Data Analytics in Virtual Power Plants: Leveraging Graphene's Unique Properties. Energy Informatics, 3(1), 29–45.

McKendry, P. (2002). Energy Production from Biomass (part 1): Overview of Biomass. Bioresource Technology, 83(1), 37–46.

Overend, R.P., Milne, T.A. (2002). Biomass Conversion Technology Principles and Practices. Kluwer Academic Publishers.

Quaschning, V. (2016). Renewable Energy and Climate Change. John Wiley & Sons.

Rawat, T., Niazi, K. R., Gupta, N., & Sharma, S. (2021). Multi-Objective Techno-Economic Operation of Smart Distribution Network Integrated with Reactive Power Support of Battery Storage Systems. Sustainable Cities and Society, 75, 103359. http://dx.doi.org/10.1016/j.scs.2021.103359.

REN21. (2023). Renewables 2023 Global Status Report. REN21 Secretariat, Paris.

Sharma, B., Singh, A., Sharma, A., Dubey, A., Gupta, V., Abaszade, R.G., Sundramoorthy, A.K., Sharma, N., Arya, S. (2024). Synthesis and Characterization of Zinc selenide/graphene Oxide (ZnSe/GO) Nanocomposites for Electrochemical Detection of Cadmium Ions. Applied Physics A, 130, 297.

Smil, V. (2017). Energy and Civilization: A History. MIT Press.

Solar Energy Industries Association (SEIA). (2022). U.S. Solar Market Insight Report. Retrieved from https://seia.org/

Soloveichik, G.L. (2011). Flow Batteries: Current Status and Trends. Chemical Reviews, 111(7), 5146–5196.

Sørensen, B. (2011). Renewable Energy: Physics, Engineering, Environmental Impacts, Economics & Planning. Academic Press.

Tester, J.W., Drake, E.M., Driscoll, M.J., Golay, M.W., Peters, W.A. (2005). Sustainable Energy: Choosing among Options. MIT Press.

The Carbon Trust (2022). The Role of Renewable Energy in Decarbonizing the Global Economy. Carbon Trust, London.

Twidell, J., Weir, T. (2015). Renewable Energy Resources. Routledge.

U.S. Energy Information Administration (EIA). (2022). Annual Energy Outlook 2022. U.S. Department of Energy, Washington, DC.

United Nations Environment Programme (UNEP). (2023). Global Trends in Renewable Energy Investment. Retrieved from https://www.unep.org/

Whittingham, M.S. (2004). Lithium Batteries and Cathode Materials. Chemical Reviews, 104(10), 4271–4302.

World Bank Group (2022). State of the Global Clean Energy Market 2022. World Bank, Washington, DC.

Zhou, Y., Wang, L. (2021). Graphene-Based Technologies in Virtual Power Plants: A New Era for Renewable Energy Systems. IEEE Transactions on Smart Grid, 12(1), 356–370.

Chapter 4

Fluid dynamics and heat transfer enhancement techniques in solar thermal renewable energy systems

Aman Kumar and Ambrish Maurya

4.1 INTRODUCTION

Renewable energy resources have gained significant momentum in the 21st century. This is because in the present time, the changes in climate, the increasing cost of fossil fuels day by day, and environmental degradation are major concerns (Ilbahar et al., 2019).

Solar thermal energy systems have emerged as a promising technology for harnessing renewable energy from the sun. These systems work by converting solar radiation into heat, which is then transferred to a working fluid that stores, transfers, and utilizes the energy. Heat obtained as an output from a solar thermal energy system is directly or indirectly used in various applications. For example, it is used in water heating, industrial process heating, and even converted into electricity (which is one of the very important applications) (Aktar et al., 2021; Kober et al., 2020). Solar thermal energy systems typically include components such as solar collectors, storage units, heat exchangers, pump and piping systems as shown in Figure 4.1.

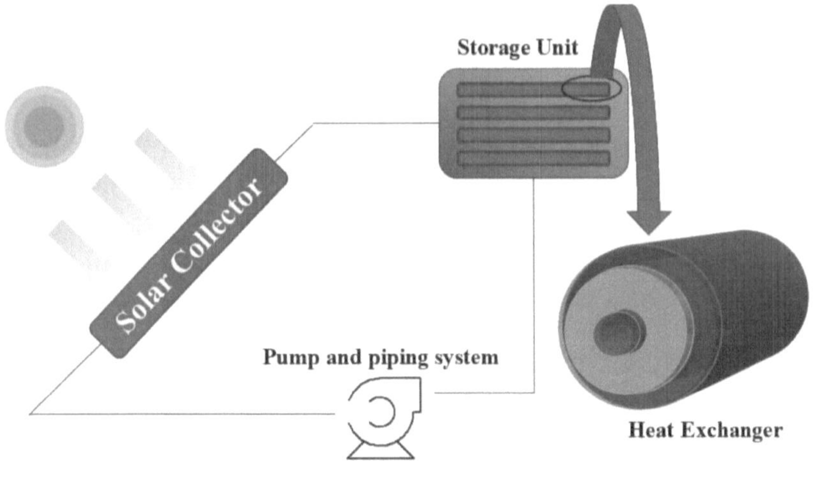

Figure 4.1 Solar thermal energy system.

DOI: 10.1201/9781003534785-4

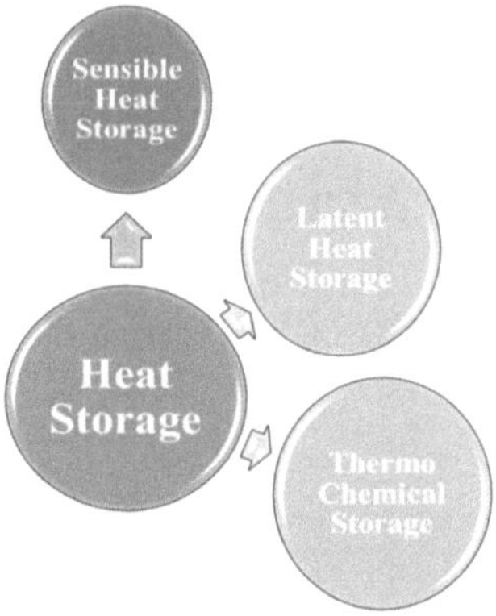

Figure 4.2 Types of heat storage.

The storage unit plays a critical role in ensuring the reliability and efficiency of the system because efficient harnessing, transferring, storage, and utilization of solar thermal energy present significant challenges. The storage medium of storage units can be selected based on the type of heat storage as shown in Figure 4.2. One of the key hurdles is the intermittent nature of solar radiation, which requires effective storage unit solutions to ensure a continuous energy supply, even during periods of low sunlight or nighttime (Mahdi et al., 2019). This is where fluid dynamics and heat transfer mechanisms come into play, offering innovative ways to improve the efficiency, reliability, and scalability of solar thermal energy systems. Understanding and optimizing the behavior of fluids within these systems is crucial for enhancing overall system performance. Whether it's the working fluid in a solar collector, the heat transfer fluid (HTF) in a heat exchanger, or the phase change material (PCM) in a storage unit, the dynamics of fluid flow and heat transfer have a profound impact on the system's efficiency.

This chapter aims to provide a detailed understanding of fluid dynamics and heat transfer enhancement techniques in solar thermal energy systems. It explores the fundamental principles of heat transfer—convection, conduction, and radiation—and examines advanced techniques such as nanofluids, PCMs, extended surfaces, and turbulence promoters to improve system efficiency. The chapter also highlights computational and experimental approaches for optimizing heat transfer, along with practical challenges in integrating these techniques into real-world applications. Additionally, it assesses the impact of these enhancements on energy efficiency and sustainability, offering insights into reducing thermal losses and enhancing system performance.

4.2 FLUID DYNAMICS IN SOLAR THERMAL ENERGY SYSTEMS

Fluid dynamics play a crucial role in the effective and efficient operation of any solar thermal energy system and in minimizing overall energy losses from the system. This is because fluid dynamics directly influences the heat transfer process happening within the system. Various fluid flow characteristics, such as velocity and turbulence mixing, determine how effectively heat is transported within the system. These are aspects of fluid dynamics, such as flow designs. The HTF in the system can have either laminar or turbulent flow, which depends on factors like the Reynolds number, the system's geometry, and the properties of the fluid. If the fluid flows turbulently, there will be good mixing, resulting in an increase in overall heat transfer. However, this will also cause a pressure drop, requiring additional pumping power. Therefore, optimizing fluid flow is necessary, and fluid dynamics plays a significant role in optimization fluid flow of solar thermal energy systems (Andreozzi et al., 2020; Salomoni et al., 2008).

The efficiency of solar thermal energy systems is also affected by the flow patterns of different substances in other various parts of the system. For instance, the HTF, which is generally water, air, or oil, circulates through the solar collector. The flow pattern of this fluid greatly impacts the system's efficiency. Even the flow patterns within the storage unit, which typically uses PCM in its molten state, influence the storage unit's performance. In this way, fluid dynamics play a crucial role in the overall efficiency of solar thermal energy systems (Migliari et al., 2016).

The primary governing equations for fluid flow includes the continuity equation, the Navier-Stokes equations, and the energy equation. These describe the conservation of mass, momentum, and energy, respectively. The interaction between fluid flow and heat transfer in a solar thermal energy system involves convection, conduction, and, in some cases, radiation.

4.2.1 Governing equations

The behavior of fluid flow in solar thermal energy systems is governed by the fundamental principles of fluid mechanics, including the continuity equation, momentum equation (Navier-Stokes equations), and energy equation (Esapour et al., 2018).

Continuity equation: Ensures the conservation of mass in the fluid flow.

$$\frac{\partial \rho}{\partial t} + \nabla \cdot (\rho \mathrm{v}) = 0$$

where ρ is the fluid density, t is time, and v is the velocity vector.

Momentum equation: Describes the conservation of momentum and includes forces such as pressure, viscous, and external body forces.

$$\rho \frac{\partial \mathrm{v}}{\partial t} + \rho (\mathrm{v} \cdot \nabla) \mathrm{v} = -\nabla p + \mu \nabla^2 \mathrm{v} + \rho g$$

where p is the pressure, μ is the dynamic viscosity, and g is the gravitational acceleration.

Energy equation: Ensures the conservation of energy, accounting for conduction, convection, and internal heat sources.

$$\rho c_p \frac{\partial T}{\partial t} + \rho c_p (\mathrm{v} \cdot \nabla T) = k\nabla^2 T + q$$

where c_p is the specific heat capacity, T is the temperature, k is the thermal conductivity, and q is the internal heat generation.

4.3 HEAT TRANSFER MECHANISMS IN SOLAR THERMAL ENERGY SYSTEMS

The heat transfer mechanism plays a very important role in the efficient working of solar thermal energy systems. The heat transfer mechanisms seen in any solar thermal energy system include conduction, convection, and radiation, which mostly impact the solar collector and the storage unit of the solar thermal energy system. To design an efficient solar thermal energy system, it is essential to understand these three mechanisms happening in the system (Ghazouani et al., 2020; Hami et al., 2012).

4.3.1 Conduction

Conduction is the transfer of heat through a solid or stationary fluid due to temperature gradients. When solar energy is absorbed by the absorber plate of the solar thermal energy system, this energy is transferred to the circulating working fluid via conduction. Conduction heat transfer is also observed in the storage unit or containment vessel of the solar thermal energy system. In the storage unit, which typically uses PCM, conduction plays an important role in internal heat transfer when the PCM is in a solid state. The more conductive the heat transfer, the better the performance of the PCM-based storage unit, and consequently, the overall performance of the solar thermal energy system improves. The thermal conductivity of the storage unit is a critical parameter influencing the rate of heat transfer. Materials with high thermal conductivity allow rapid heat distribution, enhancing the overall efficiency of the solar thermal energy system.

4.3.2 Convection

Convection is the transfer of heat through the movement of a fluid, either by natural buoyancy-driven flow (natural convection) or by forced circulation (forced convection). The HTF in a solar thermal energy system, which circulates through the solar collector or storage unit, is largely governed by natural and forced convection. For instance, the working fluid, which absorbs heat from the absorbing plate, flows through the collector and transfers its heat to the storage unit via forced convection. In systems where hot fluid (less dense) and cold fluid (denser) are involved, natural convection is observed, a phenomenon known as stratification. Natural convection also plays an important role in the performance of PCM within the storage unit, particularly when the PCM melts into its liquid form. Studies have found that natural convection significantly affects the overall performance of such systems, and some of these studies are discussed in later sections.

4.3.3 Radiation

Radiation is the transfer of heat in the form of electromagnetic waves, primarily in the infrared spectrum. In any solar thermal energy system, radiation plays an important role as solar radiation must be absorbed. However, for components like the storage unit and piping system, radiation plays a much smaller role compared to conduction and convection. But if the solar thermal energy system operates at high temperatures or has a large surface area exposed to the environment, radiative heat transfer becomes much more significant. Without controlling radiative heat transfer losses, the overall thermal efficiency of the solar thermal energy system cannot be maintained.

4.4 HEAT TRANSFER ENHANCEMENT TECHNIQUES IN SOLAR THERMAL ENERGY SYSTEMS

Enhancing heat transfer in solar thermal energy systems is vital for maximizing energy storage and retrieval efficiency and thus maximizing the overall thermal efficiency of solar thermal energy systems. Although many techniques can be used for this, often small approaches can show significant improvement in overall heat transfer. For instance, instead of using conventional fluids in various components of a solar thermal energy system, replacing them with nanofluids—fluids mixed with nanoparticles—can result in notable improvements in the system's overall heat transfer. Similarly, proper connection of the piping system in solar thermal energy systems also has a major effect on overall heat transfer. Additionally, modifying the surface roughness and geometry of these piping systems to induce turbulence can enhance overall convective heat transfer. Various other approaches, such as the use of vortex generators or swirl flow techniques, are also employed to improve convective heat transfer in solar thermal energy systems (Ferchichi et al., 2019).

Some major heat transfer enhancement techniques are also applied in the storage units of solar thermal energy systems to maximize storage and retrieval efficiency, thereby improving overall thermal efficiency. One critical component of the storage unit is the heat exchanger as shown in Figure 4.1, whose primary function is to transfer heat between the HTF and the storage medium. The design of the heat exchanger directly impacts the system's efficiency, heat transfer rate, and the pressure drop across the heat exchanger.

4.4.1 Enhancing heat exchanger performance

Researchers have applied various design strategies to enhance the performance of heat exchangers, one of which is the optimization of the heat exchanger's surface area. The larger the expanded surface area between the storage unit and the HTF, the more effectively heat will transfer between them (Lunsford, 1998). Keeping this in mind, researchers have tried different shapes and configurations of heat exchangers, including spherical, rectangular, and cylindrical shapes (Assis et al., 2007; Jellouli et al., 2007; Singh & Halder, 2023; Tan et al., 2009). Among these, cylindrical-shaped heat exchangers have proven to be the most effective and have garnered significant attention. In fact, the most commonly used heat exchanger shape in solar thermal energy systems is the cylindrical one. There are various designs of cylindrical heat exchangers, the most studied being the shell-and-tube heat exchanger (Ebadi et al., 2018). In a shell-and-tube heat exchanger, the storage medium is on one adjacent side of the HTF, while insulation is on the other side. There have been many improvements, such as the use of triple-tube-type heat exchangers, which are commonly used as heat exchangers in solar thermal energy systems. This design places storage medium on both adjacent sides of the HTF, thereby exposing more surface area between the storage medium and the HTF, which increases heat transfer. Khatibi et al. (2021) also observed this in their study, where they found that when using a triple-tube storage unit, performance improved compared to when they used a shell-and-tube storage unit Gomaa et al. (2016) investigated the thermo-fluid characteristics for two types of heat exchangers and found that triple-tube heat exchangers showed higher effectiveness and energy savings per unit length compared to double-tube heat exchangers.

To further extend the available heat transfer surface area, extended surfaces such as fins are used in heat exchangers. Finned heat exchangers are not only used in solar thermal energy systems but also in applications like air conditioning, refrigeration, and the automotive industry. Several researchers have investigated different types of fins in heat exchangers. For example, Kang et al. (1994) studied plain fins, slit fins, corrugated wavy fins, and smooth wavy fins; Yun and Lee (1999) investigated plain fins, slit fins, and louver fins; and Wang et al. (1998) examined convex-louver fins and wavy fins. Many studies have explored the use of fins in the heat transfer region of heat exchangers to enhance fluid flow. Moreover, fins are also extensively used in the storage medium region of heat exchangers, which will be discussed in detail in a later section. Another design strategy for enhancing heat exchanger performance is to increase turbulence within the heat exchanger. As discussed in previous section, when the flow of the HTF transitions from laminar to turbulent, increased mixing is observed, leading to greater heat transfer. Therefore, in some cases, mechanisms are added to the heat exchanger to achieve turbulent flow in the HTF. For example, Yildiz et al. (1996) added a propeller effect to the HTF flowing through the pipes of a heat exchanger to induce rotation, and they found that this increased the heat transfer rate by approximately

250%. El-Maghlany et al. (2012) conducted an experimental study on a rotating inner pipe in a double-pipe heat exchanger and found that increasing the rotational speed raised the Reynolds number of the fluid flow, resulting in an overall increase in heat transfer rate. Another strategy to enhance heat exchanger performance is selecting high thermal conductivity materials for the heat exchanger. Therefore, in most studies, the materials selected for heat exchangers are typically copper, aluminum, brass, etc., as these materials have high thermal conductivity. For example, Beemkumar et al. (2017) used copper, aluminum, and brass for the heat exchanger and extended surfaces, finding that during the charging process, copper fins reduced heat loss by 15%, aluminum fins by 10%, and brass fins by 15%, compared to the case without fins. Farahani et al. (2022) examined the effect of different fin materials on PCM-based storage units during the charging process and found that using copper fins enhanced the melting rate by 3% compared to aluminum fins. Moreover, the thermal conductivity of the storage medium inside the heat exchanger also significantly affects the overall heat transfer, which will be discussed in a later section.

4.4.2 Incorporation of phase change materials

Other major heat transfer enhancement technique applied in solar thermal energy system storage units is the incorporation of PCMs. Since PCM absorbs and releases heat in the form of latent heat during phase transitions, it can store and release a large amount of heat. Therefore, integrating PCM into solar thermal energy systems enhances the system's storage and retrieval capacity, allowing for a more consistent energy supply. PCM primarily plays a role as the storage medium inside the heat exchanger. Using PCM as a storage medium enables more efficient energy storage and reduces temperature fluctuations, ensuring a steady supply of heat during times when direct sunlight is not available. However, significant issues are observed with PCM in the storage medium region of heat exchangers, one of the major issues being that PCM has very low thermal conductivity. This results in a slow phase transition (referred to as the charging and discharging process), degrading the performance of the storage unit and the overall solar energy system.

To overcome the issue of low thermal conductivity in PCM, various techniques are used. One popular technique is the addition of nanoparticles to PCM. Rabienataj Darzi et al. (2016) added copper nanoparticles to PCM in a storage unit and found that this increased the melting and solidification rates. Nakhchi et al. (2021) conducted a similar study and found that the addition of nanoparticles significantly improved the melting performance, with the nanoparticles acting as heat sinks that absorb heat. Another popular technique for enhancing the thermal conductivity of PCM is embedding it in metallic foam. Esapour et al. (2018) embedded PCM in metallic foam in a storage unit and studied its behavior during the melting and solidification processes. They found improvements in both processes, though the solidification process saw greater enhancement compared to the melting process. Deng et al. (2017) studied PCM embedded in porous metal foam in a storage unit and found that the performance of the PCM was enhanced but also depended on the foam's porosity. Authors have found that for melting, porous metal foam with a smaller fractal dimension was more beneficial.

The third thermal conductivity enhancement technique for PCMs, which is popular and can be found in many research papers, is the addition of fins to the PCM region of the storage unit, meaning metallic fins are added to the PCM region of the storage unit. The concept is similar to what was discussed earlier in that the fin acts as an extended surface, achieving better heat transfer. A lot of studies have been conducted by researchers on the addition of metallic fins, such as Kirincic et al. (2021), who studied straight fins in the PCM region of the storage unit and found that when straight fins, which are the most conventional shape for fins, are added, the phase transition time improves. Apart from this conventional straight shape, researchers have also studied various other shapes of fins, such as "V" shaped fins, "Y" shaped fins, and different kinds of branched fins. Some have studied leaf-shaped branched fins, some tree-shaped branched fins, and others snowflake-shaped branched fins (Jiang & Huang, 2023; Ren et al., 2022; Sciacovelli et al., 2015; Shen et al., 2023; Shouguang et al., 2023).

So, different shapes of fins have been studied, with the same concept of using fins to increase the extended surface area, thus increasing the heat transfer area and improving heat transfer. However, fins can also be used to improve the flow dynamics of PCM in the PCM region and create turbulence, as discussed in earlier sections. Turbulence also results in better mixing and improved heat transfer. When PCM flow dynamics improve, better mixing and heat transfer are observed, which is why some innovative fin shapes have been used in research, such as curved shapes. Khedher, Mahdi, et al. (2023) demonstrated that adjusting the length and angle of arc-shaped fins in the PCM region of the storage unit significantly improved heat transfer and phase change, with optimization of fin array parameters boosting both heat transfer and melt PCM movement. In a recent study (Kumar & Maurya, 2023), we also analyzed the effects of amplitude and peak numbers of wavy-shaped curved fins, concluding that increasing amplitude notably enhanced phase change, though peak number showed no significant effect. Abed et al. (2022) evaluated the impact of arc-shaped fins in both upward and downward orientations, finding significant enhancements in PCM melting rates and thermal properties. Their results emphasized the positive role of staggered fin layouts in boosting melt PCM movement and melting efficiency. Khedher, Biswas, et al. (2023) conducted computational simulations to study various fin structures, finding that reverse arc-shaped fins resulted in the fastest PCM melting and overall superior performance. Their sensitivity analysis linked improved heat transfer to higher Reynolds numbers.

4.5 NUMERICAL MODELING AND SIMULATION

From the discussion in the previous sections, it is clear how important the role of fluid dynamics and heat transfer is for the effective performance of a solar thermal energy system. To optimize and analyze fluid dynamics and heat transfer in a solar thermal energy system, a very important and powerful tool is used, known as computational fluid dynamics (CFD) modeling. With the help of CFD, when numerical simulations are carried out, we can easily obtain detailed information such as temperature distribution, heat transfer rate, etc., in different parts of the solar thermal energy system, which could have been very difficult to obtain through experiments. This is because in CFD, what we do is solve all the governing equations of fluid dynamics and heat transfer so that we can estimate the velocity, temperature, pressure distribution, etc. CFD modeling is also a very effective tool for the storage unit because the commonly used storage medium inside the heat exchanger of the storage unit is PCM, in which phase change occurs as solidification and melting, which is a very complex phenomenon in itself. But for such complex phenomena, CFD tools like ANSYS are very useful, and these tools have been implemented by many researchers. CFD is also very useful for simulating the interaction between the HTF and the storage medium, and it is also helpful for optimizing and designing components of the storage unit, such as the design of the heat exchanger and the configuration of the HTF pipes. By the way, the most important role of CFD modeling is that it makes the parametric study of various parameters involved in a solar thermal energy system very easy. For example, there can be many parameters related to the storage unit, such as flow rate, inlet temperature, the geometry of the tubes used in the heat exchanger, their size, shape, and the properties of the PCM used in the storage region of the heat exchanger. To evaluate the performance of all these, CFD proves to be very useful and has been used in many studies (Joybari et al., 2017; Tavakoli et al., 2023).

4.6 CONCLUSION AND FUTURE SCOPE

Solar thermal energy systems represent a vital technology in harnessing renewable energy. By capturing solar radiation and converting it into heat, these systems provide a sustainable solution for applications ranging from water heating to electricity generation. The efficiency and performance of solar thermal energy systems depend on critical factors such as fluid dynamics, heat transfer mechanisms, and the design of storage units. Through the use of PCMs in storage units, it becomes possible to efficiently store heat for later use, ensuring the system's performance during non-sunlight hours. The principles of fluid dynamics—governed by the continuity, momentum, and energy equations—dictate the behavior of the HTFs within the system, influencing overall thermal performance. Additionally, various heat transfer enhancement techniques, such as the use of nanofluids, fin designs, and optimized heat exchangers, have proven effective in improving the heat transfer rate and ultimately increases the system's overall efficiency. CFD modeling tools are essential for optimizing system performance, especially in complex situations such as phase transitions in PCM.

The future development of solar thermal energy systems holds immense promise for improving global energy sustainability. Some potential areas for future research and innovation include:

- Advanced PCM materials: Developing more efficient PCMs for better heat storage and release.
- Nanoparticle-enhanced fluids: Developing and utilizing novel nanofluids to improve heat transfer in existing systems.
- Innovative heat exchanger designs: Creating new designs, like multitube or spiral exchangers, to enhance heat transfer efficiency.
- Optimization of fin configurations: Researching unconventional fin shape, size, angle, and material to improve heat distribution.
- Hybrid systems: Integrating solar thermal energy systems with other renewable technologies (e.g., PV or wind) and energy storage for more reliable power.
- Artificial intelligence (AI) in CFD: Using AI to optimize system designs and predict performance.
- Cost-effective materials: Finding affordable, sustainable materials to reduce system costs.

REFERENCES

Abed, A. M., Mouziraji, H. R., Bakhshi, J., Dulaimi, A., Mohammed, H. I., Ibrahem, R. K., Ben Khedher, N., Yaïci, W., & Mahdi, J. M. (2022). Numerical analysis of the energy-storage performance of a PCM-based triplex-tube containment system equipped with arc-shaped fins. *Frontiers in Chemistry*, *10*(December), 1–18. https://doi.org/10.3389/fchem.2022.1057196

Aktar, M. A., Alam, M. M., & Al-Amin, A. Q. (2021). Global economic crisis, energy use, CO2 emissions, and policy roadmap amid COVID-19. *Sustainable Production and Consumption*, *26*, 770–781. https://doi.org/10.1016/j.spc.2020.12.029

Andreozzi, A., Bianco, N., Iasiello, M., & Naso, V. (2020). Thermo-fluid-dynamics of a ceramic foam solar receiver: A parametric analysis. *Heat Transfer Engineering*, *41*(12), 1085–1099. https://doi.org/10.1080/01457632.2019.1600876

Assis, E., Katsman, L., Ziskind, G., & Letan, R. (2007). Numerical and experimental study of melting in a spherical shell. *International Journal of Heat and Mass Transfer*, *50*(9–10), 1790–1804. https://doi.org/10.1016/j.ijheatmasstransfer.2006.10.007

Beemkumar, N., Karthikeyan, A., & Ramachandran, S. (2017). Heat transfer enhancement of the latent heat storage system using different encapsulating materials with and without fins. *International Journal of Ambient Energy*, *38*(1), 77–84. https://doi.org/10.1080/01430750.2015.1035801

Deng, Z., Liu, X., Zhang, C., Huang, Y., & Chen, Y. (2017). Melting behaviors of PCM in porous metal foam characterized by fractal geometry. *International Journal of Heat and Mass Transfer*, *113*, 1031–1042. https://doi.org/10.1016/j.ijheatmasstransfer.2017.05.126

Ebadi, S., Tasnim, S. H., Aliabadi, A. A., & Mahmud, S. (2018). Melting of nano-PCM inside a cylindrical thermal energy storage system: Numerical study with experimental verification. *Energy Conversion and Management*, *166*(April), 241–259. https://doi.org/10.1016/j.enconman.2018.04.016

El-Maghlany, W., Eid, E., T, M., & Shahrour, I. (2012). Experimental study for double pipe heat exchanger with rotating inner pipe. *International Journal of Advanced Scientific and Technical Research*, *4*(2), 507–524. http://www.rspublication.com/ijst/index.html

Esapour, M., Hamzehnezhad, A., Rabienataj Darzi, A. A., & Jourabian, M. (2018). Melting and solidification of PCM embedded in porous metal foam in horizontal multi-tube heat storage system. *Energy Conversion and Management*, *171*(January), 398–410. https://doi.org/10.1016/j.enconman.2018.05.086

Farahani, S. D., Farahani, A. D., Hajian, E., & Öztop, H. F. (2022). Control of PCM melting process in an annular space via continuous or discontinuous fin and non-uniform magnetic field. *Journal of Energy Storage*, *55*(July). https://doi.org/10.1016/j.est.2022.105410

Ferchichi, S., Kessentini, H., Morales-Ruiz, S., Rigola, J., Bouden, C., & Oliva, A. (2019). Thermal and fluid dynamic analysis of direct steam generation parabolic trough collectors. *Energy Conversion and Management*, *196*(June), 467–483. https://doi.org/10.1016/j.enconman.2019.05.107

Ghazouani, M., Bouya, M., Benaissa, M., Anoune, K., & Ghazi, M. (2020). Thermal energy management optimization of solar thermal energy system based on small parabolic trough collectors for bitumen maintaining on heat process. *Solar Energy*, *211*(October), 1403–1421. https://doi.org/10.1016/j.solener.2020.10.074

Gomaa, A., Halim, M. A., & Elsaid, A. M. (2016). Experimental and numerical investigations of a triple concentric-tube heat exchanger. *Applied Thermal Engineering*, *99*, 1303–1315. https://doi.org/10.1016/j.applthermaleng.2015.12.053

Hami, K., Draoui, B., & Hami, O. (2012). The thermal performances of a solar wall. *Energy*, *39*(1), 11–16. https://doi.org/10.1016/j.energy.2011.10.017

Ilbahar, E., Cebi, S., & Kahraman, C. (2019). A state-of-the-art review on multi-attribute renewable energy decision making. *Energy Strategy Reviews*, *25*(September 2018), 18–33. https://doi.org/10.1016/j.esr.2019.04.014

Jellouli, Y., Chouikh, R., Guizani, A., & Belghith, A. (2007). Numerical study of the moving boundary problem during melting process in a rectangular cavity heated from below. *American Journal of Applied Sciences*, *4*(4), 251–256. https://doi.org/10.3844/ajassp.2007.251.256

Jiang, S., & Huang, Y. (2023). Investigation on the heat discharge performance of a bio-inspired latent heat exchanger with leaf-like fins. *Sustainable Energy Technologies and Assessments*, *57*, 103187. https://doi.org/10.1016/J.SETA.2023.103187

Joybari, M. M., Haghighat, F., Seddegh, S., & Al-Abidi, A. A. (2017). Heat transfer enhancement of phase change materials by fins under simultaneous charging and discharging. *Energy Conversion and Management*, *152*(September), 136–156. https://doi.org/10.1016/j.enconman.2017.09.018

Khatibi, M., Nemati-Farouji, R., Taheri, A., Kazemian, A., Ma, T., & Niazmand, H. (2021). Optimization and performance investigation of the solidification behavior of nano-enhanced phase change materials in triplex-tube and shell-and-tube energy storage units. *Journal of Energy Storage*, *33*(November 2020), 102055. https://doi.org/10.1016/j.est.2020.102055

Khedher, N. B., Biswas, N., Togun, H., Mohammed, H. I., Mahdi, J. M., Ibrahem, R. K., & Talebizadehsardari, P. (2023). Geometry modification of a vertical shell-and-tube latent heat thermal energy storage system using a framed structure with different undulated shapes for the phase change material container during the melting process. *Journal of Energy Storage*, *72*(PB), 108365. https://doi.org/10.1016/j.est.2023.108365

Khedher, N. B., Mahdi, J. M., Dulaimi, A., Chatroudi, I. S., Ebrahimnataj Tiji, M., Ibrahem, R. K., Yvaz, A., & Talebizadehsardari, P. (2023). On the application of novel arc-shaped fins in a shell-and-tube type of latent heat storage for energy charge enhancement. *Journal of Energy Storage*, *73*(PA), 108697. https://doi.org/10.1016/j.est.2023.108697

Kirincic, M., Trp, A., & Lenic, K. (2021). Numerical evaluation of the latent heat thermal energy storage performance enhancement by installing longitudinal fins. *Journal of Energy Storage*, *42*(July), 103085. https://doi.org/10.1016/j.est.2021.103085

Kober, T., Schiffer, H. W., Densing, M., & Panos, E. (2020). Global energy perspectives to 2060—WEC's world energy scenarios 2019. *Energy Strategy Reviews*, *31*(December 2019), 100523. https://doi.org/10.1016/j.esr.2020.100523

Kumar, A., & Maurya, A. (2023). Transient analysis of phase change material-based triple-tube unit with sinusoidal wavy-shaped fin. *Arabian Journal for Science and Engineering*. https://doi.org/10.1007/s13369-023-08525-x

Kang, H. J., Li, W., Li, H. Z., Xin, R. C., & Tao, W. Q. (1994). Experimental study on heat transfer and pressure drop characteristics of four types of plate fin-and-tube heat exchanger surfaces. *Journal of Thermal Science*, *3*(1), 34–42.

Lunsford, K. (1998). Increasing heat exchanger performance. *Hydrocarbon Engineering*, 1–13, 786–793.

Mahdi, J. M., Lohrasbi, S., Ganji, D. D., & Nsofor, E. C. (2019). Simultaneous energy storage and recovery in the triplex-tube heat exchanger with PCM, copper fins and Al_2O_3 nanoparticles. *Energy Conversion and Management*, *180*(November 2018), 949–961. https://doi.org/10.1016/j.enconman.2018.11.038

Migliari, L., Arena, S., Puddu, P., & Cocco, D. (2016). Thermo-fluid dynamic analysis of a CSP solar field line during transient operation. *Energy Procedia*, *101*(September), 1167–1174. https://doi.org/10.1016/j.egypro.2016.11.158

Nakhchi, M. E., Hatami, M., & Rahmati, M. (2021). A numerical study on the effects of nanoparticles and stair fins on performance improvement of phase change thermal energy storages. *Energy*, *215*, 119112. https://doi.org/10.1016/j.energy.2020.119112

Rabienataj Darzi, A. A., Jourabian, M., & Farhadi, M. (2016). Melting and solidification of PCM enhanced by radial conductive fins and nanoparticles in cylindrical annulus. *Energy Conversion and Management*, *118*, 253–263. https://doi.org/10.1016/j.enconman.2016.04.016

Ren, F., Du, J., Cai, Y., Guo, J., Liu, Y., Zhang, D., & Li, M. (2022). Study on thermal performance of a new optimized snowflake longitudinal fin in vertical latent heat storage. *Journal of Energy Storage*, *50*(February), 104165. https://doi.org/10.1016/j.est.2022.104165

Salomoni, V. A., Majorana, C. E., Giannuzzi, G. M., & Miliozzi, A. (2008). Thermal-fluid flow within innovative heat storage concrete systems for solar power plants. In *International Journal of Numerical Methods for Heat and Fluid Flow*, 18(7–8). https://doi.org/10.1108/09615530810899051

Sciacovelli, A., Gagliardi, F., & Verda, V. (2015). Maximization of performance of a PCM latent heat storage system with innovative fins. *Applied Energy*, *137*, 707–715. https://doi.org/10.1016/j.apenergy.2014.07.015

Shen, Y., Mazhar, A. R., Zhang, P., & Liu, S. (2023). Structure optimization of tree-shaped fins for improving the thermodynamic performance in latent heat storage. *International Journal of Thermal Sciences*, *184*(November 2022), 108003. https://doi.org/10.1016/j.ijthermalsci.2022.108003

Shouguang, Y. A. O., Min, Z. U. O., & Xinyu, H. (2023). Evaluation and optimization of the thermal storage performance of a triplex-tube thermal energy storage system with V-shaped fins. *Journal of Thermal Science*, 32, 2048–2064. https://doi.org/10.1007/s11630-023-1795-x

Singh, P. C., & Halder, P. (2023). Performance assessment of an air-conditioning system utilizing a PCM-based annulus cylindrical latent heat storage. *Arabian Journal for Science and Engineering*. https://doi.org/10.1007/s13369-023-08009-y

Tan, F. L., Hosseinizadeh, S. F., Khodadadi, J. M., & Fan, L. (2009). Experimental and computational study of constrained melting of phase change materials (PCM) inside a spherical capsule. *International Journal of Heat and Mass Transfer*, *52*(15–16), 3464–3472. https://doi.org/10.1016/j.ijheatmasstransfer.2009.02.043

Tavakoli, A., Farzaneh-Gord, M., & Ebrahimi-Moghadam, A. (2023). Using internal sinusoidal fins and phase change material for performance enhancement of thermal energy storage systems: Heat transfer and entropy generation analyses. *Renewable Energy*, *205*(October 2022), 222–237. https://doi.org/10.1016/j.renene.2023.01.074

Wang, C. C., Tsai, Y. M., & Lu, D. C. (1998). Comprehensive study of convex-louver and wavy fin-and-tube heat exchangers. *Journal of Thermophysics and Heat Transfer*, *12*(3), 423–430. https://doi.org/10.2514/2.6354

Yildiz, C., Bíçer, Y., & Pehlívan, D. (1996). Influence of fluid rotation on the heat transfer and pressure drop in double-pipe heat exchangers. *Applied Energy*, *54*(1), 49–56. https://doi.org/10.1016/0306-2619(95)00070-4

Yun, J. Y., & Lee, K. S. (1999). Investigation of heat transfer characteristics on various kinds of fin-and-tube heat exchangers with interrupted surfaces. *International Journal of Heat and Mass Transfer*, *42*(13), 2375–2385. https://doi.org/10.1016/S0017-9310(98)00310-X

Chapter 5

Efficiency enhancement of twin pipe heat exchangers using twisted tape inserts

Shanmukh Sudhir Arasavelli, Govinda Rao Budda, and Anand Kumar Solanki

5.1 INTRODUCTION

Heat exchangers are devices that are used to interchange heat between unmixed fluids. They are differentiated on the basis of application, geometry, fluid flow, etc. They can be constructed into tubular type, extended surface type, etc. In tubular type, twin pipe heat exchanger [TPHE] is made up of concentric pipes to swap heat with finite difference in temperature. It is a leading heat exchanger for cooling or heating in high-temperature applications. Some of the unique applications are solar collectors, food pasteurization, waste heat recovery, etc. TPHE is preferred due to its outstanding effectiveness with lower capital cost. They have lower maintenance cost and occupy less space in comparison with other tubular types. They are also simple in design and easy to install in industries (Kakac and Liu, 2009).

Effectiveness of TPHE can be magnified by using enhancement techniques. These techniques are classified into passive and active. Passive techniques do not require inward power, whereas active techniques require inward power. Appropriate passive techniques for TPHE are twisted tapes, wired coils, extended surfaces, etc. (Omidi *et al.*, 2017). Numerical and experimental works are reviewed on various tape inserts. Majority of experimental works established correlations to find Nusselt number. Few numerical works on twisted tapes are also discussed (Varun *et al.*, 2016). Temperature gradients and wall shear in friction coefficient are altered for better heat transfer. Expressions of convective coefficient and friction factor are generalized for a broad range of Reynolds numbers, i.e., laminar, laminar to turbulent and turbulent regimes (Sarma *et al.*, 2005).

The effect of hollow tape inserts on heat transfer in a tube under laminar regime is simulated. Cross-hollow tapes performed better than other tapes with different widths (Li *et al.*, 2015). A new tape is modeled with a parallelogram winglet to insert into a tube for intensification in heat transfer. Nusselt number is maximized by 191% approximately and corresponding performance factor is also increased by 1.85 times (Zhi-Min et al., 2017). Numerical work is performed on heat transfer in a cross tape inserted tube for laminar flow. The increase in convective coefficient is indicated due to reduction in the boundary layer (Liu *et al.*, 2018).

 DOI: 10.1201/9781003534785-5

Three dimensional simulations are carried out to investigate both pressure drop and heat transfer from air with eight-fold tape and spiral tape inserts. Nusselt number for eight-fold insert found to be 10% higher than the corresponding value of spiral insert (Cui and Tian, 2010). Space ratio, rotated angle and twist ratio of short length tape are optimized in an air flowing tube under a turbulent zone. Numerical study specified that higher rotation with lower twists on tape insert gives greater heat transfer and flow resistance (Wang *et al.*, 2011). Square cut tapes are created to improve thermal performance from a tube numerically. Maximum performance is obtained from square tape with smaller perforated length to width ratio and higher perforated width to width ratio (Saysroy and Eiamsa-Ard, 2017).

The effect of classic, jagged and butterfly inserts on the heat transfer from tubes of air cooler are analyzed. A 90° inclined butterfly insert showed better performance than the other two different inserts in the experimental range of Reynolds numbers. Computational fluid dynamics (CFD) results on performance of three different tape inserts are observed to be in line with corresponding experimental results (Shabanian *et al.*, 2011). Influence of twisted tubes on pressure drop and Nusselt number in a heat exchanger are tested. Experimental study revealed that augmentation in heat transfer from twisted tubes also enhances its pressure drop. Numerical study showed that both convective coefficient and friction factor increase with increase in axis ratio and also decrease with increase in pitch length (Tan *et al.*, 2012). Behavior of pressure drop and convective coefficient are examined in a space tape inserted heat exchanger. Their experimental results indicated that performance in heat transfer increased with decrease in both space ratio and twist ratio (Eiamsaard *et al.*, 2014).

So far, the numerical work on both friction factor and Nusselt number in a TPHE with tape inserts for turbulent conditions has not been studied. Therefore, present work is focused to realize the behavior of swirl flow inside the inner pipe fitted with tape inserts numerically. Influence of twist ratios on fluid flow and heat transfer characteristics for different flow rates of hot water are also examined in TPHE.

5.2 CFD MODELING

5.2.1 Governing equations

The incompressible fluid considered in this numerical model is water and their properties are assumed to be constant with respect to temperature. In the present analysis, flow is turbulent inside TPHE and it is confirmed by calculating Reynolds number using Eq. (5.1). Fluid flow with convective heat transfer is governed by Navier stokes equations for continuity, momentum, and energy at steady state. Reynolds-averaged Navier-Stokes (RANS) method is employed to render Navier stokes equations. The RANS approach is widely adopted for practical engineering applications to compute time-dependent flows. Equations (5.2–5.4) are RANS-applied governing equations in tensor form which describes the fluid flow with heat transfer (Oni and Paul, 2016).

Reynolds number

$$\mathrm{Re} = \frac{ud}{v} \tag{5.1}$$

Continuity equation

$$\frac{\partial}{\partial x_i}(\rho u_i) = 0 \tag{5.2}$$

Momentum equation

$$\frac{\partial}{\partial x_j}(\rho u_i u_j) = -\frac{\partial P}{\partial x_i} + \frac{\partial}{\partial x_j}\left[\mu\left(\frac{\partial u_i}{\partial x_j} + \frac{\partial u_j}{\partial x_i}\right)\right] + \frac{\partial}{\partial x_j}\left(-\rho \underline{u_i' u_j'}\right) \tag{5.3}$$

Energy equation

$$\frac{\partial}{\partial x_i}(\rho u_i T) = \frac{\partial}{\partial x_j}\left[\left(\frac{\mu}{Pr} + \frac{\mu_t}{Pr}\right)\frac{\partial T}{\partial x_j}\right] \tag{5.4}$$

Different models of transport equations are compared and suggested that renormalization group (RNG) $k-\varepsilon$ model is superior to study heat transfer and pressure drop (Lim *et al.*, 2018). The RNG $k-\varepsilon$ model is a two-equation model with one transport equation for turbulence kinetic energy (k) and another transport equation for its dissipation rate (ε). Equations (5.5) and (5.6) are two transport equations (Oni and Paul, 2016).

$$\frac{\partial(\rho k u_i)}{\partial x_i} = \frac{\partial}{\partial x_j}\left[\left(\mu + \frac{\mu_t}{\sigma_k}\right)\frac{\partial k}{\partial x_j}\right] + \Gamma_k - \rho\varepsilon \tag{5.5}$$

$$\frac{\partial(\rho \varepsilon u_i)}{\partial x_i} = \frac{\partial}{\partial x_j}\left[\left(\mu + \frac{\mu_t}{\sigma_\varepsilon}\right)\frac{\partial \varepsilon}{\partial x_j}\right] + C_{1\varepsilon}\frac{\varepsilon}{k}\Gamma_k + C_{2\varepsilon}\rho\frac{\varepsilon^2}{k} \tag{5.6}$$

Model constants are $C_{1\varepsilon} = 1.44$, $C_{2\varepsilon} = 1.90$, $\sigma_k = 1.0$, $\sigma_\varepsilon = 1.2$.

5.2.2 Convective coefficient

Along with turbulence model, i.e., RNG $k-\varepsilon$ model, energy equation is enabled for applying boundary conditions of fluid inlet temperature and pipe wall temperature. Using the heat balance, wall heat transfer coefficient (h)

can be evaluated from fluid outlet temperature using Eq. (5.7) (Tian and Barigou, 2015).

$$mC_p(T_{out} - T_{in}) = hA(\Delta T)_{LMTD} \tag{5.7}$$

Log mean temperature difference $(\Delta T)_{LMTD}$, mass flow rate (m) and surface area (A) can be obtained from corresponding equations

$$(\Delta T)_{LMTD} = \frac{(T_{h,i} - T_{c,o}) - (T_{h,o} - T_{c,i})}{\ln\left(\frac{(T_{h,i} - T_{c,o})}{(T_{h,o} - T_{c,i})}\right)}, m = \rho A u \text{ and } A = \pi \text{ d L}$$

5.2.3 CFD simulations

Three dimension simulations were conducted to simulate and compare both steady flow and swirl flow using software package ANSYS 2022 R1. Pre-processing, solving, and post-processing are performed in Fluent 2022. Dimensions of copper inner pipe, steel outer pipe, aluminum twisted tapes that are used in TPHE are given in Table 5.1 (Arasavelli *et al.*, 2021). Using design modeler, flow geometry is created with a combination of outer pipe, inner pipe, tape insert and two fluid zones. Three different inserts, inner pipe, outer pipe and one combination of TPHE are illustrated in Figure 5.1. Named selections are hot fluid, inner pipe, cold fluid and outer pipe. Three distinct tape inserts are named as TTR 07, TTR 12 and TTR 17 (where TTR is tape twist ratio). Outer surface of the outer pipe is designated as an adiabatic wall to avoid heat losses from TPHE.

Meshing is a vital step in which geometry will be divided into a number of control volumes, i.e., cells. Meshing in the current heat exchanger is performed by using polyhedral cells as depicted in Figure 5.2. A polyhedral cell

Table 5.1 Dimensions for models in TPHE

S. No	*Parameter*	*Value*
1	Inner pipe internal diameter	15 mm
2	Inner pipe thickness	2 mm
3	Inner pipe length	2000 mm
4	Outer pipe internal diameter	29 mm
5	Outer pipe thickness	3 mm
6	Outer pipe length	2000 mm
7	Tape thickness	2 mm
8	Tape width (w)	14 mm
9	Tape length	2000 mm
10	Tape pitches (p)	100, 170 and 240
11	Tape twist ratios (TTR = p/w)	7, 12 and 17

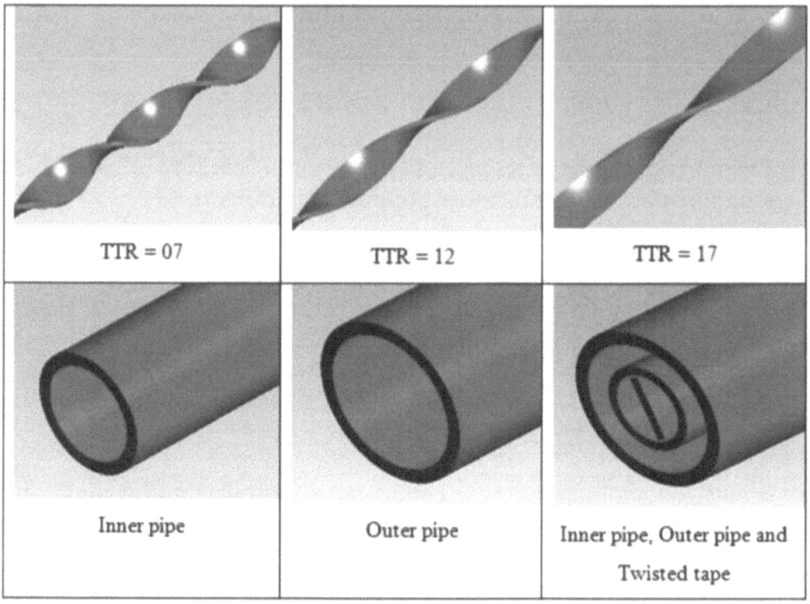

Figure 5.1 Geometric models of TPHE.

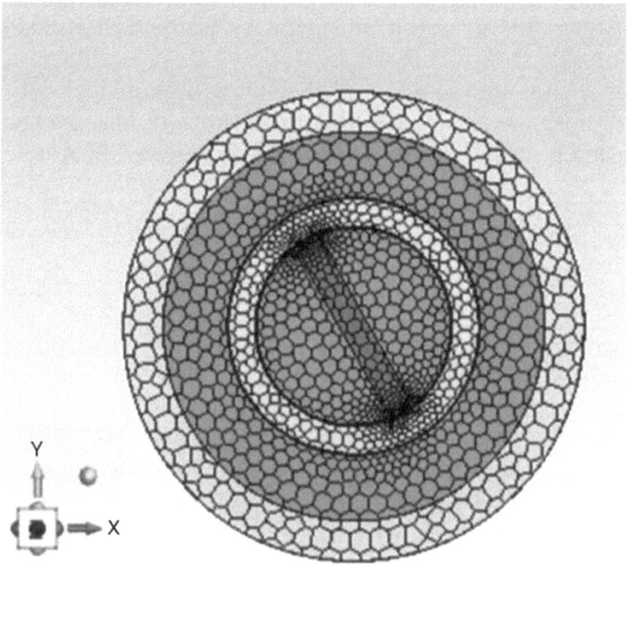

Figure 5.2 Polyhedral meshing in TPHE.

Table 5.2 Selected material properties in fluent database

S. No	*Fluid material*	*Density (kg/m³)*	*Specific heat [J/kg K]*	*Thermal conductivity [W/m K]*
1	Water liquid	998.2	4182	0.6
2	Copper	8978	381	387.6

combines the benefits of both hexahedral and tetrahedral cells from unstructured mesh (Sosnowski *et al.*, 2018). Selection of solver, model specifications, material properties followed by zone conditions are set for simulation after meshing. Heat transfer is enabled by switching on the energy equation. Along with energy equation, RNG $k-\varepsilon$ model is selected from the model list to create turbulent flow in TPHE. Water liquid from fluids and copper from solids are created to perform simulation. Allocated properties of fluid and solid materials are listed in Table 5.2.

Boundary type is selected for named selections that are given in the geometry section of CFD simulation. Boundary values are entered according to the type of selected boundary. Boundary conditions with respect to boundary types and values are mentioned in Table 5.3 for present simulations.

Numerical investigations are performed using a fluent package for diverse cases of TTR and mass flow rates. Reynolds numbers are calculated from mass flow rates of hot water using Eq. (5.1). TTR in TPHE is varied from 7 to 17, whereas Reynolds numbers in the same heat exchanger are also alerted from 10,710 to 23,703. Each of the four cases, i.e., without TTR, TTR = 17, 12 and 7 are further divided into four sub-cases of Re = 10,710, 14,994, 19,278 and 23,703. So, numerical simulations are conducted for a total number of 16 cases in present work. Independence study on mesh is done by performing a number of simulations with various mesh sizes for four cases, i.e., three cases with different TTRs and one case without TTR. It is necessary to optimize mesh size until simulation results are independent of outcomes.

Five inflation layers are created on outer and inner surfaces of the inner pipe to capture changes in temperature profiles near the wall. First layer thickness is further expanded by a growth rate of 20% to enhance mesh resolution near the wall. Numbers of cells for each centimeter along the heat exchanger length and TPHE cross-section are found to be 244,980 and 14,410, respectively, for

Table 5.3 Boundary conditions applied in TPHE

S. No	*Named selections*	*Boundary type*	*Boundary value*
1	Adiabatic wall	Heat flux	0 W/m²
2	Hot inlet	Mass flow inlet	0.083, 0.12, 0.15 and 0.18 kg/s
		Temperature	313 K
3	Cold outlet	Outflow	Static pressure of 1 atmosphere
4	Cold inlet	Mass flow inlet	0.13 kg/s
		Temperature	301 K
5	Hot outlet	Outflow	Static pressure of 1 atmosphere

Table 5.4 Independence test on mesh size

S. No	*No. of cells*	*Nu*	$\frac{Nu^{i+1} - Nu^{i}}{Nu^{i+1}}$	*f*	$\frac{f^{i+1} - f^{i}}{f^{i+1}}$
1	70,21,600	86.28	-	0.0142	-
2	75,28,884	92.67	0.068	0.0175	0.189
3	80,20,168	97.23	0.046	0.0201	0.129
4	85,12,693	100.1	0.028	0.0223	0.099
5	90,28,190	101.62	0.014	0.0238	0.063
6	95,01,496	101.78	0.001	0.0240	0.008

one of the cases. Changes in Nusselt number and friction factor in correspondence with the number of cells for a Reynolds number of 10,710 and TTR of 17 are represented in Table 5.4 (Nakhchi and Esfahani, 2019).

Pressure and velocity are coupled through a SIMPLE scheme in solution methods. Default selections are retained in the spatial discretization. Solution is initialized through standard initialization. Option of all zones is chosen from the drop list of compute. In residual monitors, absolute criteria are changed to 10^{-5} to ensure accuracy in steady state conditions.

5.2.4 Validation of CFD model

Figure 5.3(a) showed authentication of the numerical model with experimental results in the absence of twisted tapes (Arasavelli et al., 2020). Friction factors from the numerical model of TPHE are compared with experimental values of friction factor at different Reynolds number from 10,710 to 23,703. Numerical values of friction factor are in good agreement with corresponding experimental values. Numerical friction factor had maximum deviation of 8% with experimental friction factor at a Reynolds number of 19,278, which is nominal.

Figure 5.3(b) indicates the comparison of Nusselt number for present study with experimental results for the case of without inserts in TPHE (Arasavelli et al., 2020). Numerical results of Nusselt number are differentiated with respective experimental results in the turbulent range of Reynolds numbers. It is apparent from Figure 5.3(b) that there is a fine consistency between numerical and experimental Nusselt numbers. Present results of Nusselt number had the highest divergence of 6% with experimental results of Nusselt number.

5.3 RESULTS AND DISCUSSION

5.3.1 Temperature contours

Figure 5.4 demonstrates the contour diagrams for temperature at different Reynolds numbers and TTRs. Figure 5.4(a–d) represents temperature contours for plain TPHE at four different Reynolds numbers from 23,703 to 10,710. Temperature contours for a TTR = 17 in the same range of Reynolds

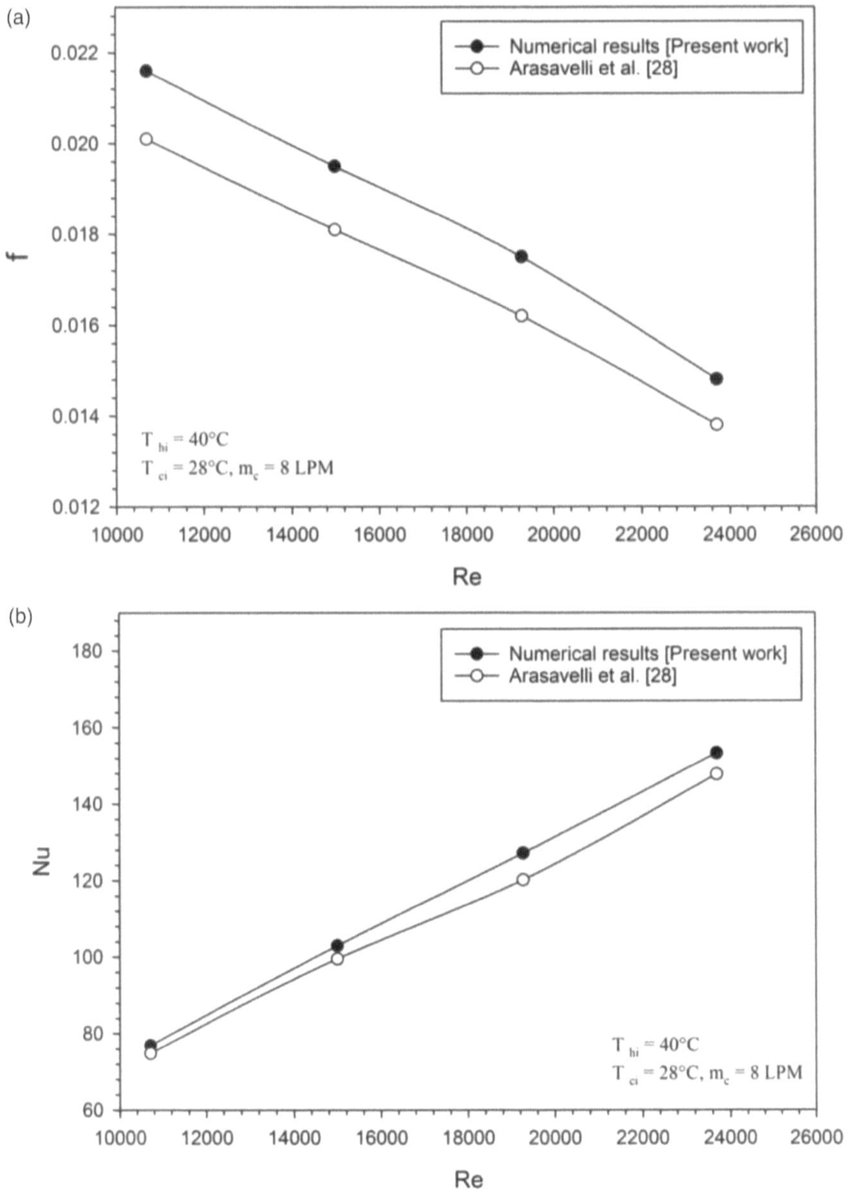

Figure 5.3 (a) Authentication of numerical results with experimental results. (b) Validation of numerical values with experimental values.

number are depicted in Figure 5.4 (e–h). Another set of contours in Figure 5.4 (i–l) illustrates temperature changes for TPHE with 12 TTR and four distinct Reynolds numbers. Last four diagrams in Figure 5.4(m–p) showed the temperature contours for twisted tape insert TPHE (TTR = 07) at the same Reynolds numbers.

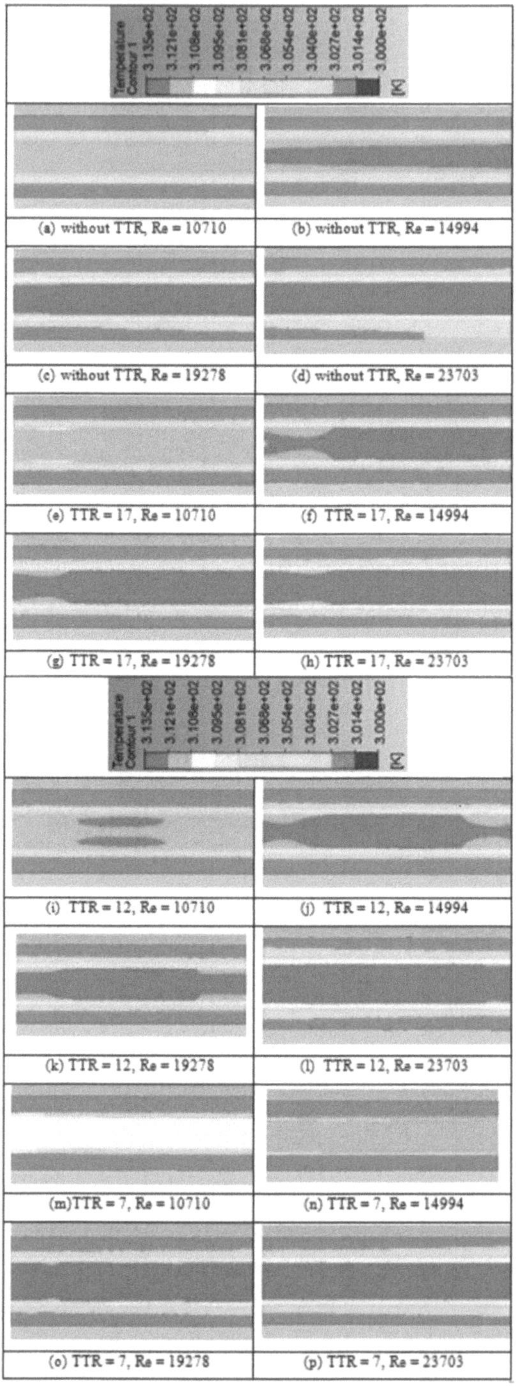

Figure 5.4 Contour diagrams of temperature for different TTR and Re.

TPHE with tape inserts of all three twist ratios provided higher distribution in temperature than corresponding distribution in plain TPHE for all Reynolds numbers. This is credited to effective swirl flow, which happens inside the heat exchanger with tape inserts. At any Reynolds number, the lowest TTR has better distribution than the other two TTRs. This is endorsed to higher swirl flow due to increase in the number of tape turns. Reduction in the temperature distribution for other two TTRs can be seen in both axial and radial directions. Hence, TTR with smallest value, i.e., 7 had consistent distribution in temperature when compared with other two TTRs of 12 and 17.

5.3.2 LMTD

Figure 5.5 shows the variation in log mean temperature difference (LMTD) with Reynolds number for plain TPHE and TPHE with three different TTRs. It is observed that LMTD slides down with a drop in Reynolds number. It is also clear that LMTD reduces further with decrease in the value of TTR at each value of Reynolds number. Decline in LMTD is because of the swirl flow created by tape insert in the tube side of TPHE. Tape inserts will continuously generate changes in the flow path of hot water. They improve both residence time and surface area of the fluid. Frequent modifications in fluid direction also generate turbulence for better convection and subsequently lead to higher heat exchange (Arasavelli *et al.*, 2021). Results specified that the LMTD for TTR of 7 is 11% lower than the LMTD in plain TPHE at the

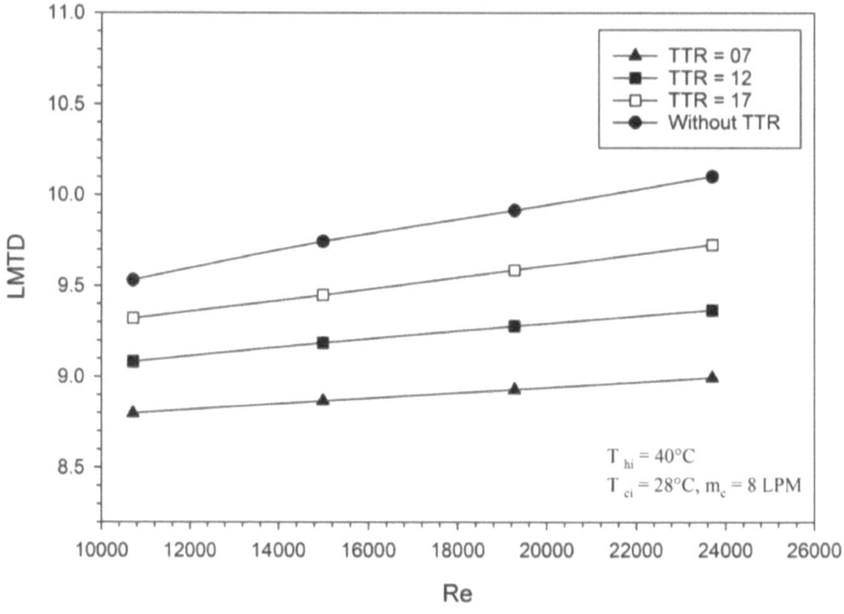

Figure 5.5 Log mean temperature as a function of Reynolds number.

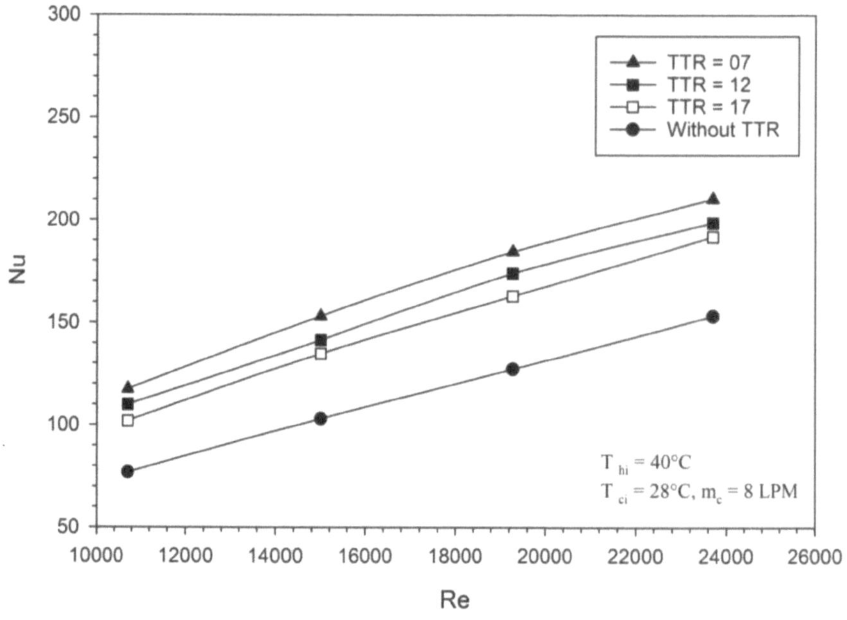

Figure 5.6 The effect of different twist ratios on Nusselt number.

highest Reynolds number. LMTD is affected by both twist ratio and Reynolds number. Best LMTD is achieved at minimum TR and Reynolds number, i.e., 7 and 10710, respectively.

5.3.3 Twist ratio

Change in numerical value of Nusselt number with respect to Reynolds numbers and twist ratios are presented in Figure 5.6. Results showed that there is a rise in Nusselt number when the Reynolds number is increased and the same Nusselt number grows further when the twist ratio is dropped to 7. Consequently, Nusselt number at TTR = 7 is greater than Nusselt number at the other two twist ratios for any Reynolds number. Lower TTR induces higher swirl flow to break the formation of boundary layer regularly. Swirl flow also alters energy between fluid layers for convenient transfer of thermal energy. This leads to better convective coefficient between fluid zone and pipe walls in TPHE (Liu and Sakr, 2013). Quantitative results in Figure 5.6 illustrated that Nusselt number with TTR of 7, 12 and 17 are 53%, 42% and 32% higher than the Nusselt number of plain TPHE at Re of 10,710.

5.3.4 Friction factor

Figure 5.7 presents the variations in friction factor values with different Reynolds numbers for both cases of with and without TTR in TPHE. Friction

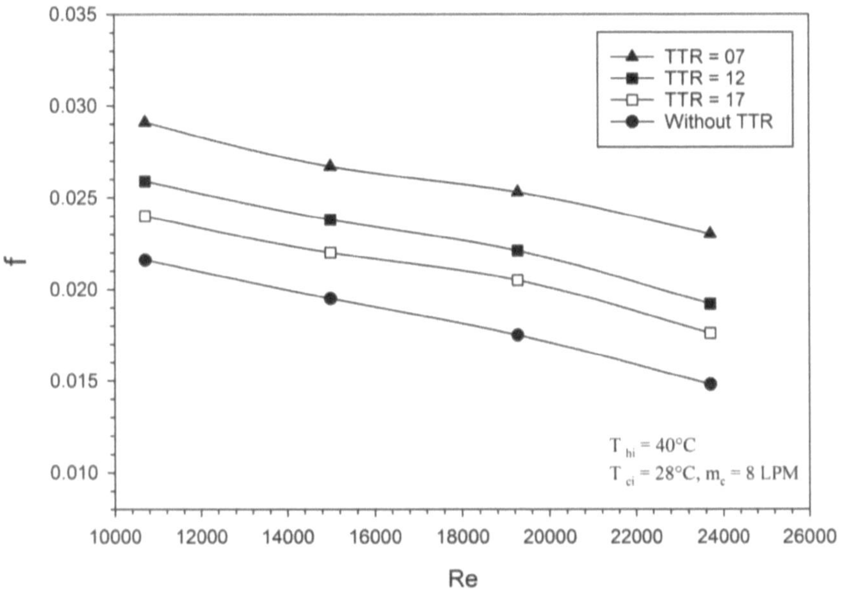

Figure 5.7 Friction factor results for various twist ratios.

factor value slides down with increase in the value of Reynolds number for all four cases in Figure 5.7. This downslide is due to decrease in viscous force from the walls of inner pipe with corresponding increase in the flow rate. It can also be observed that the friction factor with any value of TTR is more than the friction factor without TTR. Friction factor was found to be stepped up further by decreasing TTR from 17 to 7. By decreasing TTR, more swirls are created in the hot water. These swirls will enhance viscous force between hot water and surrounding walls of inner pipe (Sivashanmugam, 2009). At Reynolds number of 10,710, friction factor of TTR = 7, 12 and 17 are 35%, 20% and 11% more than plain friction factors in present TPHE.

5.3.5 Pressure drop

Pressure drops are displayed for distinct Reynolds numbers at different TTR of tape inserts in Figure 5.8. Pressure drop is raised in upward direction when Reynolds number is stepped up for any of the TTR. Similar trend of increase is followed by pressure drop for the case of plain TPHE. Pressure drop for any one of the three TTRs are greater than pressure drop without TTR at all Reynolds numbers. TPHE with smaller TTR had higher pressure drop when compared to other TTRs of 12 and 17 in same heat exchanger. Increase in pressure drop is due to increase in flow blockage by twisted tapes. Smaller TTR will have more number of turns and leads to higher congestion to the fluid flow. Higher blockage in the fluid flow is responsible for higher-pressure

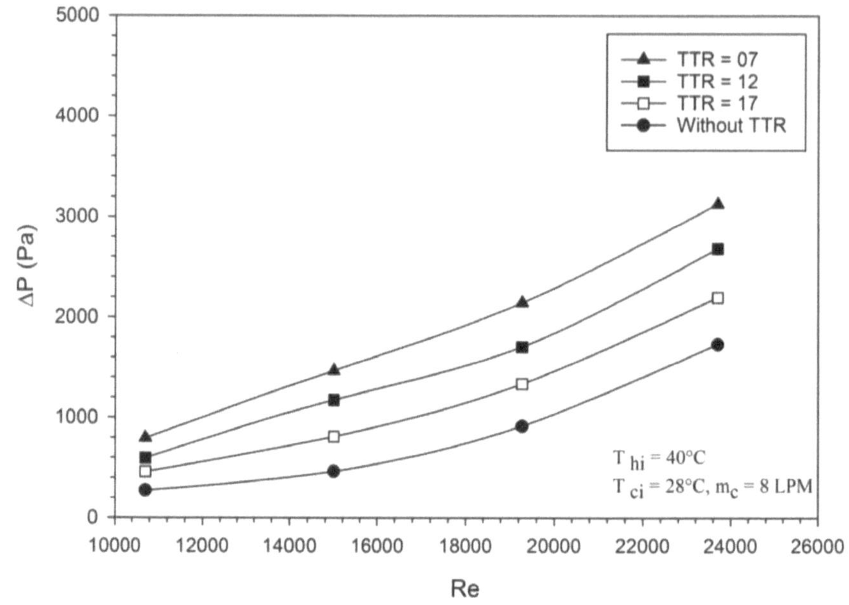

Figure 5.8 Evaluation of pressure drop against distinct twist ratios.

drop at smaller TTR and bigger Reynolds number (Varun *et al.*, 2016). Pressure drop with TTR of 7, 12 and 17 are, respectively, around 38%, 23% and 14% higher than that of plain TPHE at a Reynolds number of 10,710.

5.3.6 Enhancement efficiency

Fifty-three percent of enhancement in Nusselt number leads to 35% of increase in friction factor at TTR = 7 and Re = 10,710. Feasibility between Nusselt number increase and friction factor increase is checked through enhancement efficiency (η). It is defined as the relation between Nusselt number ratio and friction factor ratio. The key parameter for the design of heat exchanger is enhancement efficiency (Wang *et al.*, 2011). Efficiencies for inserted tapes in TPHE at three different TTRs are calculated from Eq. (5.8) and showcased in Figure 5.9.

$$\eta = \frac{\frac{Nu}{Nu_o}}{\frac{f}{f_o}} \tag{5.8}$$

Enhancement efficiency dropped with increase in Reynolds number at all three TTRs in Figure 5.9. It is reduced further by increasing the value of TTR in the Reynolds number range of turbulent flow. Figure 5.9 shows that the

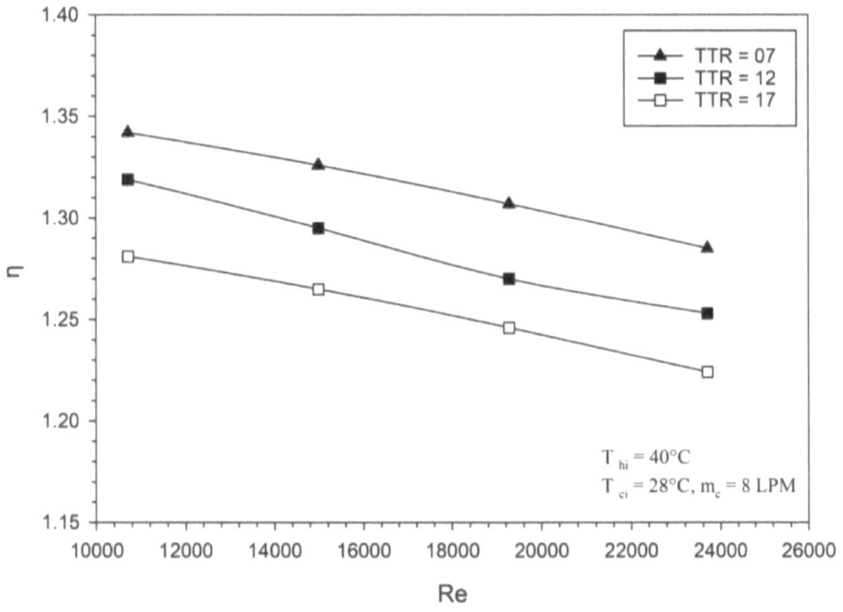

Figure 5.9 The influence of twist ratios on enhancement efficiency.

value of enhancement efficiency is high at the lower value of Reynolds number and TTR. Highest efficiency for using TTR = 7, 12 and 17 in TPHE are found to be 1.34, 1.32 and 1.28, respectively.

5.4 CONCLUSIONS AND FUTURE SCOPE

Simulations are performed on current heat exchanger with different tape inserts to study its Nusselt number as heat transfer parameter, friction factor as pressure drop parameter and enhancement efficiency as feasibility parameter. Simulation results of friction factor and Nusselt number are validated with experimental results from existing literature. Temperature contours are illustrated for distinct combinations of twist ratio and Reynolds number. Major findings are summarized into following conclusions:

1. Results from numerical simulations show that Nusselt number with twist ratios from 7 to 17 are 1.53 to 1.32 times more than the same parameter in TPHE without tape inserts at the minimum value of Reynolds number.
2. Friction factors of TPHE fitted with the same set of twist ratios are, respectively, around 1.35 to 1.11 times greater than the friction factor in TPHE without twisted tapes at the same value of Reynolds number, i.e., 10,710.

3. Enhancement efficiencies for current twist ratios are higher than 1 in the Reynolds number range of 10,710 to 23,703, which implies that the tape-inserted TPHEs are appropriate for improvement in convective coefficient.

Current work can be expanded to the following future works:

1. Heat transfer aspects and pressure drop parameters can be examined numerically with a wide range of tape inserts.
2. Nanofluids can be passed through a current heat exchanger to examine its heat transfer, pressure drop and enhancement efficiency.

REFERENCES

Arasavelli, S. S., Konijeti, R., and Budda, G. R. (2020). Effect of vibrator position and frequency on heat transfer enhancement in counter flow concentric pipe heat exchanger. World Journal of Engineering, 17(5), 751–760.

Arasavelli, S. S., Konijeti, R., and Budda, G. R. (2021). Positive impact of vibration on heat transfer in twisted tape inserted heat exchanger. Journal of Thermal Engineering, 7(7), 1731–1742.

Cui, Y.-Z., and Tian, M.-C. (2010). Three-dimensional numerical simulation of thermal-hydraulic performance of a circular tube with edgefold-twisted-tape inserts. Journal of Hydrodynamics, 22(5), 662–670.

Eiamsa-Ard, P., Piriyarungroj, N., Thianpong, C., and Eiamsa-Ard, S. (2014). A case study on thermal performance assessment of a heat exchanger tube equipped with regularly-spaced twisted tapes as swirl generators. Case Studies in Thermal Engineering, 3, 86–102.

Kakac, S., and Liu, H. (2009). Heat exchangers: Selection, rating and thermal design. CRC Press.

Li, P., Liu, Z., Liu, W., and Chen, G. (2015). Numerical study on heat transfer enhancement characteristics of tube inserted with centrally hollow narrow twisted tapes. International Journal of Heat and Mass Transfer, 88, 481–491.

Lim, D. C., Al-Kayiem, H. H., and Kurnia, J. C. (2018). Comparison of different turbulence models in pipe flow of various Reynolds numbers. In AIP Conference Proceedings (Vol. 2035, No. 1). AIP Publishing.

Liu, S., and Sakr, M. (2013). A comprehensive review on passive heat transfer enhancements in pipe exchangers. Renewable and Sustainable Energy Reviews, 19, 64–81.

Liu, X., Li, C., Cao, X., Yan, C., and Ding, M. (2018). Numerical analysis on enhanced performance of new coaxial cross twisted tapes for laminar convective heat transfer. International Journal of Heat and Mass Transfer, 121, 1125–1136.

Nakhchi, M. E., and Esfahani, J. A. (2019). Numerical investigation of rectangular-cut twisted tape insert on performance improvement of heat exchangers. International Journal of Thermal Sciences, 138, 75–83.

Omidi, M., Farhadi, M., and Jafari, M. (2017). A comprehensive review on double pipe heat exchangers. Applied Thermal Engineering, 110, 1075–1090.

Oni, T. O., and Paul, M. C. (2016). Numerical investigation of heat transfer and fluid flow of water through a circular tube induced with divers' tape inserts. Applied Thermal Engineering, 98, 157–168.

Sarma, P. K., Kishore, P. S., Rao, V. Dharma, and Subrahmanyam, T. (2005). A combined approach to predict friction coefficients and convective heat transfer

characteristics in a tube with twisted tape inserts for a wide range of Re and Pr. International Journal of Thermal Sciences, 44(4), 393–398.
Saysroy, A., and Eiamsa-Ard, S. (2017). Periodically fully-developed heat and fluid flow behaviors in a turbulent tube flow with square-cut twisted tape inserts. Applied Thermal Engineering, 112, 895–910.
Shabanian, S. R., Rahimi, M., Shahhosseini, M., and Alsairafi, A. A. (2011). CFD and experimental studies on heat transfer enhancement in an air cooler equipped with different tube inserts. International Communications in Heat and Mass Transfer, 38(3), 383–390.
Sivashanmugam, P. (2009). CFD modeling of heat transfer augmentation in a circular tube fitted with right-left helical inserts in turbulent flow. International Journal of Food Engineering, 5(1).
Sosnowski, M., Krzywanski, J., Grabowska, K., and Gnatowska, R. (2018). Polyhedral meshing in numerical analysis of conjugate heat transfer. EPJ Web of Conferences, 180, 02096.
Tan, X.-H., Zhu, D.-S., Zhou, G.-Y., and Zeng, L.-D. (2012). Experimental and numerical study of convective heat transfer and fluid flow in twisted oval tubes. International Journal of Heat and Mass Transfer, 55(17–18), 4701–4710.
Tian, S., and Barigou, M. (2015). An improved vibration technique for enhancing temperature uniformity and heat transfer in viscous fluid flow. Chemical Engineering Science, 123, 609–619.
Varun, Garg, M. O., Nautiyal, H., Khurana, S., and Shukla, M. K. (2016). Heat transfer augmentation using twisted tape inserts: A review. Renewable and Sustainable Energy Reviews, 63, 193–225.
Wang, Y., Hou, M., Deng, X., Li, L., Huang, C., Huang, H., Zhang, G., Chen, C., and Huang, W. (2011). Configuration optimization of regularly spaced short-length twisted tape in a circular tube to enhance turbulent heat transfer using CFD modeling. Applied Thermal Engineering, 31(6–7), 1141–1149.
Zhi-Min, L., Liang-Bi, W., Mei, L., Wei, D., and Yong-Heng, Z. (2017). Numerical study of the laminar flow and heat transfer characteristics in a tube inserting a twisted tape having parallelogram winglet vortex generators. Applied Thermal Engineering, 115, 644–658.

Chapter 6

Optimization of microwave heated conversion of algae bio-oil into biodiesel using RSM

Bioenergy conversion

Divyangana Sharma, Vivek Goel, and Sunil Kumar

6.1 INTRODUCTION

6.1.1 Optimization

Finding the ideal situation under the circumstances is the very method or approach of optimization. Designers have to arrive at multiple organizational and technical choices at different stages of the creation, construction, as well as operation of any type of technological framework. The goal to maximize the prospective result or minimize the required labour is the primary intent of all such assessments. Identifying the settings that cause a function to have its maximum or minimum value can be the process of optimization, as the amount of work required or the intended result in any practical scenario can be described as a function of certain criteria for decision-making.

"Selecting $a \in A$ where $\forall_{x\in A} f(a) \geq f(x)$ is the process of optimization, where A is an ordered collection of variables and f is the total order on A." It indicates that an ideal x is one way that $\forall_{x\in A} f(a) \geq f(x)$. A model is represented by set A, which is a space that holds all of the potential objects x of the framework; the component x represents specific roots, or literal states, of the system. All pairs of elements in A are comparable because the total order f is an anti-symmetric, reflexive, transitive, and severely associated relationship that evaluates and arranges the components to optimized items that have the potential to be launched (Birkhoff, 1940). Almost all mathematical optimizations fall under our concept of optimization. The goal of operations research, a branch of mathematics, is to apply scientific methods to identify the best answers to decision-making issues. The development of linear programming techniques began during World War II when mathematicians assisted the British military in efficiently allocating manpower (Rao, 2009).

6.1.2 Innovative technologies for optimization

In recent years, strong and popular approaches for tackling complicated engineering optimization problems have emerged: these methods are known as modern optimization methods, or sometimes non-traditional optimization

 DOI: 10.1201/9781003534785-6

methods. Fuzzy maximizing efficiency, optimization with neural networks, ant colony optimization, optimizing particle swarm, genetic algorithms, and that imitates scorching are some of these techniques. The principles of inherent selection as well as innate genetics serve as the basis for digital discovery, including algorithms of optimization, which are called genetic algorithms. One researcher, namely John Holland, first presented the idea of genetic algorithms in 1975. The behaviour of social creatures such as colonies or crowds of parasites (such as termites, ants, wasps, and honeybees), a cluster of fauna, and even an ensemble of aquatic creatures is replicated by the particle swarm optimization approach. Two researchers first proposed the algorithm in 1995. The partnership behaviour of the ant crowd, which can evaluate the quickest pathway to their destination place of residence for a food supply, is the foundation of ant colony optimization.

In 1992, one investigator created the approach for the very initial moment. The neural network approaches rely on the enormous nervous system computational capacity, which allows it to process information in parallel and resolve perceptual issues even in the face of vast amounts of sensory information. Two scientists implemented the optimization method for the initial instance in 1985. Fuzzy optimization techniques were created to address optimization issues involving uncertain and linguistically described design inputs, objective functions, and restrictions. Rao (2009) introduced fuzzy techniques for both single- and multiple-objectify optimization in engineering planning back in 1986.

6.2 THE FRAMEWORK OF AN OPTIMIZATION PROBLEM

Objective function—An objective function is a mathematical expression that defines what is being optimized. It could be a cost function to minimize or a utility function to maximize.

Decision variable—These are the variables that we can control or manipulate to optimize the objective function. They represent the quantities we are trying to determine.

Constraints—Constraints are conditions that must be satisfied for a solution to be feasible. Constraints can be equality constraints.

The relationship between machine learning (ML) and mathematical optimization has gained attention recently and has been extensively covered in the literature (Figure 6.1). A recent study about optimization techniques from the point of view of ML is given by the researchers (Sun et al., 2020). Basic ML tasks like adversarial learning, regression, clustering, and classification, among other techniques, are presented as problems of optimization by Gambella et al. (2021) in another review. The ML application to assist in the formulation and resolution of optimization problems is receiving increasing interest in the optimization discipline. Large and challenging

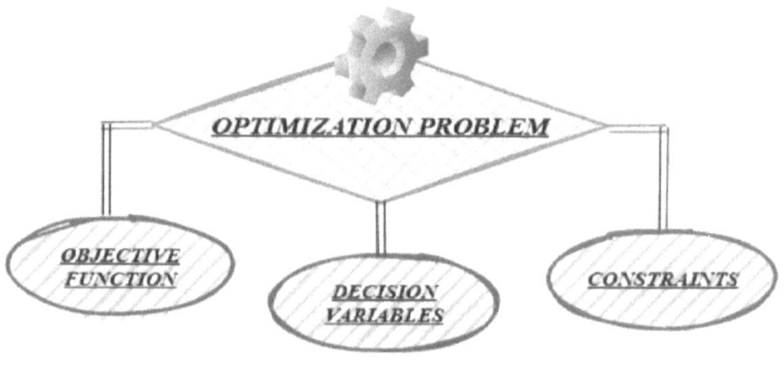

Figure 6.1 Optimization problem framework.

practical optimization problems can be solved, and an increasing number of authors are taking advantage of ML's flexibility and utility to help solve challenging practical optimization concerns (Abbasi et al., 2020; Fischetti and Fraccaro, 2019).

Apart from the conventional ML methods like neural networks, linear regression in addition to the regression tree methods (Verwer et al., 2017; Villarrubia et al., 2018). Alternative strategies like programming genetic as well as linear programming using mixed integers have also been employed to generate constraints for issues related to optimization (Pawlak and Krawiec, 2019, 2017). Optimization as a mathematical challenge is a paradigm that has been closely associated with management theory and operations research. Since the introduction of linear programming in the middle of the 20th century, these fields have grown rapidly. A variety of fields, including econometrics, statistics, ML, and optimal control, are using this type of offline optimization nowadays. One of the main pillars of contemporary engineering study and practice is mathematical optimization. Nonetheless, mathematical optimization is typically regarded as a numerical subject across all application domains.

The development of contextual optimization techniques for deciding what to do when faced with ambiguity is examined in this survey. It divides the literature into three categories: decision rule optimization, sequential learning and optimization, and integrated learning and optimization. It enumerates current study domains and presents a standard syntax for examining approaches (Sadana et al., 2024). Significant value dispersion occurs during grinding processes, which are impacted by sporadic disruptions. Tool characteristics alter over time as grinding progresses, affecting tool durability. The lifespan of a tool is the shortest time that doesn't go over bounds. Maintaining the condition of tools enhances their use, lowers expenses, and boosts productivity (Kacalak et al., 2024). The chapter defines optimization, talks about its drawbacks, and looks at how it affects our understanding of people and communities. Future work will try to elucidate these limitations with the use of actual cases (Carissimo and Korecki, 2024).

6.3 RESPONSE SURFACE METHODOLOGY

An approach to statistics known as the response surface methodology (RSM) is employed to plan experiments as well as maximize the influence of the factors of the method. It is founded on the ideas of design of experiments (DOE), an applied statistics area that evaluates factors influencing parameter values by designing, carrying out, evaluating, and interpreting controlled tests (Veza et al., 2023). On the contrary, the statistical planning of experiments (DOE) tool can be used to pre design series of test runs for engines, including the number of trials and crucial operational circumstances. Prediction and optimization might thereafter be carried out using the experiment data. This is where biofuel study and development can benefit greatly from the use of a DOE method like RSM. RSM can significantly reduce exhaust emissions from biofuel-fuelled engines while optimizing their performance and combustion characteristics. The utilization of RSM for biodiesel has been covered in a sizable number of published papers (Gupta et al., 2022; Nazarpour et al., 2022; Tamoradi et al., 2021). A CaO-catalysed glycerolysis reaction was modelled and optimized using RSM by researchers under the effect factors (Selemani and Kombe, 2022). Simbi et al. (2022) used RSM to determine the effects of operation-related factors on the ultimate yield of the biodiesel produced from sunflowers. To examine the effects of catalysts being loaded, temperature, reaction time span, and the methanol-to-oil ratio, Etim optimized transformation into different biodiesel utilizing the Taguchi design of RSM (Etim et al., 2022).

RSM is a quantitative and numerical procedure for creating an experiment set. Optimizing the reaction that is influenced by several independent variables is the aim. An equation for a second-order polynomial is portrayed in Equation (6.1), which is employed to anticipate the resultant response by accounting for the input components.

$$\mathrm{R} = p_0 + \sum_{i}^{k} p_i x_i + \sum_{i}^{k} p_{ii} x_i^{\,2} + \sum_{i,j=1,j\neq i}^{k} p_{ij} x_i x_j \tag{6.1}$$

Here, R displayed the result, i and j portrayed the quadratic in addition to the linear coefficients, p_0 shown an average of the results, and p_i, p_{ii}, and p_{ij} are the coefficients of the result.

The range of lengths of RSM parameters will be ascertained by experimental design. It ought to be selected per the quantity of optimization factors and the necessary degree of accuracy. For instance, the quadratic model in a two-variable framework has to be resembled by not fewer than four runs. A three-variable design requires a minimum of seven runs. More simulations are usually recommended for a more accurate model and to correct for any potential inaccuracies. As such, it is imperative to carefully select the variety of variables to ensure that the optimal subset of the surface of response is sufficiently evaluated.

6.4 SOFTWARE FOR RSM

Several computer software programs fall into one of two categories: those that are generally statistical or those that are expressly devoted to experimental design. Commercial software packages Design-Expert (State-Ease, Inc.) and Minitab (Minitab, Inc.) are frequently utilized for doing RSM. R studio, on the other hand, is open-source and free of charge means software for non-commercial use. In addition, there are alternative RSM software programs to MATLAB (Mathworks) and Statistica (Stat Soft), such as Chemoface and Develve.

6.5 DESIGN OF SURFACE OF RESPONSE IN RSM

A collection of sophisticated DOE strategies used to maximize the response is called a response surface design. The next crucial level involves constructing the trials by determining the points at which responds of interest might have accurately anticipated as well as assessed after the factors and responses have been established. A response surface can be obtained by developing some DOEs after the number of considerations has been determined. The key properties of these response surface designs—such as variance and the total number of experiments—are used to designate them. Because of this, choosing an appropriate design strategy is crucial to the process as a whole. It should be noted, therefore, that the majority of previously published publications only provide a cursory explanation of the selection criteria for response surface design rather than a thorough analysis. Full factorial design (FFD), CCD which stands as a central composite design, and including Box-Behnken design (BBD) displayed the three main categories of response surface design.

A factor in experimental design is a variable that can affect how a system or process responds. The many values or settings that a factor is pre-determined to throughout the experiment are referred to as its levels. As it enables scholars to methodically examine the connection between variables along with responses under a variety of settings, the concept of orders is significant. The particular objectives of the study and the range across which the parameter is anticipated to have an impact on the response will determine which values are appropriate for each factor. By selecting these particular levels, the research project can look at how temperature affects the respond factor on various dots along the area of focus. To guarantee that the experiment gathers pertinent data and offers insightful information, it is crucial to choose the right

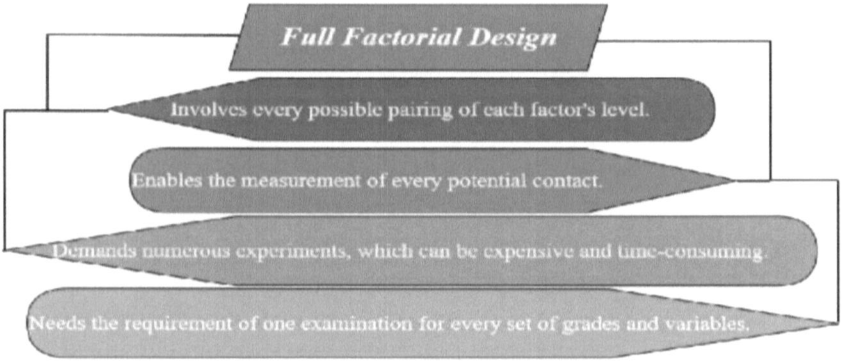

Figure 6.2 Summary of full factorial design (FFD).

levels for each component. The levels ought to cover an assortment that is indicative of the actual working environment or the area of interest. The accuracy and precision of the experiment outcomes can also be affected by how many categories are selected for each factor. To effectively capture the response variable's behaviour over the range of factors, a sufficient number of layers should be chosen.

6.6 RSM FOR FULL FACTORIAL DESIGN FFD

Full factorial design (FFD) is a technique that records each interaction and calculates the number of trials or repeats in an experiment by taking into account every potential combination of variables (Figure 6.2). Large repetitions can lessen random variation and enhance the evaluation of the effects of components. Two-level FFD is useful for trials with fewer components. FFD divides input factors into two or three levels.

The three-level FFD, in which the factors are capable of handling three values—minimum, centre, and maximum—is a more popular type of FFD. Consequently, a total of 3^k experiments were conducted to examine k components at the three level. Nonetheless, a significant disadvantage of three-level FFD is the need for multiple trial runs, which frequently result in needless high-order interactions (Karimifard and Alavi Moghaddam, 2018). According to the researcher, the three-level FFD is more beneficial when there are no more than five components (Sakkas et al., 2010).

6.7 RSM FOR CENTRAL COMPOSITE DESIGN (CCD)

A CCD response surface design is the most often used. Incorporating axial and central points, CCD may become accustomed to expanding on previous factorial trials, which makes it very helpful in chronological experimental

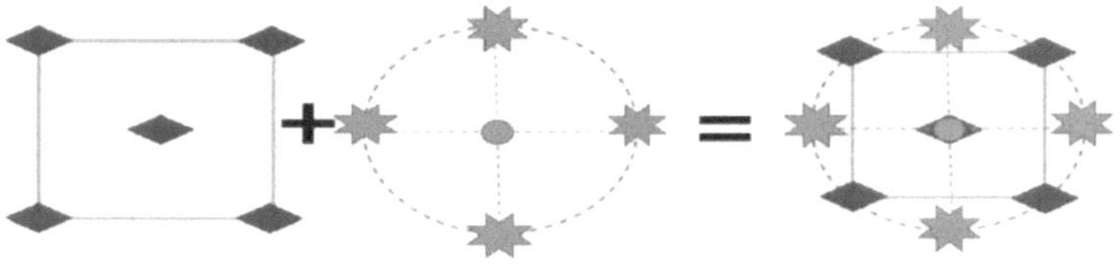

Figure 6.3 Graphical presentation of CCD.

work. A CCD might be utilized to efficiently approximate expressions of the first and second orders and method a respond variable utilizing contours by including axial points and the factorial design's centre point (Figure 6.3).

The locations midway through the investigation domain including the "star" exterior of it allow for the estimation of the response surface curvature. Factorial designs have point levels of ±1, whereas "star" designs have point levels of $\pm\alpha$, where $|\alpha| \geq 1$. CCD can be categorized into three groups (Ait-Amir et al., 2020). The value of the α parameter is determined by the computational capacities as well as the necessary precision of a surface response computation. The position of the spots has an impact on the prediction performance. It is significant to remember that the trial number at the domain's centre and the α value setting have an impact on the estimation precision.

6.8 RSM FOR BOX-BEHNKEN DESIGN (BBD)

Although BBD is a longer and less expensive response surface design than CCD, it is not suitable for chronological experimental activity since it does not include an incorporated factorial design (Figure 6.4). BBD is more effective and adaptable than other designs, producing a more complicated response with fewer runs. Together, BBD and CCD preserve the higher-order surface. Each factor demands three levels of BBD, which is rotatable. The BBD can adjust to the entire polynomial structure of responses to surface design, just like CCD can (Ferreira et al., 2007). The treatment combinations inside the BBD are situated at the midpoints of the cube's edges as well as its centre. When doing experimental operations with more than two variables and where the optimum is anticipated to lie somewhere within the factor ranges, consideration should be given to this response surface design. The researcher optimized dual-fuel engines powered by waste-derived biogas and applied ML to algae bio diesel and BBD (Sharma et al., 2023).

This is crucial to evaluate the precision of the model, exactness, and dependability via arrange of quantitative markers when contrasting the RSM conclusions to accurate information. Among the most important metrics to

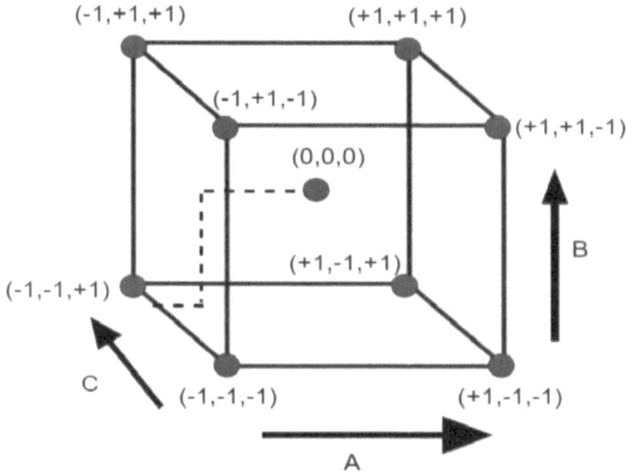

Figure 6.4 Graphical presentation of Box-Behnken design (BBD).

assess are coefficient of determination (R^2), adjusted R^2, mean-squared error (MSE), root mean squared error (RMSE), mean absolute error (MAE), and residual analysis. Lower values indicate a better model fit. The mean absolute error computes the mean absolute difference between the anticipated and actual outcomes of the RSM. In contrast, the mean-squared error (MSE) measures the mean-squared disparity between authentic and projected outcomes. Higher error rates are R^2 prioritized and showed sensitivity to variances. A higher model fit is indicated by lower MSE values. It is simpler to read the squared root of MSE, or RMSE, which provides an estimation of errors in prediction in exactly the identical units as the response variable. Improved model fit is displayed by lower RMSE values. R^2, which has integers that vary from 0 to 1, indicates the percentage of the variation of the dependent variable that the RSM framework discusses.

6.9 GENERATION AND SOURCES OF BIODIESEL

A broad spectrum of feed stocks are able to be employed for constructing biodiesel, and among them are edible fats (first generation) comprising soybean, canola, sunflower, rapeseed, coconut, corn, etc.; edible oils (second generation) including as lard (pork), chicken, rubber, caster, linseed, etc.; animal fats (third generation) such as tallow (rendered beef or mutton fat), lard (pork), poultry, etc.; recycled cooking oil (used frying oils from restaurants); waste vegetable oil (from food processing industries); and microbial oils (fourth generation) such as fungi, microalgae, *Chlorella vulgaris*, Nostoc, etc. (Table 6.1) (Athar and Zaidi, 2020; Chhandama et al., 2021; Singh et al., 2020; Topare et al., 2022).

Table 6.1 Substrate applied to the fabrication of biodiesel

First generation (Edible oils' stock)	*Second generation (Non-edible oils' stock)*	*Third generation (Waste or leftover oils' stock)*	*Fourth generation (Micro-organism oils' stock)*
Coconut	Karanja	Waste	Microalgae
Sunflower	Jatropha	Chicken	Fungi
Peanut	Jatropha	Pork lard	Nostoc
Corn	Coffee	Beef tallow	*Chlorella vulgaris*
Soya bean	Linseed	Poultry	Isochrysis

6.9.1 The transesterification process of collecting the biodiesel

Transesterification represents an interaction linking alcohol and lipids that ends during the production of alkyl esters of fatty acids. Alcohol and triglycerides are transformed to yield glycerol and FAAE. In the initial platform, alcohols and triglycerides integrate giving rise to di-glycerides, which are further broken down becoming into glycerol and mono-glycerides, each leading to an alkyl ester (Thangaraj et al., 2019). Transesterification seems a mechanism by which triglycerides transform di-glycerides, monoglycerides, and subsequently glycerol. This technique is how biodiesel originates. Usually, just one phase within this process includes alkali catalysts. Nevertheless, if the feedstock possesses enormous quantities of water or free fatty acids (FFAs), a two-stage procedure may become necessary. The initial procedure in transforming FFAs into fatty acid esters is recognized as acid-catalysed alcohol lysis or esters of fatty acid. The next level is the transesterification process, which converts the esters of fatty acid into biodiesel (Farouk et al., 2024).

Organic acids, as per Wyse-Masonetal and Beckles (2012), developed catalysts to produce biodiesel, the vegetable oils are simultaneously esterified and trans-esterified to produce a lower volume of biodiesel. The researcher evaluated a variety of the use of heterogeneous catalysts for the generation of biodiesel (Di Serio et al., 2008). The investigator reported optimizing the transesterification process under alkaline conditions to produce biodiesel from canola oil (Issariyakul and Dalai, 2010). Marta conducted a rheological assessment of the biodiesel made from castor oil (Conceição et al., 2005). According to Trotter in the last few years' study on the significance of ionic solutions and the DESs for the manufacture of biodiesel, DESs function as a co-solvent and catalyst by generating an insoluble matrix with methanol, reducing the amount of direct contact between the base catalyst and FAME and preventing hydrolysis. DEES effectively removes glycerol and non-reacted catalysts (Shahbaz et al., 2011; Troter et al., 2016).

Algae are plant-like creatures that are powered by photosynthetic processes. They are made up of carbohydrates, proteins, lipids, and specific pigments called carotenoids. Microalgal species have rapid rates of growth and high lipid material like Scenedesmus sp., Chlorella sp., and Chlorococcum sp. Microalgae development and cultivation are influenced by several factors, including salinity, light intensity, temperature, pH, aeration, and photoperiod (Brindhadevi et al., 2021).

The independent input parameters like alcohol to bio-oil proportion, temperature at which the reaction occurs, reaction time, and catalyst quantity must be chosen to optimize the process. To analyse the effects of the modification in input parameters, one or more output parameters are finally chosen. A straightforward experimental design (TR) was employed in earlier research on the generation of microalgae biodiesel (Hasanudin et al., 2022; Jiménez Callejón et al., 2014; Li et al., 2007; Miao and Wu, 2006). This methodology necessitates a significant number of tests because it entails modifying one variable while maintaining the other factors constant. This method has been used to optimize several microalgae strains for the generation of biodiesel, particularly *Chlorella protothecoides*, *Nannochloropsis gaditana*, and *C. vulgaris* (Al-Humairi et al., 2022; Hasanudin et al., 2022; Jiménez Callejón et al., 2014; Li et al., 2007; Miao and Wu, 2006). However, because a large number of tests are needed, this method demands an extended period as well as resources (Table 6.2).

The speed at which biodiesel from sunflower oil by exploiting CaO to be a heterogeneous catalyst was examined by the researchers (Vujicic et al., 2010). Before the process, catalyst activation was conducted at 900°C. At a reaction process temperature of 80°C, a 91% fatty acid methyl ester yield was attained under ideal circumstances of a 6:1 methanol/oil proportion, 1% weight catalyst concentration, and 5.5 hours of reaction time. Gold-silver core-shell nanoparticles were employed through the investigators as a catalyst to pivot the sunflower transforming oil into biodiesel (Banerjee et al., 2014). The ideal parameters were a catalyst concentration of 5% and a reaction temperature of 65°C for 2 hours, with an alcohol-to-oil molar ratio of 5:1. Studying the kinetics of sunflower oil transesterification, Reyero achieved a high yield of 99% at the ideal catalyst concentration (NaOH) of 0.3 weight percent, reaction temperature of 50°C, and ethanol to oil ratio of 24:1 (Reyero et al., 2015). An investigator used KOH/ZSM_5 as an agent to change sunflower oil into biodiesel (Saba et al., 2016). The catalyst's dosage of 60% wt. and the methanol/oil combination of 18:1 at the temperature of 65°C for the reaction period of 5 h were found to be the ideal parameters. A yield of 97.7% biodiesel was achieved. Ejected catalyst K_2CO_3/Al_2O_3 in various configurations in the production of biodiesel production from sunflower oil was investigated by Silveira Junior et al. (2019). The optimal combination, according to the data, was 35% K_2CO_3/Al_2O_3 of 65%, which produced a FAME production of 99.3% at an action 80°C temperature and a 1:12 oil/methanol molar ratio for the time of 4 hour and 5% of catalyst concentration. According to a researcher, ($Ca600/Mg_4Al_2HT$) is an excellent catalyst for the synthesis of biodiesel from sunflower oil (Dahdah et al., 2020).

Table 6.2 Establishment of experiments to maximize the generation of diverse vegetable oils executed to make biodiesel

Feedstock	*Design of experiments*	*Independent parameters*	*Outputs*	*References*
Nannochloropsis gaditana	TR	Temperature (°C), time (h), and alcohol ratio (mL/g)	Conversion rate (%)	Jiménez Callejón et al. (2014)
Chlorella protothecoides	TR	Temperature (°C), time (h), and alcohol ratio (mL/g)	Conversion rate (%)	Li et al. (2007)
Rice bran	TR	Alcohol ratio (ml), Catalyst (g),	Conversion rate (%)	Prabu et al. (2021)
Chlorella vulgaris	TR	Temperature (°C), duration (minute), catalyst (%), and alcohol ratio (mL)	FAME yield (%)	Hasanudin et al. (2022)
C. vulgaris	TR	Alcohol ratio (molar); temperature (°C), time (minute)	FAME yield (%)	Al-Humairi et al. (2022)
Jatropha curcas Elaeisguineensis (palm)	RSM	Catalyst (%) and alcohol ratio (molar)	FAME/FAEE content (%)	Arce Saavedra et al. (2022)
Sunflower oil	RSM	Alcohol ratio (molar), temperature (°C), catalyst (g)	Viscosity (cSt)	Calero et al. (2020)
Refined soybean oil (RSO)	RSM	Alcohol ratio (molar), temperature (°C), catalyst (%), ultrasound power (W)	Conversion rate (%)	Zou et al. (2023)
Pistacia lentiscus	RSM	Alcohol ratio (molar), temperature (°C), catalyst (%)	FAME yield (%)	Khiari et al. (2019)
Refined palm oil suitable for food (RPO)	RSM-CCD	Alcohol ratio (molar), temperature (°C) time (h), catalyst (%)	Conversion rate (%)	Thoai et al. (2017)
C. vulgaris +castor oil	RSM-CCD	Alcohol ratio (mL), temperature (°C), catalyst (%)	Conversion rate (%)	Beyene et al. (2022)
Chlorella protothecoides	RSM-CCD	Alcohol ratio (mol/mol), temperature (°C), residence time (minute), pressure (bar), water content (wt%)	FAME yield (%)	Nan et al. (2015)
Chlorella variabilis	RSM-CCD	Alcohol ratio (molar), temperature (°C), time (minute), catalyst (wt%)	FAME yield (%)	Nirmala and Dawn (2021)
Palm oil	RSM-BBD	EFI (V/cm), time (minute)	Conversion rate (%)	Sangsawang et al. (2021)
Palm oil	RSM-BBD	Alcohol ratio (molar), temperature (°C), time (minute)	FAME yield (%)	Latchubugata et al. (2018)
N. gaditana	RSM-BBD	Alcohol ratio (mL/ g), catalyst (%), hexane/SL ratio (mL/g)	FAME yield (%)	Macías-Sánchez et al. (2018)

6.10 FUNDAMENTAL COMPONENTS AND APPROACHES

Liable Essential Oil Products Private Limited, Noida, Uttar Pradesh based company, was the algal oil provider. Purchases of chemicals such as potassium hydroxide (KOH) and methanol can be made at Naveen Traders, Civil Lines market. The substance was manufactured using 99% exceptional class methanol and 99.0% extra pure KOH in compact form. The transesterification process requires the following four variables: time (5–7 minutes), catalyst (0.5–1.5 grams), methanol/oil (30–40%), and microwave power (250–350 watts), in the defined order.

6.11 SYNTHESIS OF BIODIESEL

The procedure of transesterification was employed to produce biodiesel (Figure 6.5). Triglycerides and alcohol—typically methanol—interact during the transesterification process to produce glycerine and esters. Three consecutive reversible reactions usually make up a broad method. In the initial step, triglycerides are modified into di-glycerides; then in the very second step, di-glycerides are changed into mono-glycerides; and in the final level, ester as well as glycerol are extracted. The equipment used to produce biodiesel and the components with the method variables consists of a chemical reaction vial, condenser, thermometer, and motor operated by electricity to create agitation. This glass reactor is distinguished by its single-litter capability, including three exposed points (one each for the stirrer and condenser).

Figure 6.5 A separation picture of glycerol and algae biodiesel with its outcome.

6.12 OBSERVATIONS AND DISCUSSION

The above-discussed operational circumstances cover a wide range of situations where biodiesel is produced through transesterification. As a result, optimizing the operating state will be a significant task. The impacts of the variables on the amount of biodiesel produced (%) in this investigation were examined in the section that followed, utilizing the Box-Behnken experiment design. The experimental outputs for the biodiesel produced (in percentage) at design locations are compiled in Table 6.3.

Table 6.3 The final result for the process of transesterification of algae oil

Run	*A—Time (minutes)*	*B—Catalyst (grams)*	*C—Methanol/oil (percentage %)*	*D—Power (Watt)*	*Experimental data*	*Predicted data via RSM*
1	0	0	0	0	98.54	98.95
2	0	+1	0	−1	91.08	94.63
3	0	0	+1	+1	91.15	87.25
4	0	0	+1	−1	85.23	86.96
5	0	+1	+1	0	78.05	84.43
6	0	−1	+1	0	99.81	96.12
7	−1	0	−1	0	92.65	92.25
8	+1	0	−1	0	95.10	92.25
9	0	+1	0	+1	91.80	92.25
10	+1	0	0	−1	99.73	93.59
6	0	−1	−1	0	80.19	81.83
12	0	0	0	0	84.29	90.03
13	+1	0	+1	0	88.57	80.87
14	0	0	0	0	75.55	76.31
15	0	0	−1	−1	79.92	86.69
16	0	0	−1	+1	98.31	96.41
17	−1	0	+1	0	91.62	92.25
18	0	−1	0	+1	96.91	96.86
19	0	0	0	0	90.32	88.38
20	0	−1	0	−1	92.12	89.79
21	−1	0	0	+1	95.09	99.01
22	−1	+1	0	0	96.21	94.71
23	+1	0	0	+1	98.91	101.86
24	0	+1	−1	0	99.27	97.53
25	+1	+1	0	0	86.39	92.25
26	0	0	0	0	74.63	76.86
27	+1	−1	0	0	94.84	96.20
28	−1	0	0	−1	96.84	92.85
29	−1	−1	0	0	88.79	85.55

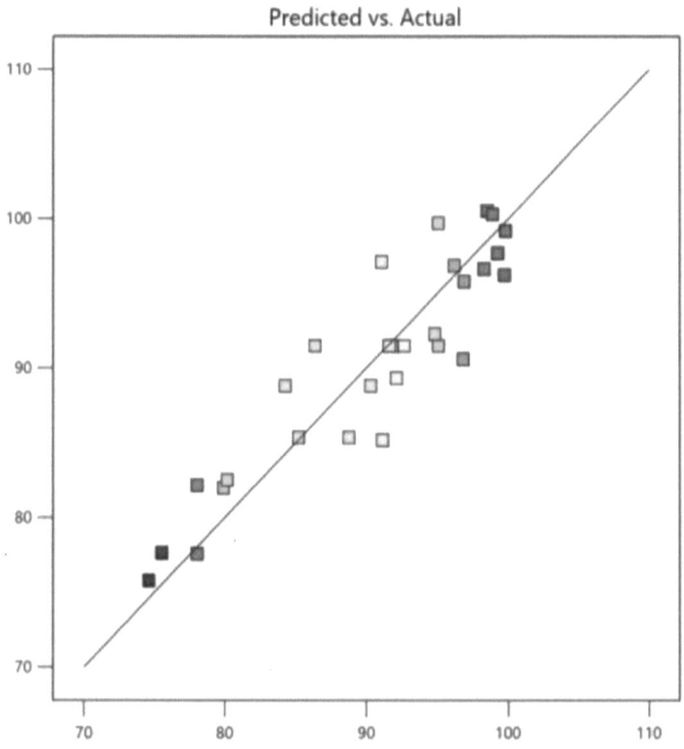

Figure 6.6 Representation of graph between the expected and real data.

The relationship between the expected and genuine integrating performance (%) graph demonstrates how closely the predicted values using the RSM match the experimental values, confirming the improved model's ability to reliably create a relationship between the yield of the algae oil and the variable process in Figure 6.6.

The contour curve is depicted in Figure 6.7. The straightforward correlation in-between the two separate factors and how they affect the response variable is shown by the contour. The yield of biodiesel reduces with rising catalyst concentration. This could be due to a higher amount of catalyst, which aids in triglyceride side reactions. Furthermore, excessive catalysts

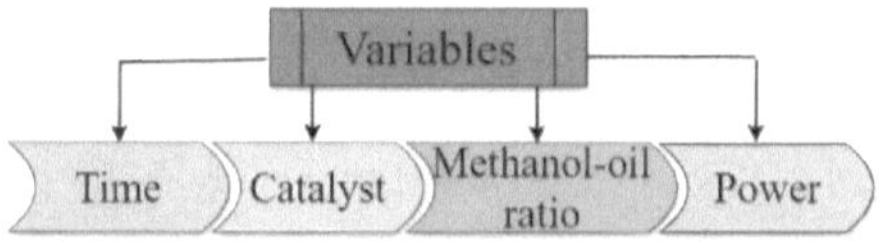

Figure 6.7 Experimental constructed variables.

Table 6.4 Variables utilized for Box-Behnken during transesterification of algae oil

	+1	*0*	*−1*
A—Time	5	6	7
B—Catalyst	0.5	1	1.5
C—Methanol/oil	30	40	50
D—Power	250	300	350

cause the creation of an emulsion, which increases viscosity. The optimal methanol/oil ratios, when combined with other reaction parameters, have a significant impact on whole biodiesel yield (Table 6.4). The maximum molar ratio boosts the yield. Transesterification is a bidirectional process; hence, an excessive amount of alcohol must be consumed to preserve the reaction moving ahead.

The data from the experiment runs and the result of the biodiesel synthesis is shown by the vertical bars on the organized bar graph. The blue thin lines show the experimental results of 29 inspections, whereas the red thin bands show the outcomes determined by the software. On the graph, the horizontal axis displayed the quantity of experimental runs, and the vertical axis displayed the yield.

6.13 ENHANCEMENT VIA OPTIMIZATION

The optimization procedure for every response was performed based on corresponding evolved mathematical equations in order to optimize the production of algae biodiesel. Table 6.5 shows the optimal input process parameter value.

Experiments were carried out with optimized process parameter values for validation reasons, and Table 6.6 demonstrates that the projected response matches well with the experimental data, with an error of 1.865%. With methanol/oil (36%) and KOH as catalyst (1.1 grams), an algal biodiesel yield of 97.975% was achieved in 5 minutes at the power of 250 watts.

Table 6.5 Optimizing for optimal output of algal biodiesel

Variables	*Objective*	*Optimum value*
Time	Within the range	5.000
Catalyst quantity	Within the range	1.103
Methanol/Oil	Within the range	35.627
Power	Within the range	250.089
Biodiesel Production	To maximize	99.840

Table 6.6 The findings of validation under optimal circumstances

Serial number	*Time*	*Catalyst quantity*	*Methanol/oil*	*Power*	*Predicted production*	*Laboratory production*	*Error %*
1	5 minutes	1.1 grams	36 %	250 watts	99.840 %	97.975 %	1.865%

Table 6.7 Distinctive features of algae oil biodiesel

Serial number	*Characteristics (units)*	*Data*	*Standards*
1.	Density (g/cm3)	0.863	ASTM-D-4052
2.	Pour Point (°C)	−7	ASTM-D-2500
3.	Calorific value (MJ/kg)	44	IS:1350(P-2) 1970, ASTM-D
4.	Cloud point (°C)	−2	ASTM-D-2500
5.	Flash Point (°C)	66	ASTM-D-93
6.	Viscosity at 40°C	4.00	ASTM-D-45
7.	FFA (%)	<1	ASTM-D5555-95
8.	Acidic value (mg KOH/g)	0.73	ASTM D664

6.14 CHARACTERISTICS OF BIODIESEL MANUFACTURED

The features of the biodiesel samples made from algae oil are examined. The majority of the physicochemical parameters of algae biodiesel are matched with standard values, as shown in Table 6.7.

6.15 CONCLUSION

The ultimate objective of this endeavour is to optimize the synthesis of biodiesel from algal oil by utilizing the transesterification process technique. Four process variables were optimized BBD with the aid of Design-Expert software to achieve the best results: the methanol/oil (30–50%), the reaction time (5–7 minutes), the catalyst concentration (KOH; 0.5–1.5 grams), and the power (250–350 watts). At the power of 250 watts, 5 minutes were needed to get a 97.975% biodiesel production with the percentage of oil and methanol (36%) and KOH as a catalyst (1.1 grams). The yield of algae oil was shown to be heavily influenced by the reaction duration and catalyst concentration, power, and methanol/oil volume proportion. This investigation unequivocally demonstrates that the RSM was a suitable means of optimizing the operating circumstances to enhance the methyl esters' synthesis. The actual collected experimental data was compared with the expected responses obtained from the RSM. The outcome demonstrated that the instruments provided accurate forecasts because of the R-square values (i.e., 0.8065 for RSM). The models' ability to fit data was evaluated. To find the ideal point, graphic response curves were employed. The findings of the current study were determined to have potential significance regarding the numerical simulation effectiveness of RSM for the biodiesel synthesizing mechanism and the evaluation of treatment methods.

REFERENCES

Abbasi, B., Babaei, T., Hosseinifard, Z., Smith-Miles, K., Dehghani, M., 2020. Predicting solutions of large-scale optimization problems via machine learning: A case study in blood supply chain management. Computers & Operations Research 119, 104941. https://doi.org/10.1016/j.cor.2020.104941

Bouzid Ait-Amir, Philippe Pougnet, Abdelkhalak El Hami, 6 - Meta-Model Development, Editor(s): Abdelkhalak El Hami, Philippe Pougnet, Embedded Mechatronic Systems 2 (Second Edition), ISTE, 2020, Pages 157–187, ISBN 9781785481901,https://doi.org/10.1016/B978-1-78548-190-1.50006-2(https://www.sciencedirect.com/science/article/pii/B9781785481901500062)

Al-Humairi, S.T., Lee, J.G.M., Salihu, M., Harvey, A.P., 2022. Biodiesel production through acid catalyst in situ reactive extraction of *Chlorella vulgaris* foamate. Energies 15, 4482. https://doi.org/10.3390/en15124482

Arce Saavedra, T., Bueno-Borges, L.B., Sangaletti-Gerhard, N., de Alencar, S.M., Regitano-d'Arce, M.A.B., 2022. Optimized conventional and ultrasound-assisted ethyl transesterification of jatropha (*Jatropha curcas*) and palm (*Elaeis guineensis*) oil mixtures. Chemical Engineering Communications 209, 1482–1495. https://doi.org/10.1080/00986445.2021.1977925

Athar, M., Zaidi, S., 2020. A review of the feedstocks, catalysts, and intensification techniques for sustainable biodiesel production. Journal of Environmental Chemical Engineering 8, 104523. https://doi.org/10.1016/j.jece.2020.104523

Banerjee, M., Dey, B., Talukdar, J., Chandra Kalita, M., 2014. Production of biodiesel from sunflower oil using highly catalytic bimetallic gold–silver core–shell nanoparticle. Energy 69, 695–699. https://doi.org/10.1016/j.energy.2014.03.065

Beyene, D., Abdulkadir, M., Befekadu, A., 2022. Production of biodiesel from mixed castor seed and microalgae oils: Optimization of the production and fuel quality assessment. International Journal of Chemical Engineering 2022, 1536160. https://doi.org/10.1155/2022/1536160

Birkhoff, G., 1940. Lattice Theory. American Mathematical Soc.

Brindhadevi, K., Mathimani, T., Rene, E.R., Shanmugam, S., Chi, N.T.L., Pugazhendhi, A., 2021. Impact of cultivation conditions on the biomass and lipid in microalgae with an emphasis on biodiesel. Fuel 284, 119058. https://doi.org/10.1016/j.fuel.2020.119058

Calero, J., Luna, D., Luna, C., Bautista, F.M., Romero, A.A., Posadillo, A., Estevez, R., 2020. Optimization by response surface methodology of the reaction conditions in 1,3-selective transesterification of sunflower oil, by using CaO as heterogeneous catalyst. Molecular Catalysis 484, 110804. https://doi.org/10.1016/j.mcat.2020.110804

Carissimo, C., Korecki, M., 2024. Limits of optimization. Minds & Machines 34, 117–137. https://doi.org/10.1007/s11023-023-09633-1

Chhandama, M.V.L., Satyan, K.B., Changmai, B., Vanlalveni, C., Rokhum, S.L., 2021. Microalgae as a feedstock for the production of biodiesel: A review. Bioresource Technology Reports 15, 100771. https://doi.org/10.1016/j.biteb.2021.100771

Conceição, M.M., Candeia, R.A., Dantas, H.J., Soledade, L.E.B., Fernandes, Valter J., Souza, A.G., 2005. Rheological behavior of castor oil biodiesel. Energy Fuels 19, 2185–2188. https://doi.org/10.1021/ef050016g

Dahdah, E., Estephane, J., Haydar, R., Youssef, Y., El Khoury, B., Gennequin, C., Aboukaïs, A., Abi-Aad, E., Aouad, S., 2020. Biodiesel production from refined sunflower oil over Ca–Mg–Al catalysts: Effect of the composition and the thermal treatment. Renewable Energy 146, 1242–1248. https://doi.org/10.1016/j.renene.2019.06.171

Di Serio, M., Tesser, R., Pengmei, L., Santacesaria, E., 2008. Heterogeneous catalysts for biodiesel production. Energy Fuels 22, 207–217. https://doi.org/10.1021/ef700250g

Etim, A.O., Musonge, P., Eloka-Eboka, A.C., 2022. A green process synthesis of biocomposite heterogeneous catalyst for the transesterification of linseed-marula bi-oil methyl ester. Results in Engineering 16, 100645. https://doi.org/10.1016/j.rineng.2022.100645

Farouk, S.M., Tayeb, A.M., Abdel-Hamid, S.M.S., Osman, R.M., 2024. Recent advances in transesterification for sustainable biodiesel production, challenges, and prospects: A comprehensive review. Environ Sci Pollut Res 31, 12722–12747. https://doi.org/10.1007/s11356-024-32027-4

Fischetti, M., Fraccaro, M., 2019. Machine learning meets mathematical optimization to predict the optimal production of offshore wind parks. Computers & Operations Research 106, 289–297. https://doi.org/10.1016/j.cor.2018.04.006

Gambella, C., Ghaddar, B., Naoum-Sawaya, J., 2021. Optimization problems for machine learning: A survey. European Journal of Operational Research 290, 807–828. https://doi.org/10.1016/j.ejor.2020.08.045

Gupta, S., Patel, P., Mondal, P., 2022. Biofuels production from pine needles via pyrolysis: Process parameters modeling and optimization through combined RSM and ANN based approach. Fuel 310, 122230. https://doi.org/10.1016/j.fuel.2021.122230

Hasanudin, N., Abd Ghani, N., Ab Rahim, A.H., Azman, N.S., Rosdi, N.A., Masri, A.N., 2022. Optimization and kinetic studies on biodiesel conversion from *Chlorella vulgaris* microalgae using pyrrolidinium-based ionic liquids as a catalyst. Catalysts 12, 277. https://doi.org/10.3390/catal12030277

Ferreira, S.C., Bruns, R.E., Ferreira, H.S., Matos, G.D., David, J.M., Brandão, G.C., ... Dos Santos, W.N.L., 2015. Box-Behnken design: An alternative for the optimization of analytical methods. Analytica Chimica Acta 597(2), 179–186.

Issariyakul, T., Dalai, A.K., 2010. Biodiesel production from greenseed canola oil. Energy Fuels 24, 4652–4658. https://doi.org/10.1021/ef901202b

Jiménez Callejón, M.J., Robles Medina, A., Macías Sánchez, M.D., Hita Peña, E., Esteban Cerdán, L., González Moreno, P.A., Molina Grima, E., 2014. Extraction of saponifiable lipids from wet microalgal biomass for biodiesel production. Bioresource Technology 169, 198–205. https://doi.org/10.1016/j.biortech.2014.06.106

Kacalak, W., Lipiński, D., Szafraniec, F., 2024. Selected aspects of precision grinding processes optimization. Materials 17, 607. https://doi.org/10.3390/ma17030607

Karimifard, S., Alavi Moghaddam, M.R., 2018. Application of response surface methodology in physicochemical removal of dyes from wastewater: A critical review. Science of The Total Environment 640–641, 772–797. https://doi.org/10.1016/j.scitotenv.2018.05.355

Khiari, K., Tarabet, L., Awad, S., Loubar, K., Mahmoud, R., Tazerout, M., 2019. Optimization of *Pistacia lentiscus* oil transesterification process using central composite design. Waste Biomass Valor 10, 2575–2581. https://doi.org/10.1007/s12649-018-0257-2

Latchubugata, C.S., Kondapaneni, R.V., Patluri, K.K., Virendra, U., Vedantam, S., 2018. Kinetics and optimization studies using response surface methodology in biodiesel production using heterogeneous catalyst. Chemical Engineering Research and Design 135, 129–139. https://doi.org/10.1016/j.cherd.2018.05.022

Li, X., Xu, H., Wu, Q., 2007. Large-scale biodiesel production from microalga *Chlorella protothecoides* through heterotrophic cultivation in bioreactors. Biotechnology and Bioengineering 98, 764–771. https://doi.org/10.1002/bit.21489

Macías-Sánchez, M.D., Robles-Medina, A., Jiménez-Callejón, M.J., Hita-Peña, E., Estéban-Cerdán, L., González-Moreno, P.A., Navarro-López, E., Molina-Grima, E., 2018. Optimization of biodiesel production from wet microalgal biomass by direct transesterification using the surface response methodology. Renewable Energy 129, 141–149. https://doi.org/10.1016/j.renene.2018.06.001

Miao, X., Wu, Q., 2006. Biodiesel production from heterotrophic microalgal oil. Bioresource Technology 97, 841–846. https://doi.org/10.1016/j.biortech.2005.04.008

Nan, Y., Liu, J., Lin, R., Tavlarides, L.L., 2015. Production of biodiesel from microalgae oil (*Chlorella protothecoides*) by non-catalytic transesterification in supercritical methanol and ethanol: Process optimization. The Journal of Supercritical Fluids 97, 174–182. https://doi.org/10.1016/j.supflu.2014.08.025

Nazarpour, M., Taghizadeh-Alisaraei, A., Asghari, A., Abbaszadeh-Mayvan, A., Tatari, A., 2022. Optimization of biohydrogen production from microalgae by response surface methodology (RSM). Energy 253, 124059. https://doi.org/10.1016/j.energy.2022.124059

Nirmala, N., Dawn, S.S., 2021. Optimization of *Chlorella variabilis*. MK039712.1 lipid transesterification using response surface methodology and analytical characterization of biodiesel. Renewable Energy 179, 1663–1673. https://doi.org/10.1016/j.renene.2021.07.123

Pawlak, T.P., Krawiec, K., 2017. Automatic synthesis of constraints from examples using mixed integer linear programming. European Journal of Operational Research 261, 1141–1157. https://doi.org/10.1016/j.ejor.2017.02.034

Pawlak, T.P., Krawiec, K., 2019. Synthesis of constraints for mathematical programming with one-class genetic programming. IEEE Transactions on Evolutionary Computation 23, 117–129. https://doi.org/10.1109/TEVC.2018.2835565

Prabu, A., Premkumar, I., Pradeep, A., 2021. Production of rice bran biodiesel by the bubble wash method. International Journal of Ambient Energy 42, 981–984. https://doi.org/10.1080/01430750.2019.1583128

Rao, S.S., 2009. Engineering optimization: theory and practice (4th ed.). John Wiley & Sons, Hoboken, N.J.

Reyero, I., Arzamendi, G., Zabala, S., Gandía, L.M., 2015. Kinetics of the NaOH-catalyzed transesterification of sunflower oil with ethanol to produce biodiesel. Fuel Processing Technology 129, 147–155. https://doi.org/10.1016/j.fuproc.2014.09.008

Saba, T., Estephane, J., El Khoury, B., El Khoury, M., Khazma, M., El Zakhem, H., Aouad, S., 2016. Biodiesel production from refined sunflower vegetable oil over KOH/ZSM5 catalysts. Renewable Energy 90, 301–306. https://doi.org/10.1016/j.renene.2016.01.009

Sadana, U., Chenreddy, A., Delage, E., Forel, A., Frejinger, E., Vidal, T., 2024. A survey of contextual optimization methods for decision-making under uncertainty. European Journal of Operational Research. https://doi.org/10.1016/j.ejor.2024.03.020

Sakkas, V.A., Islam, Md.A., Stalikas, C., Albanis, T.A., 2010. Photocatalytic degradation using design of experiments: A review and example of the Congo red degradation. Journal of Hazardous Materials 175, 33–44. https://doi.org/10.1016/j.jhazmat.2009.10.050

Sangsawang, T., Rongrat, N., Yothayuth, A., 2021. Optimization for biodiesel production by transesterification with electric fields. Engineering Journal 25, 261–268. https://doi.org/10.4186/ej.2021.25.2.261

Selemani, A., Kombe, G.G., 2022. Glycerolysis of high free fatty acid oil by heterogeneous catalyst for biodiesel production. Results in Engineering 16, 100602. https://doi.org/10.1016/j.rineng.2022.100602

Shahbaz, K., Mjalli, F.S., Hashim, M.A., AlNashef, I.M., 2011. Using deep eutectic solvents based on methyl triphenyl phosphonium bromide for the removal of glycerol from palm-oil-based biodiesel. Energy Fuels 25, 2671–2678. https://doi.org/10.1021/ef2004943

Sharma, P., Sahoo, B.B., Said, Z., Hadiyanto, H., Nguyen, X.P., Nižetić, S., Huang, Z., Hoang, A.T., Li, C., 2023. Application of machine learning and Box-Behnken design in optimizing engine characteristics operated with a dual-fuel mode of algal biodiesel and waste-derived biogas. International Journal of Hydrogen Energy, The 5th International Conference on Alternative Fuel, Energy & Environment (ICAFEE 2021) 48, 6738–6760. https://doi.org/10.1016/j.ijhydene.2022.04.152

Silveira Junior, E.G., Perez, V.H., Reyero, I., Serrano-Lotina, A., Justo, O.R., 2019. Biodiesel production from heterogeneous catalysts based K_2CO_3 supported on extruded γ-Al_2O_3. Fuel 241, 311–318. https://doi.org/10.1016/j.fuel.2018.12.074

Simbi, I., Aigbe, U.O., Oyekola, O., Osibote, O.A., 2022. Optimization of biodiesel produced from waste sunflower cooking oil over bi-functional catalyst. Results in Engineering 13, 100374. https://doi.org/10.1016/j.rineng.2022.100374

Singh, D., Sharma, D., Soni, S.L., Sharma, S., Kumar Sharma, P., Jhalani, A., 2020. A review on feedstocks, production processes, and yield for different generations of biodiesel. Fuel 262, 116553. https://doi.org/10.1016/j.fuel.2019.116553

Sun, S., Cao, Z., Zhu, H., Zhao, J., 2020. A survey of optimization methods from a machine learning perspective. IEEE Transactions on Cybernetics 50, 3668–3681. https://doi.org/10.1109/TCYB.2019.2950779

Tamoradi, T., Kiasat, A.R., Veisi, H., Nobakht, V., Besharati, Z., Karmakar, B., 2021. MgO doped magnetic graphene derivative as a competent heterogeneous catalyst producing biofuels via transesterification: Process optimization through response surface methodology (RSM). Journal of Environmental Chemical Engineering 9, 106009. https://doi.org/10.1016/j.jece.2021.106009

Thangaraj, B., Solomon, P.R., Muniyandi, B., Ranganathan, S., Lin, L., 2019. Catalysis in biodiesel production—A review. Clean Energy 3, 2–23. https://doi.org/10.1093/ce/zky020

Thoai, D.N., Tongurai, C., Prasertsit, K., Kumar, A., 2017. A novel two-step transesterification process catalyzed by homogeneous base catalyst in the first step and heterogeneous acid catalyst in the second step. Fuel Processing Technology 168, 97–104. https://doi.org/10.1016/j.fuproc.2017.08.014

Topare, N.S., Jogdand, R.I., Shinde, H.P., More, R.S., Khan, A., Asiri, A.M., 2022. A short review on approach for biodiesel production: Feedstock's, properties, process parameters and environmental sustainability. Materials Today: Proceedings, International Chemical Engineering Conference 2021 (100 Glorious Years of Chemical Engineering & Technology) 57, 1605–1612. https://doi.org/10.1016/j.matpr.2021.12.216

Troter, D.Z., Todorović, Z.B., Đokić-Stojanović, D.R., Stamenković, O.S., Veljković, V.B., 2016. Application of ionic liquids and deep eutectic solvents in biodiesel production: A review. Renewable and Sustainable Energy Reviews 61, 473–500. https://doi.org/10.1016/j.rser.2016.04.011

Verwer, S., Zhang, Y., Ye, Q.C., 2017. Auction optimization using regression trees and linear models as integer programs. Artificial Intelligence, Combining Constraint Solving with Mining and Learning 244, 368–395. https://doi.org/10.1016/j.artint.2015.05.004

Veza, I., Spraggon, M., Fattah, I.M.R., Idris, M., 2023. Response surface methodology (RSM) for optimizing engine performance and emissions fueled with biofuel: Review of RSM for sustainability energy transition. Results in Engineering 18, 101213. https://doi.org/10.1016/j.rineng.2023.101213

Villarrubia, G., De Paz, J.F., Chamoso, P., la Prieta, F.D., 2018. Artificial neural networks used in optimization problems. Neurocomputing 272, 10–16. https://doi.org/10.1016/j.neucom.2017.04.075
Vujicic, Dj., Comic, D., Zarubica, A., Micic, R., Boskovic, G., 2010. Kinetics of biodiesel synthesis from sunflower oil over CaO heterogeneous catalyst. Fuel 89, 2054–2061. https://doi.org/10.1016/j.fuel.2009.11.043
Wyse-Mason, R.R., Beckles, D.M., 2012. An investigation of restaurant waste oil characteristics for biodiesel production in Trinidad and Tobago. Energy for Sustainable Development 16, 515–519. https://doi.org/10.1016/j.esd.2012.05.005
Zou, S., Zhang, H., Wang, J., 2023. Ultrasound-assisted pickering interfacial catalysis for transesterification: Optimization of biodiesel yield by response surface methodology. Journal of Oleo Science 72, 233–243. https://doi.org/10.5650/jos.ess22340

Chapter 7

Desiccant rotary regenerators for sustainable cooling in buildings

D.B. Jani

7.1 INTRODUCTION

The requirement of electrical energy especially in many different building cooling applications has substantially increased due to constant increase in growing global population. As the growth rate increases, a significant amount of energy is required to run more machines for enhanced cooling requirement of people and industrial applications for daily use. The expectation that multiple physical tasks be completed shortly and simultaneously has motivated to a busy lifestyle. When people are sweaty and tired, they automatically think to turn on the room air conditioners to relieve their discomfort by maintaining necessary comfort by controlled cooling and dehumidification. As a result, it is crucial to evaluate the cooling system's energy performance while keeping indoor design conditions in mind for effective cooling and dehumidification [1–3]. To maintain necessary indoor thermal comfort in terms of temperature and humidity, mostly occupants can make use of vapor compression based conventional air conditioners [4]. One of the problems is that air conditioners won't be able to rely on fossil fuels as their supply will decrease with over pricing in the next few decades [5–9]. This could be probably one of the reasons why the sustainable energy performance of the traditionally used vapor compression based air conditioners is about to decline because it uses of high-grade electrical energy for its operation. This is because the vapor compression refrigeration (VCR) only uses energy to cool the environment through coolants; it was unable to lower the cooling load through dehumidification due to low due point temperature of cooling re-circulating room air [10–12]. In addition, the vapor compression based traditional cooling system has played a significant role in the release of greenhouse gases like as carbon dioxide (CO_2) and substances that are mainly responsible for to deplete the ozone layer which saves mother earth from harmful ultraviolet rays, many other harmful gases like as hydro chlorofluorocarbons (HCFCs) and chlorinated fluorocarbons (CFCs). Elevated concentrations of these harmful coolants in indoor closed space have increased the potential threat to degrade indoor air quality (IAQ) due to poor ventilation provided in conventional VCR cooling. Because the traditionally used vapor

DOI: 10.1201/9781003534785-7

compression based air conditioners causes the furniture in the vicinity to rot, especially when the surrounding relative humidity surpasses 60–70%, it has partially resulted in significant losses for air conditioner users [13–16]. This occurs because the system was unable to independently regulate the temperature and humidity of the indoor room space. Increased amounts of water vapor in the air due to high humidity encourage growth of intensive mold due to excessive indoor moisture retention on furniture and lead to the rotting of the user's costlier furniture because the traditionally used VCR-based air conditioners were unable to independently regulate the temperature and humidity inside of the room space to control effectively the latent heat load part of the cooling load. Solid desiccant materials must be used in conjunction with rotary regenerators in order to absorb interior moisture and effective control over interior humidity through the dehumidification process [17–21]. Dehumidified air can reduce the risk of indoor furniture rotting while simultaneously providing dry, cool air to the interior environment for necessary indoor thermal comfort. Due to the fact that desiccant-assisted cooling can judiciously be coupled to freely available renewable energy sources like as solar energy or low cost industrial waste heat will also be able to reduce the energy cost of air conditioners because, as it removes moisture from the air, it will also be able to cool the air around it to a certain extent, reducing the need for cooling because the indoor environment will be more comfortable than usual [22–24].

Desiccant-assisted rotary regenerator based dehumidification and air conditioning systems can also boost air ventilation to improve IAQ, get rid of unwanted odors and inborn air pollutants in room re-circulated air, and enhance the quality of the air inside buildings by air cleaning through reactivation process [25–27]. These findings mainly came from an economic analysis based on payback period and comparison between a conventional VCR system and a desiccant-assisted rotary regenerator based dehumidification and novel air conditioning system. Freely available solar heat and surplus industrial waste heat from cogeneration system can be used to regenerate the desiccant dehumidifier. Due to the system's ability to handle temperature and humidity independently, it can function on a broad range of sensible humidity ratios [28–29].

7.2 PRINCIPAL OF WORKING OF ROTARY REGENERATORS

A substance or chemical known as a desiccant is one that absorbs moisture by having natural affinity of water vapor from humid air and leaves the surrounding area dry (adsorption). In industry, this chemical is typically used to extract moisture from damp air, remove it, and maintain the items' dryness to prevent corrosion. These chemicals are placed in product containers during transportation to prevent moisture damage to the transported goods in presence of water vapor. Drips also prevent condensation

Table 7.1 Properties of important desiccant materials used in rotary regenerators

Sr. no.	*Type of desiccant materials*	*Moisture removal capacity (gm/kg)*	*Regeneration temperature (°C)*	*Applications*
1	Silica gel	0.7	60–80	Dehumidification
2	Metal organic frame works	1.7	70–80	Humidity pump
3	Molecular sieve	0.43	60–70	Cooling
4	High amylose starch	0.78	50–70	ERVs (energy recovery ventilators)
5	SG/LiBr	0.4	80–100	Cooling
6	SG/$MgCl_2$	0.3	70–90	Cooling
7	Vermiculite/$LiNO_3$	0.5	55	Adsorption chiller
8	Activated carbon	0.19	115	Dehumidification
9	Basolite	0.65	85	Adsorption heat pump

from forming between window panes [30–33]. A wide range of industries, including manufacturing, agriculture, food and beverages, electronics, packaging, and super stores, makes use of desiccants extensively. They are available in a variety of solid or liquid forms. There are two primary categories of desiccants, each with specific chemical properties and purposes as shown in Table 7.1. These groups include desiccants that are liquid as well as solid.

The rotary regenerator wheel is a rotary-type regenerative energy wheel that is coated with solid desiccant materials like silica gel and molecular sieves and is made of hundreds of flow channels constituted by layers of desiccant-coated aluminum sheets in a matrix. As the wheel rotates, the heat and moisture from one process, air-stream, are stored in the substrate and transferred to the other, reactivation air-stream. During winter operations, energy from warm and humid air from the building exhaust will be transferred to cool and dry outdoor air [34–37]. The various parts of desiccant cooling systems include hot air for desiccant regeneration, processed air, revolving desiccant wheels in rotary regenerator forms, and sensible cooling systems to control supply design temperature of indoor space for comfort cooling and maintaining thermally cooled atmosphere. Desiccant wheel integrated rotary regenerators in air conditioning systems fall into one of the following categories, depending on the type of desiccant material utilized in the matrix of the dehumidifier: Desiccant dehumidification integrated innovative cooling and air conditioning systems that come in two varieties: liquid and solid. Desiccant-assisted rotary regenerator integrated dehumidification and sustainable innovative cooling systems help save unnecessary repairs from costing hundreds of rupees by reducing the moisture content in the right places to effective humidity control for the product or processes for air conditioning [38–42]. The adsorption of moisture takes place in rotary regenerator commonly known as a desiccant

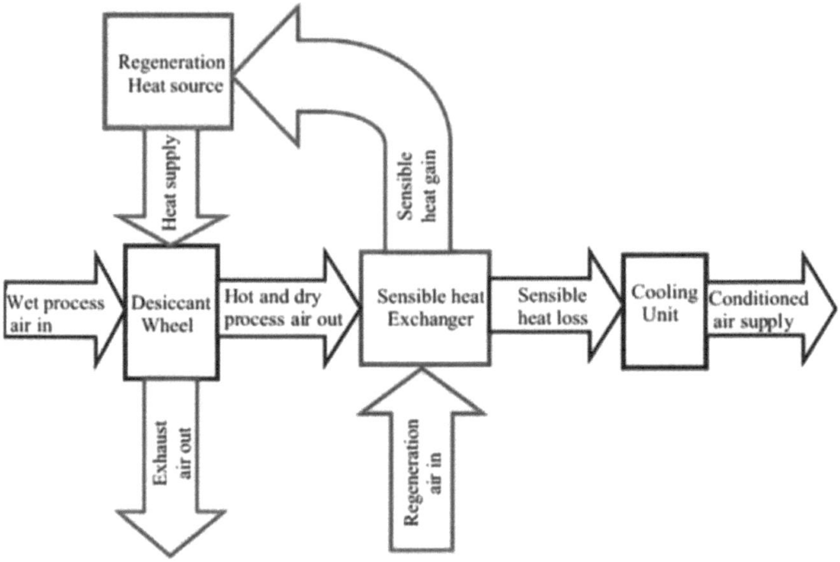

Figure 7.1 Working principle of desiccant rotary regenerator based hybrid cooling systems.

wheel (Figure 7.1) that uses silica gel or other available absorptive materials for maximum adsorption power [43].

7.3 SYSTEM OPERATION

Depending on how it is configured, desiccant cooling can be either an open cycle or a closed cycle. It uses a desiccant wheel and a thermal wheel to accomplish both cooling and dehumidification. Reactivating the desiccant can be accomplished through waste heat or solar thermal energy. Similar to the thermal wheel, which recovers heat energy by rotating and reactivating the desiccant, the desiccant wheel also functions by reactivating the desiccant. It is placed between the supply and exhaust air streams. Figure 7.2 illustrates the cycle's basic idea [44–45]. There are several variations based on the humidity control requirements, air change requirements, and the state of the outside ambient air. It can be used in conjunction with an evaporative cooling system or a traditional vapor compression cycle, but it has a limited operating range and requires large air flows. As outside air passes through the desiccant wheel in the air cycle, heat from the adsorption of water vapor dissipates, drying and intensifying the air (process 2–3). The air is then forced into a regenerative evaporative cooler (REC), where it is cooled in sensible heat exchangers (SHX) by three to four degrees. After that, the heated air is exhausted and the refrigerant (R-410A) is condensed by feeding it into the evaporative condenser (process 4–5).

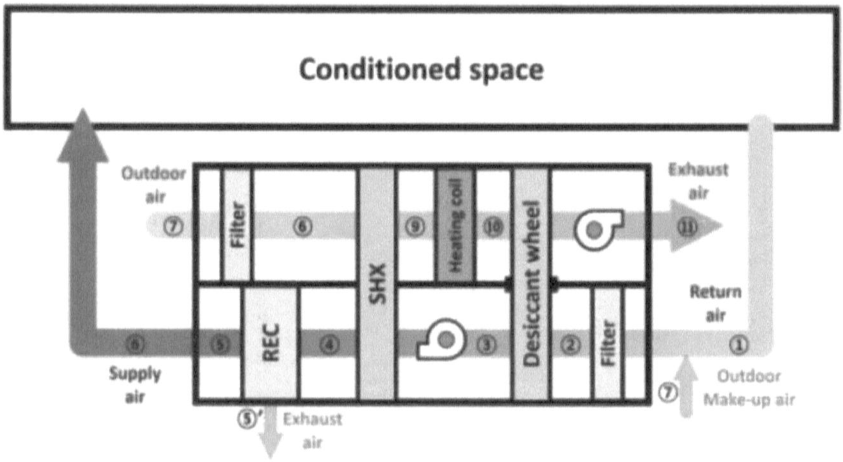

Figure 7.2 Schematic layout of desiccant rotary regenerator based hybrid cooling system.

When air from the outside is heated to the regeneration temperature (process 9–10) at the heating coil, it enters the regeneration channel and absorbs moisture from supplied process air in rotary regenerator and at the end, exhaust it to open atmosphere. With reference to Figure 7.3, as air from outside passes through the desiccant wheel, its temperature rises and its moisture content decreases. It gains some moisture as it travels through an evaporative cooler after being sensibly cooled by the thermal wheel to the necessary supply temperature [46]. After passing through an evaporative cooler, the return

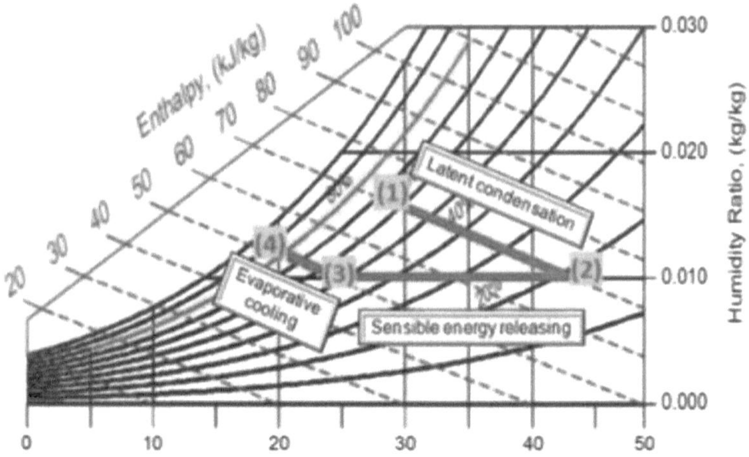

Figure 7.3 Psychrometric representation of adsorption and desorption processes in rotary regenerator.

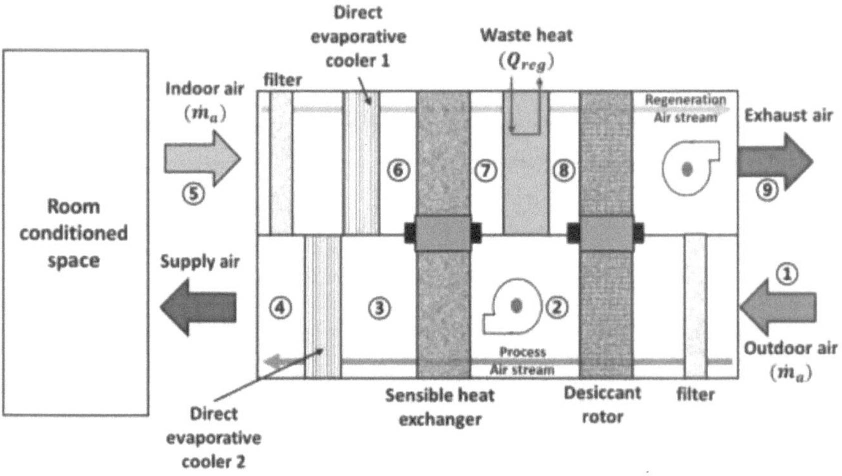

Figure 7.4 Different components of rotary regenerator based hybrid cooling system.

air from indoors—which is, let's say, 25°C—enters the thermal wheel at a lower temperature and a higher moisture content. It is sensibly heated as it travels through the thermal wheel and is heated even more by the heater to allow the desiccant to reactivate before exhausting. It is possible to control the by-pass to prevent the application of extra heat. Figure 7.4 shows an illustration of the thermal wheel, i.e., SHX.

7.4 USE OF DIVERSIFIED REACTIVATION HEAT SOURCES FOR SUSTAINABLE COOLING DEVELOPMENT

Almost negligible electrical power is needed in desiccant regenerator based sustainable cooling systems to run the fans in preparation of process air for desorption and reactivation air for regeneration. Since the desiccant system can run on low heat, solar thermal systems or any other condenser or industrial waste heat source from power plants, chimneys, etc., can supply heat energy needed for desiccant regeneration [47]. The desiccant-based rotary generators were commonly used in hybrid cooling where rotary dehumidifiers can be regenerated using heat recovery from waste heat rejected by the VCR (vapor compression refrigeration) condenser, as shown in Figure 7.5. Consequently, the solid desiccant-based rotary regenerators integrated hybrid air conditioning system operates better and the condenser cools more efficiently. Thus, in solid desiccant-based vapor compression integrated innovative indoor space cooling, the exhaust heat from the VCR condenser can also be utilized to meet the reactivation heat requirement of the dehumidifier; in the absence of this, the heat is released into the open atmosphere. This shows that the system is economically feasible as running cost lowers when waste

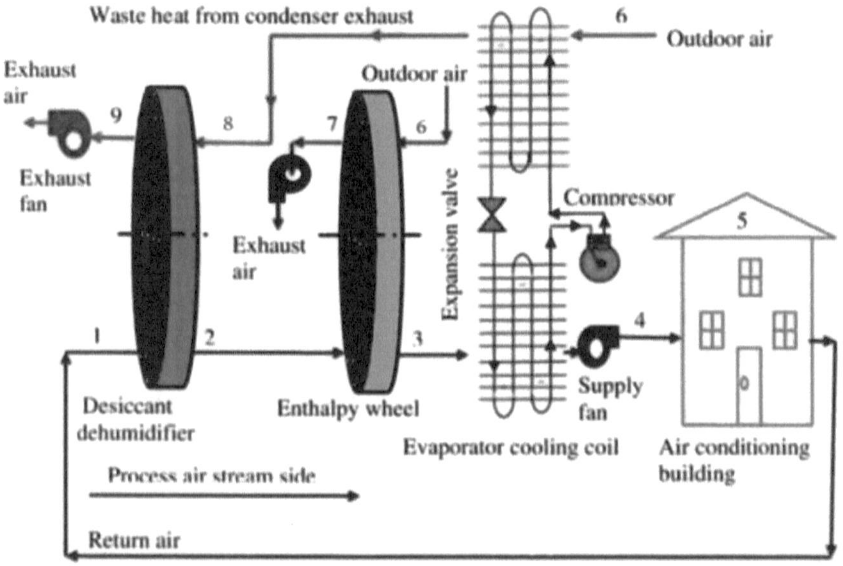

Figure 7.5 Use of condenser waste heat in rotary regenerator's reactivation in hybrid cooling.

heat is used for dehumidifier reactivation because it not only leads to higher energy savings but also slightly lower system operating costs resulting in sustainable cooling [48–51]. Desiccant regenerator based innovative cooling and dehumidification systems, however, may become inoperable if these sources become temporarily unavailable due to cloudy weather, in which case they would need to rely on electricity or other supplemental or auxiliary power heating sources to function. Storing thermal energy as regenerating desiccant dehumidifier and using it for dehumidification in the absence of thermal energy is one way to solve this issue.

The drawback of desiccant cooling is that, in order to desorb saturated desiccant rotary regenerator during a portion of the exchange process, a significant amount of reactivation heat is to be added by freely available renewable energy source like solar or wind for better energy conservation [52–54]. The system will be effective if this heat can be produced at no cost using renewable energy sources like solar power. Sun radiation and the need for cooling also increase at the same time. By regenerating the heat produced by solar heat, various solar collectors (Figure 7.6) can easily completely regenerate the rotary regenerator. Some time it is observed that the required solar radiation is insufficient to generate the necessary reactivation heat because of the heating system used during the winter or other seasonal weather. Since solar energy has been found to be the most abundant at the continental places and is also freely available, using it to regenerate the rotary regenerated based desiccant dehumidifier is an intriguing way to meet the needs of moist spaces for effective drying. Two strategies can be used to capture solar energy. First,

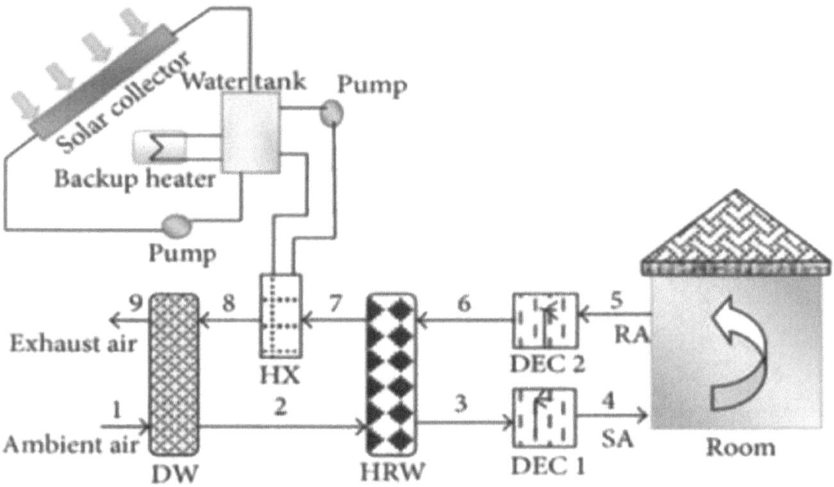

Figure 7.6 Regenerating the desiccant rotary regenerators by using renewable solar heat.

a photovoltaic system that uses solar energy to generate electricity for the fans and drive motors. Second, the weak desiccant regenerators that can be renewed by using the heat energy from the sun. Prior to passing the reactivation heat energy, the desiccant regenerator needed to be reactivated, it is directly heated using a solar collector and regenerator combination. Water is used in another kind of solar collector to transfer heat from the collector to the desiccant matrix available in regenerator. Since the desiccant regenerator receives all of the heat energy absorbed by the collector, the earlier type of solar collector device is more efficient. The latent heat component of cooling load can be extracted from the moist or humid room air using desiccant regenerator based dehumidification and cooling systems. However, sensible heat cannot be removed from the air by a desiccant regenerator integrated innovative cooling system. In order to eliminate latent and sensible heat before introducing it into the comport room space, the desiccant-based rotary regenerator dehumidifier systems are frequently used in conjunction with contact or non-contact type evaporative cooling systems, vapor compression based conventional cooling systems or vapor absorption systems.

The solid desiccant-assisted rotary regenerator integrated hybrid cooling system powered by waste energy received from engine exhaust allows for the resupply of heat to the rotating dehumidifier for its effective desorption through the use of either the micro-CHP (where CHP is combined heat and power) unit or the engine exhaust of the DG set (Figure 7.7).

This enhanced engine thermal efficiency by lowering the exhaust sink temperature and fuel efficiency were accompanied by a desiccant regenerator integrated innovative evaporative cooling HVAC system, and better overall system cooling performance by the desiccant-powered dehumidification and exhaust cooling system [55, 56].

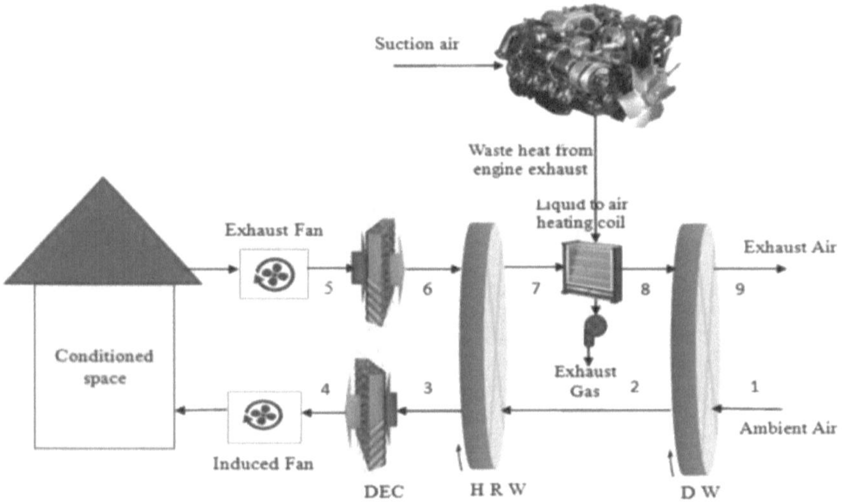

Figure 7.7 Use of engine exhaust heat as regeneration heat sources of desiccant regenerator in solid desiccant integrated hybrid dehumidification and cooling system.

7.5 PERFORMANCE ASSESSMENT OF DESICCANT-BASED SUSTAINBALE COOLING

A small seating cabin of dimensions about (3 m ~3 m ~ 3 m) has been chosen for the investigation. The cooling loads that are sensible and latent are evaluated as per cooling load factor method obtained as 1.36 kW and 0.38 kW, respectively. Ratio of sensible heat (SHR) based on calculated sensible and latent heat load is computed to be 0.76. Process air stream and regeneration flow rate measurements can be obtained as 121.5 m^3/h and 364 m^3/h, respectively. Requirements for indoor comfort as per standard indoor room design supply conditions can be taken as 50% relative humidity and 26°C dry bulb temperature (DBT) of indoor space. Figure 7.8 depicts the schematic diagram of the solid desiccant regenerated integrated hybrid air conditioning system. In the return room air (condition 1) process, air side experiences a significant amount of adsorption of moisture as it travels through the revolving solid desiccant regenerator wheel, becoming drier and hotter as it exits the rotary regenerator wheel. In the air-to-air sensible heat recovery wheel (condition 2), sensible cooling of this air steam is accomplished between states condition 2 and 3, concurrently preheating the surrounding air in the reactivation side. As a result, while heating the reactivation air in the heater to the necessary regeneration temperature required for the desiccant material to desorb, useful energy would be saved. The air is conveyed to the indoor conditioned room after being further chilled between states conditions 3 and 4 by passing over the traditional vapor compression unit of air conditioning having sensible cooling evaporator coil

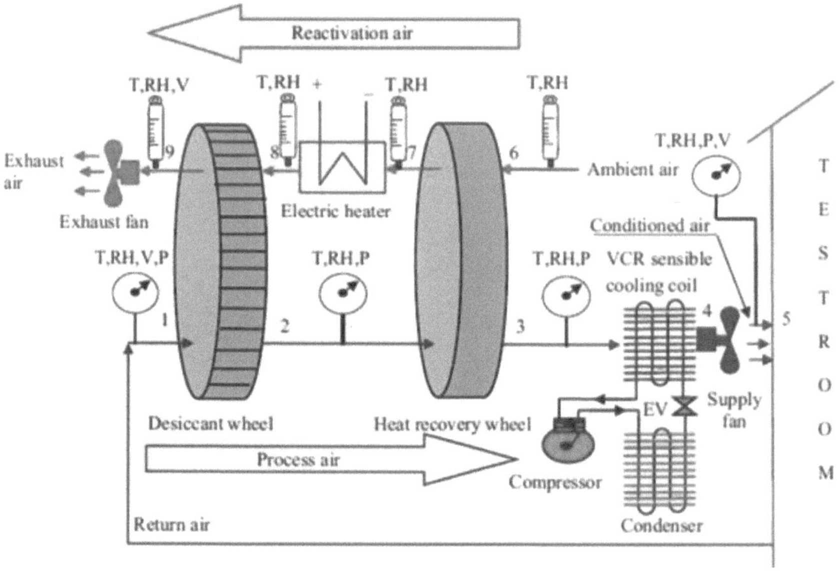

Figure 7.8 Different locations of measurement devices in schematic layout of rotary regenerator integrated hybrid cooling system.

[57]. After sensibly cooled process, dried air available from open outdoor fresh air in the reactivation air side is slightly preheated from state 6 to state 7 as it passes through the air to air sensibly heat recovery wheel. With the help of an external regeneration heat source to produce required reactivation temperature for effective desiccant regeneration, this preheated air is further heated to final desorption state point condition 8, or the regeneration temperature required for the desiccant material to desorb the stored moisture within rotary regenerator matrix. At state point 9, the desiccant wheel is completely recharged for next cycle operation and the air is at the last exhausted to the atmosphere from rotary regenerator. In the different locations of experimental set up, the measurement devices were located for concurrent measurements of process and reactivation of air dry bulb temperature, moist air relative humidity, flow of both air streams, pressure drop across dehumidifier, etc.

At significant junctures in the rotary regenerator running cycle, the relative humidity and pressure drop variations have been demonstrated in Figure 7.9. It illustrates how a specific cooling process parameter like relative humidity and pressure drop across dehumidifier changes over the course of a typical measuring duration. Simultaneously, one can also observe variations in one parameter in relation to the other parameters, depicting effect of one over another. This allows for the prediction of overall system performance at any given time duration of cycle operation [58]. The start and stop of control in

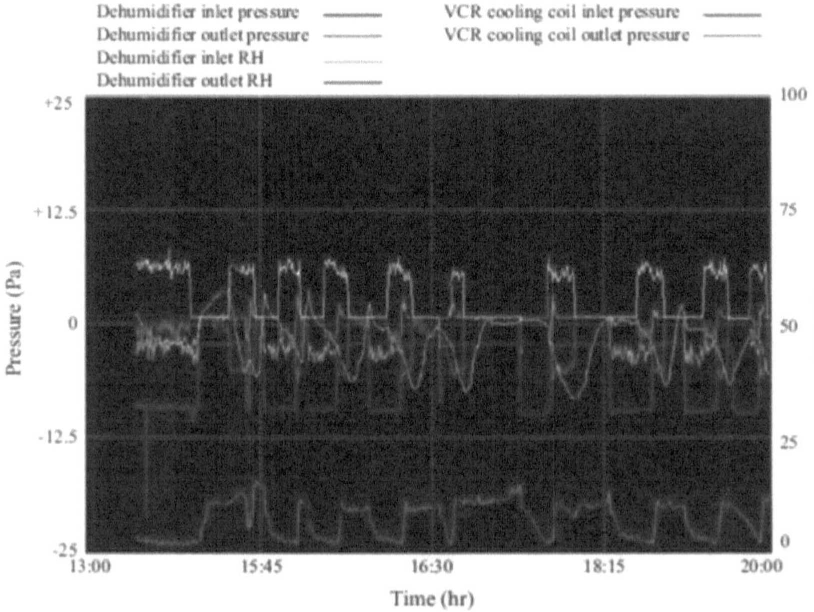

Figure 7.9 Deviations recorded for relative humidity and pressure drop during measurements.

attaining the intended design supply room conditions for producing necessary comfort conditions within the room result in lines or interruptions of varying degrees among different parameters. This can also be used to determine the individual parameter performance for better control strategy.

Figure 7.10 illustrates how the effectiveness of a rotary regenerator dehumidifier varies with variations in the surrounding relative humidity at various outdoor temperatures changes. It has been noted that greater relative humidity found especially in the surrounding air increases the effectiveness of desiccant-integrated rotary regenerator wheels. This is because, in the presence of higher water vapor content in form of excessive moisture in the supplied process room air, there is major deviations found in the vapor pressure of moisture in process air mixture had been large in case of the process air and the desiccant felt, which leads to higher diffusion of the water molecules transfer from the warm and dehumidified process air supplied from rotary regenerator dehumidifier to the surface of the desiccant material. This is caused by the process air inlet humidity ratio of the desiccant wheel increasing due to higher indoor relative humidity due to room moisture additions from large number of occupants [59]. As a result, higher interior relative humidity comparatively increases the humidity ratio of the dehumidified supplied to the process air inlet of the desiccant regenerator wheel, which in turn increases the amount of total moisture contained in air which is effectively cycled

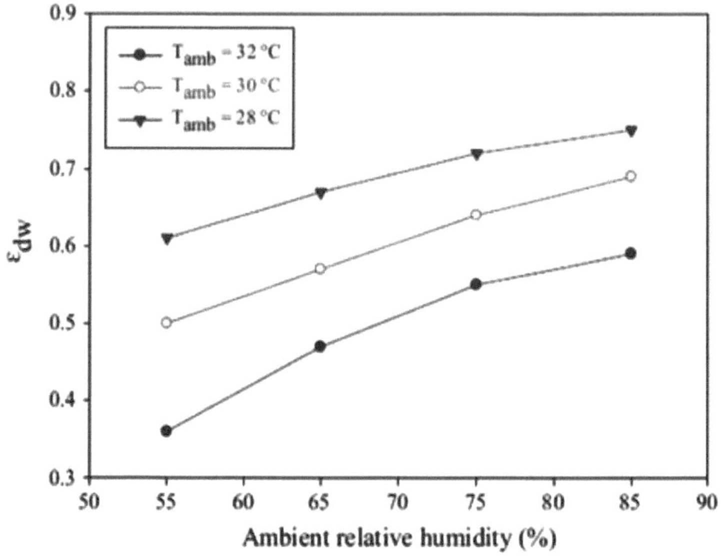

Figure 7.10 Effects of fresh air humidity on rotary regenerator effectiveness according to changes in outdoor temperature.

within the rotary dehumidifier and boosts the rotary regenerator dehumidifier's efficacy [60–61].

7.6 CONCLUSIONS

The recent development of desiccant rotary regenerators integrated novel dehumidification and air conditioning systems as an alternative to traditional vapor compression systems is accelerating due to the latter's high energy consumption and negative environmental effects mainly caused by vapor compression based traditional cooling systems. Desiccant integrated novel cooling systems therefore become more important due to the ongoing rise in energy demand, rising costs, and climate change due to detrimental effects from global warming. Furthermore, they don't use any environmentally hazardous refrigerants. Desiccant-integrated novel cooling systems are therefore more environmentally friendly due to sustainable cooling because freely available renewable solar energy or low-grade industrial waste heat is mainly used for desiccant regeneration. They therefore require less energy. The desiccant wheel regeneration process requires no additional energy, such as gas or electricity. Microorganisms in the air are eliminated as they go through reactivation part of the system to open atmosphere in form of exhaust. Due to its better ability to control moisture level of the re-circulated room air to meet required indoor comfort, there is no dew, there is no mildew or mold. As a result, these systems can stop illnesses linked to buildings occupants. Desiccant regenerators based innovative and sustainable cooling systems can also be used in a variety of climates.

REFERENCES

1. Bejan, A. (1980). Second law analysis in heat transfer. Energy, 5(8–9), 720–732.
2. San, J. Y., Worek, W. M., & Lavan, Z. (1987). Entropy generation in combined heat and mass transfer. International Journal of Heat and Mass Transfer, 30(7), 1359–1369.
3. Van den Bulck, E. S. A. K., Klein, S., & Mitchell, J. W. (1988). Second law analysis of solid desiccant rotary dehumidifiers. Journal of Solar Energy Engineering, 110(1), 2–9. https://doi.org/10.1115/1.3268233
4. Ahmed, C. K., Gandhidasan, P., Zubair, S. M., & Al-Farayedhi, A. A. (1998). Exergy analysis of a liquid-desiccant-based, hybrid air-conditioning system. Energy, 23(1), 51–59.
5. Camargo, J. R., Ebinuma, C. D., & Silveira, J. L. (2003). Thermoeconomic analysis of an evaporative desiccant air conditioning system. Applied Thermal Engineering, 23(12), 1537–1549.
6. Dincer, I., & Sahin, A. Z. (2004). A new model for thermodynamic analysis of a drying process. International Journal of Heat and Mass Transfer, 47(4), 645–652.
7. Rosen, M. A., & Dincer, I. (2004). Effect of varying dead-state properties on energy and exergy analyses of thermal systems. International Journal of Thermal Sciences, 43(2), 121–133.
8. Caliskan, H., Dincer, I., & Hepbasli, A. (2011). Exergetic and sustainability performance comparison of novel and conventional air cooling systems for building applications. Energy and Buildings, 43(6), 761–772.
9. Hürdoğan, E., Buyükalaca, O., Hepbasli, A., & Yılmaz, T. (2011). Exergetic modeling and experimental performance assessment of a novel desiccant cooling system. Energy and Buildings, 43(6), 789–798.
10. Koronaki, I. P., Rogdakis, E., & Kakatsiou, T. (2011). Thermodynamic analysis and neural network model of open cycle desiccant cooling systems with silica gel. International Review of Mechanical Engineering (IREME), 5(2), 298–304.
11. Parameshwaran, R., Kalaiselvam, S., Harikrishnan, S., & Elayaperumal, A. (2012). Sustainable thermal energy storage technologies for buildings: A review. Renewable and Sustainable Energy Reviews, 16(5), 2394–2433.
12. Enteria, N., Yoshino, H., Takaki, R., Yonekura, H., Satake, A., & Mochida, A. (2013). First and second law analyses of the developed solar-desiccant air-conditioning system (SDACS) operation during the summer day. Energy and Buildings, 60, 239–251.
13. Jani, D. B., Gwalwanshi, M., Mishra, M., & Sahoo, P. K. (2013, February). Solid desiccant cooling—An overview. In Proceedings of the International Conference on Advances in Chemical Engineering (pp. 22–24), Roorkee, India.
14. Koronaki, I. P., Rogdakis, E., & Kakatsiou, T. (2013). Experimental assessment and thermodynamic analysis of a solar desiccant cooling system. International Journal of Sustainable Energy, 32(2), 121–136.
15. Kwong, Q. J., Adam, N. M., & Sahari, B. B. (2014). Thermal comfort assessment and potential for energy efficiency enhancement in modern tropical buildings: A review. Energy and Buildings, 68, 547–557.
16. Rafique, M. M., Gandhidasan, P., Al-Hadhrami, L. M., & Rehman, S. (2016). Energy, exergy and anergy analysis of a solar desiccant cooling system. Journal of Clean Energy Technologies, 4(1), 78–83.

17. Jani, D. B., Mishra, M., & Sahoo, P. K. (2016). Solid desiccant air conditioning–A state of the art review. Renewable and Sustainable Energy Reviews, 60, 751–769.
18. Uçkan, İ, Yılmaz, T., Hürdoğan, E., & Büyükalaca, O. (2014). Exergy analysis of a novel configuration of desiccant based evaporative air conditioning system. Energy Conversion and Management, 84, 524–532.
19. Jani, D. B., Mishra, M., & Sahoo, P. K. (2015). Experimental investigations on hybrid solid desiccant–vapor compression air-conditioning system for Indian climate. In The proceedings of the 24th IIR international congress of refrigeration (pp. 16–22), Yokohama, Japan.
20. Zendehboudi, A. (2016). Implementation of GA-LSSVM modelling approach for estimating the performance of solid desiccant wheels. Energy Conversion and Management, 127, 245–255.
21. Jani, D. B., Mishra, M., & Sahoo, P. K. (2017). Application of artificial neural network for predicting performance of solid desiccant cooling systems–A review. Renewable and Sustainable Energy Reviews, 80, 352–366.
22. Jani, D. B., Deep, L., & Soham, P. (2018). A critical review on evaporative desiccant cooling. International Journal of Innovative and Emerging Research in Engineering, 5(1), 24–29.
23. Dadi, M., & Jani, D. B. (2019). TRNSYS simulation of an evacuated tube solar collector and parabolic trough solar collector for hot climate of Ahmedabad. Available at SSRN 3367102. https://ssrn.com/abstract=3367102 or http://dx.doi.org/10.2139/ssrn.3367102
24. Jani, D. B. (2019). An overview on desiccant assisted evaporative cooling in hot and humid climates. Algerian Journal of Engineering and Technology, 1, 1–7.
25. Peng, D., Zhou, J., & Luo, D. (2017). Exergy analysis of a liquid desiccant evaporative cooling system. International Journal of Refrigeration, 82, 495–508.
26. Batukray, J. D. (2019). Application of renewable solar energy in liquid desiccant powered dehumidification and cooling. Journal of Environment Protection and Sustainable Development, 5, 1–6.
27. Soares, N., Bastos, J., Pereira, L. D., Soares, A., Amaral, A. R., Asadi, E., Rodrigues, E., Lamas, F. B., Monteiro, H., Lopes, M. A. R., & Gaspar, A. R. (2017). A review on current advances in the energy and environmental performance of buildings towards a more sustainable built environment. Renewable and Sustainable Energy Reviews, 77, 845–860.
28. Jani, D. B., Mishra, M., & Sahoo, P. K. (2016). Exergy analysis of solid desiccant-vapour compression hybrid air conditioning system. International Journal of Exergy, 20(4), 517–535.
29. Jani, D. B. (2019). Liquid desiccants applications in cooling and dehumidification–an overview. Industrial & Engineering Chemistry Research, 201, 1–17.
30. Bachrun, A. S., Ming, T. Z., & Cinthya, A. (2019). building envelope component to control thermal indoor environment in sustainable building: A review. Sinergi, 23(2), 79–98.
31. Guan, B., Liu, X., & Zhang, T. (2020). Analytical solutions for the optimal cooling and heating source temperatures in liquid desiccant air-conditioning system based on exergy analysis. Energy, 203, 117860.
32. Jani, D. B., Bhabhor, K., Dadi, M., Doshi, S., Jotaniya, P. V., Ravat, H., & Bhatt, K. (2020). A review on use of TRNSYS as simulation tool in performance prediction of desiccant cooling cycle. Journal of Thermal Analysis and Calorimetry, 70(5), 2011–2031.
33. Jani, D. B. (2019). An overview on recent development in desiccant materials. International Journal of Advanced Materials Research, 5(2), 31–37.
34. Jani, D. B., Mishra, M., & Sahoo, P. K. (2018). Performance analysis of a solid desiccant assisted hybrid space cooling system using TRNSYS. Journal of Building Engineering, 19, 26–35.

35. Jani, D. B., Mishra, M., & Sahoo, P. K. (2018). Solar assisted solid desiccant—Vapor compression hybrid air-conditioning system. Applications of Solar Energy, 233–250. https://doi.org/10.1007/978-981-10-7206-2_12
36. Bhabhor, K. K., & Jani, D. B. (2019). Progressive development in solid desiccant cooling: A review. International Journal of Ambient Energy, 43(1), 992–1015.
37. Jani, D. B., Mishra, M., & Sahoo, P. K. (2017). A critical review on solid desiccant-based hybrid cooling systems. International Journal of Air-Conditioning and Refrigeration, 25(03), 1730002.
38. Ge, F., & Wang, C. (2020). Exergy analysis of dehumidification systems: A comparison between the condensing dehumidification and the desiccant wheel dehumidification. Energy Conversion and Management, 224, 113343.
39. Jani, D. B., Mishra, M., & Sahoo, P. K. (2018). Investigations on effect of operational conditions on performance of solid desiccant based hybrid cooling system in hot and humid climate. Thermal Science and Engineering Progress, 7, 76–86.
40. Vyas, V., Jani, D. B., & Brahmbhatt, P. K., (2016, April). A comprehensive study on application of renewable solar energy in thermal power generation. In National Conference on Emerging Research Trends in Engineering (NCERTE-2016), VGEC Chandkheda, Institute for Plasma Research (IPR) and CTE Gandhinagar (pp. 4–6), Ahmedabad.
41. Ahmad, T., & Chen, H. (2018). Short and medium-term forecasting of cooling and heating load demand in building environment with data-mining based approaches. Energy and Buildings, 166, 460–476.
42. Jani, D. B., Mishra, M., & Sahoo, P. K. (2016). Performance prediction of rotary solid desiccant dehumidifier in hybrid air-conditioning system using artificial neural network. Applied Thermal Engineering, 98, 1091–1103.
43. Bhabhor, K. K., & Jani, D. B. (2021). Performance analysis of desiccant dehumidifier with different channel geometry using CFD. Journal of Building Engineering, 44, 103021.
44. Dadi, M., & Jani, D. B. (2019). Solar energy as a regeneration heat source in hybrid solid desiccant – Vapor compression cooling system: A review. Journal of Emerging Technologies and Innovative Research, 6(5), 421–425.
45. Jani, D. B. (2022). TRNSYS simulation of desiccant powered evaporative cooling systems in hot and humid climate. Journal of Mechanical Engineering, 1(1), 1–6.
46. Zhang, Q., Liu, X., Zhang, T., & Xie, Y. (2020). Performance optimization of a heat pump driven liquid desiccant dehumidification system using exergy analysis. Energy, 204, 117891.
47. Jani, D. B., Mishra, M., & Sahoo, P. K. (2022). TRNSYS simulation of desiccant-assisted HVAC system. Air Conditioning and Refrigeration Journal, 24(6), 28–33.
48. Jia, C. X., Dai, Y. J., Wu, J. Y., & Wang, R. Z. (2007). Use of compound desiccant to develop high performance desiccant cooling system. International Journal of Refrigeration, 30(2), 345–353.
49. Hua, L., & Wang, R. (2022). An exergy analysis and parameter optimization of solid desiccant heat pumps recovering the condensation heat for desiccant regeneration and heat transfer enhancement. Energy, 238, 121811.
50. Jani, D. B. (2021). A recent development in desiccant assisted HVAC system. Air Conditioning and Refrigeration Journal, 24(2), 26–52.
51. Jani, D. B. (2021). Use of artificial neural network in performance prediction of solid desiccant powered vapor compression air conditioning systems. Instant Journal of Mechanical Engineering, 3(3), 5–19.
52. Jani, D. B. (2019). Desiccant cooling as an alternative to traditional air conditioners in green cooling technology. Instant Journal of Mechanical Engineering, 1(1), 1–13.
53. Halliday, S. P., Beggs, C. B., & Sleigh, P. A. (2002). The use of solar desiccant cooling in the UK: A feasibility study. Applied Thermal Engineering, 22(12), 1327–1338.

54. Kanoğlu, M., Çarpınlıoğlu, M., & Yıldırım, M. (2004). Energy and exergy analyses of an experimental open-cycle desiccant cooling system. Applied Thermal Engineering, 24(5-6), 919–932.
55. Jani, D. B., Vadera, N. J., & Patel, M. R. (2022). A progressive review on solid desiccant cooling systems. International Journal of Advanced Research in Science Communication and Technology, 2(1), 703–709.
56. Açıkkalp, E., Caliskan, H., Hong, H., Piao, H., & Seung, D. (2022). Extended exergy analysis of a photovoltaic-thermal (PVT) module based desiccant air cooling system for buildings. Applied Energy, 323, 119581.
57. Jani, D. B., Mishra, M., & Sahoo, P. K. (2022). Experimental analysis on desiccant assisted hybrid HVAC system. Air Conditioning and Refrigeration Journal, 25(5), 68–72.
58. Jani, D. B. (2022). Mathematical modeling of rotary desiccant dehumidifier. International Virtual Conference on Recent Trends in Applied Mathematics (ICRTAM - 2022) on 07th October 2022, organized by Department of Science and Humanities (Mathematics), ANNA University, Perur Chettipalayam, Coimbatore, Tamil Nadu, India.
59. Prajapti, T., Bhabhor, K. K., & Jani, D. B. (2023). Performance analysis of desiccant dehumidifier using ANSYS fluent. International Journal of Advanced Research in Science Communication and Technology, 3(5), 421–437.
60. Gorai, V., Singh, S. K., & Jani, D. B. (2024). A comprehensive review on solid desiccant-assisted novel dehumidification and its advanced regeneration methods. Journal of Thermal Analysis and Calorimetry. https://doi.org/10.1007/s10973-024-13479-9.
61. Jani, D. B., Mishra, M., & Sahoo, P. K. (2025). Emerging technologies for HVAC system efficiency. In Thermal Evaluation of Indoor Climate and Energy Storage in Buildings (pp. 249–265). https://doi.org/10.1201/9781003415695

Chapter 8

Application of heat exchangers in electronics cooling and thermal management

Kaustav Pradhan and Mukesh Kumar

8.1 INTRODUCTION

Improvement in heat transfer technologies has become crucial in addressing the growing demand for efficient cooling systems in various devices with tremendously increased volumetric heat generation [1]. Among the innovative approaches to improve heat transfer, the utilization of nanofluids as cooling agents in the thermal systems has gained significance lately [2]. On the other hand, the incorporation of various inserts in heat exchanger pipes to induce turbulence and thereby enhance heat transfer, while using the conventional coolants, has also been extensively explored [3]. Here, we explore the recent developments, experimental findings, and numerical investigations involving the use of nanofluids in along with various types of inserts for heat transfer enhancement [4]. Apart from this, application of magnetic and electric fields on the nanofluids in heat exchanger pipes has also been explored to some extent.

Nanofluids are defined as fluids consisting of metallic or other high thermal conductivity nanoparticles dispersed in conventional fluids (such as water or organic solvents) as the base fluid. As compared to the traditional coolants, nanofluids have demonstrated remarkably improved cooling capabilities owing to the significantly enhanced effective thermal conductivity. Although nanofluids provide a convenient means of improving thermal performance, researchers and industry continuously sought after techniques to improve heat transfer characteristics of engineering systems. They are typically categorized as accessories that create swirl flow and are part of the broader field of controlled swirling flows [5]. The helical twist of fluid motion extends the effective flow path, while the twist of the tape generates secondary fluid motion, enhancing convective heat transfer through swirl mixing. These collective effects contribute to an overall improvement in heat transfer efficiency and effectiveness of heat exchange [6].

Miniaturization of electronic component has resulted in medium-scale integration (MSI) in the 1960s with up to 1000 components per chip, large-scale integration (LSI) in the 1970s with up to 100000 components per chip, and very large-scale integration (VLSI) in the 1980s with up to 10000000

DOI: 10.1201/9781003534785-8

components per chip [7]. Even in 2006, chips were already manufactured that contained up to 100 million transistors per square centimetre and now the most powerful processors have about 5.5 billion transistors [8]. Pedram and Nazarian [9] found that more than 50% of all integrated circuit (IC) failures are related to thermal issues. Thus, a rule of thumb is that the failure rate of electronic components can be halved for each 10°C reduction in their junction temperature and the cooler the devices operate, the more reliable they are [7].

For many semiconductor technologies, the reliability of individual transistor is exponentially dependent on the operating temperature and the median time to failure in hours (MTF) can be estimated by the well-known Black's correlation [10], which has the form:

$$MTF = \frac{1}{AJ^2} \exp\left(\frac{E_A}{k_B T}\right) \tag{8.1}$$

where A is a constant, J is the current density (/cm^2), E_A is the activation energy (in eV) which is approximately 0.68eV for typical silicon failures, k_B is the Boltzmann constant and T is the absolute operating temperature (K). Equation 8.1 demonstrates that a modest increase in operating temperature could significantly increase the device failure rate. For example, Fallah and Pedram [11] demonstrated that 40% or even a higher percentage of the total power consumption in 90 nm process technology is due to the leakage of transistors. Over the last few years, two complementary ways of enhancing the effectiveness of cooling such modules have emerged [12]. While the first method aims to reduce the thermal resistance by removing certain layers, the second method focuses on increasing the efficiency of the cold plate through elevation of the heat transfer from the heat-sink body to the coolant.

It has been seen that most of the cooling techniques cannot achieve the required cooling performance due to the limitations in heat transfer capabilities of traditional/conventional coolants such as air, oil, and water, which are inherently poor heat transfer mediums with low thermal conductivity and convective heat transfer coefficient. As an example, an effective heat transfer coefficient of about 2×10^4 W/m^{2}K is required to accommodate a heat flux of 100 W/cm^2 at a temperature difference of 50°C; such a high value of heat transfer coefficient is usually not possible through free and forced convections of the above-mentioned coolants [13]. Thus, there is a desperate requirement of non-traditional cooling fluids with superior heat transfer capability. Consequently, some new type of fluids has emerged which can potentially be used as advanced coolants in those electronic devices. One such fluid category is nanofluids—a new class of heat transfer fluids, which are the suspensions of nanometer-sized particles (typically < 100 nm) in base fluids like water (W), ethylene glycol (EG) and oils. Nanofluids were found to possess considerably higher thermal properties particularly thermal conductivity and convective as well as boiling heat transfer compared to their base conventional fluids [14]. Besides the desirable high thermal properties, these modified fluids offer immense benefits and applications in a wide range of areas including electronics [15–18]. Nanofluids are believed to be capable of meeting the cooling demands and challenges of modern day electronic devices and appliances [19, 20].

8.2 TRADITIONAL COOLING TECHNIQUES

Figure 8.1 demonstrated the classification of cooling technologies based on different, e.g. radiation, free, forced convection and liquid evaporation [21].

Based on coolant, the cooling technologies can also be classified into three general categories, e.g. air, water and refrigerant cooling, which are shown in the Figure 8.2 [21].

Figure 8.3 illustrates the effectiveness range of various traditional cooling technologies as a function of heat flux. It is concluded from the figure that natural and free convection is effective at low heat flux values, around 0.1 to 1 kW/m^2; forced convection-air is suitable for moderate heat fluxes, approximately between 1 and 10 kW/m^2; forced convection-liquid provides cooling for a higher range, approximately from 10 to 1000 kW/m^2 and evaporation-liquid is effective in managing very high heat fluxes, from around 100 to over 10,000 kW/m^2.

Table 8.1 presents typical ranges of heat transfer coefficients of most commonly used coolants (i.e., air and water) in all cooling modes. As well known, water possesses better heat transfer coefficient resulting in higher cooling performance compared to air.

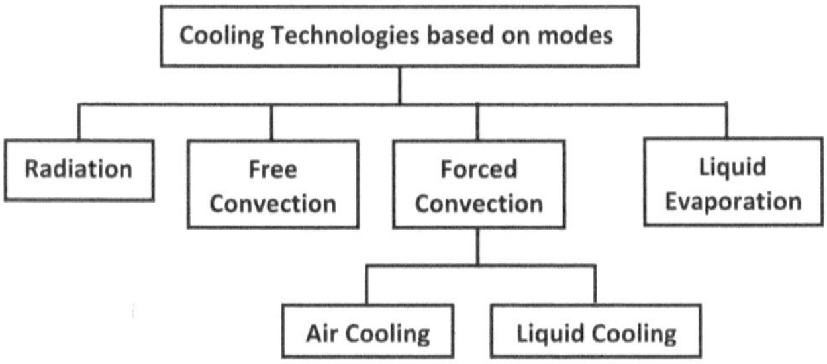

Figure 8.1 Different cooling technologies.

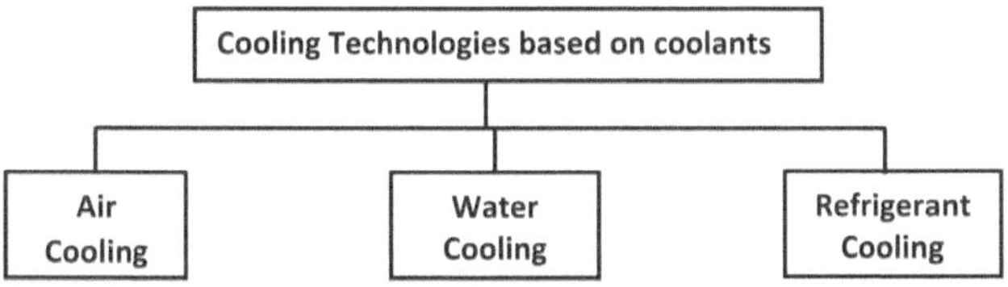

Figure 8.2 Cooling technologies based on coolants.

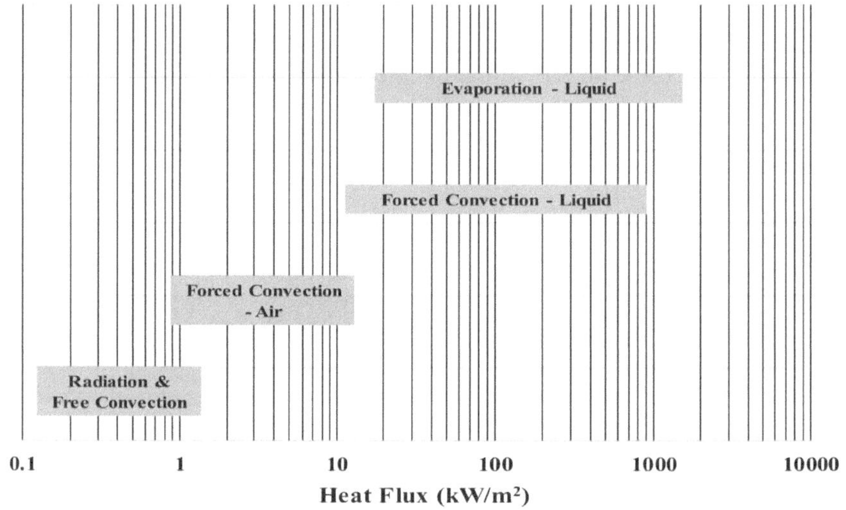

Figure 8.3 Illustration showing the effectiveness range of traditional cooling technologies based on heat loads.

Table 8.1 Range of values of the heat transfer coefficient of air and water for different cooling modes

	Cooling mode		
	Free convection	*Forced convection*	*Boiling*
Heat transfer coefficient of air (W/m^2K)	5–10	10–350	
Heat transfer coefficient of air (W/m^2K)	100–1200	500–3000	3000–100000

Direct liquid immersion (DLI) cooling is another technique that provides a high heat transfer coefficient, and hence reduces the temperature rise of the chip surface above the liquid coolant temperature. Some of the DLI cooling methods are very well established and widely used in numerous thermal management systems [22, 23]. More details about thermal management of electronics using these cooling methods as well as various emerging ones can be found elsewhere [24].

8.3 PASSIVE TECHNIQUES OF HEAT TRANSFER ENHANCEMENT

Passive heat transfer enhancement techniques refer to those that aim to improve heat transfer rates without incurring any external energy input. These techniques rely on modifications to the geometry, surface properties and flow conditions of a system to enhance the heat transfer. Passive methods are comparatively simpler and more cost-effective than the other techniques used for heat transfer enhancement. There are five passive techniques that are commonly used for heat transfer augmentation. They are as follows.

8.3.1 Heat transfer surface modification

This technique involves altering the physical characteristics of heat exchange surfaces to enhance heat transfer efficiency. This can include the generation of roughened surfaces, addition of fins and grooves, and employing coatings of high thermal conductivity material. These modifications are broadly used in heat exchangers, air-cooled condensers and electronic cooling systems to enhance overall the thermal performance.

8.3.2 PCM (phase change material) as heat transfer medium

Phase change materials (PCMs) are substances capable of storing and releasing large amounts of heat during phase change process. PCMs absorb heat from the system and change their phase from solid to liquid. Subsequently, they release the heat to the surroundings and change back to solid from the liquid phase. PCMs are widely used in thermal energy storage systems, building insulation, battery thermal management systems, etc.

8.3.3 Swirl flow inducers

Figure 8.4 illustrates a swirl inducer, a device designed to impart rotational motion or "swirl" to fluid flow within a system. These inducers typically consist of specialized inserts positioned within the flow path, which enhance heat transfer by promoting fluid mixing, disrupting boundary layers, and

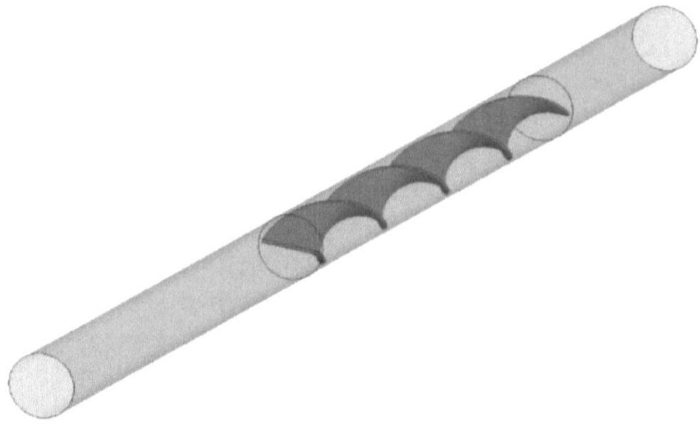

Figure 8.4 Insertion of twisted tapes in heat transfer tubes to induce swirl.

increasing the convective heat transfer coefficient. Swirl inducers are frequently used in heat exchangers, boilers, and refrigeration systems to improve thermo-hydraulic performance.

8.3.4 Nanofluids

Nanofluids are a specially formulated colloid suspensions of high thermal conductivity nanoparticles dispersed in a base fluid like water or EG. These nanoparticles are generally of the order of 1–100 nanometres. These fluids have been utilized in many heat transfer applications, including heat exchangers, solar collectors and electronic cooling systems.

One of the most popular passive techniques that has been used over the years for achieving improved performance in existing heat exchangers is the insertion of twisted tapes inside the heat transfer tubes [25]. These twisted tapes are easy to fabricate and provide a reasonably cost-effective solution to the up-gradation of existing heat exchangers. This is why they such arrangements are being increasingly contemplated for compact heat exchangers [26, 27]. Many industries that use heat exchangers with highly viscous fluids can significantly benefit by the incorporation of twisted tapes. However, the implementation of twisted tapes can result in a tremendous increase in the pumping power requirements, especially for the highly viscous fluids and small dimensions of the cooling system channels inside electronic devices [28–31].

Since twisted tapes increase both the heat transfer (h) and the pumping power requirement (f), it is difficult to quantitatively state the usefulness of twisted tapes. Ideally, the heat transfer should be improved while the pumping power should not be increased significantly. To overcome this problem of quantification, one non-dimensional parameter called thermo-hydraulic performance factor (THPF) has been used [32]. It is given by the following expression:

$$THPF = \frac{\left(\overline{Nu}/\overline{Nu_s}\right)}{\left(f/f_s\right)^{1/3}} \tag{8.2}$$

Here, $\overline{Nu}$ and $\overline{Nu_s}$ represent the Nusselt number values for the pipe with and without twisted tape, respectively, while f and f_s represent the corresponding friction factors. The Nusselt number varies directly with the mass flow rate, while the pumping power varies as the one-third power of the mass flow rate. Accordingly, the non-dimensional parameter has been defined in a number of previously published work [33, 34]. It is to be noted that $^{Nu}/_{Nu_s}$ is always greater than unity, which means that a twisted tape always improves the heat transfer. At the same time $^{f}/_{f_s}$ is also always greater than unity. THPF > 1 represents an improved performance of the device. This means that the enhancement in heat transfer is more pronounced than the increased pumping power. On the other hand, THPF < 1 indicates a fall in the performance of the device characterized by greater increase in pumping power requirements as compared to the improved heat transfer.

8.4 ACTIVE TECHNIQUES OF HEAT TRANSFER ENHANCEMENT

Active techniques refer to those that require some external energy input to bring about the desired effect of heat transfer enhancement. Using this technique, heat transfer can be augmented beyond the passive technique. This technique plays a crucial role when high heat transfer augmentation is required. There are four active techniques broadly used for heat transfer augmentation. These four techniques are discussed below.

8.4.1 Electric field

Electro-hydrodynamics (EHD) refers to the study of the effects of an electric field on the flow field and vice versa. Using the concepts of EHD, heat transfer augmentation may be achieved by applying an electric field across the fluid flow occurring in the system. When the electric field is applied across a conducting fluid, ions present in the fluid create an electrostatic force that results in an additional driving force for the fluid flow. Hence, the flow is augmented, which in turn augments the heat transfer.

8.4.2 Magnetic field

This technique involves the application of a magnetic field onto an existing flow field. The magnetic field alters the flow field and vice versa; this branch of science is referred to as magneto hydrodynamics (MHD). When a magnetic fluid (such as a nanofluid with magnetic nanoparticles) moves across a magnetic field, an electric current is generated, leading to the formation of Lorentz forces that influence fluid motion. This alteration in the flow field leads to an improvement in the heat transfer performance of the cooling system. It is commonly employed in advanced electronic cooling systems, fusion reactors and metallurgical processes.

8.4.3 Ultrasonic vibration

Ultrasonic waves are used to agitate fluids in motion, disrupt boundary layers and promote mixing. As a result, heat transfer improvement takes place due to the rupture of the boundary layer flow near the heat transfer surface. It is commonly used in various applications, including heat exchangers, chemical reactors and biomedical devices.

8.4.4 Pulsating flow technique

Pulsating flow technique involves the external periodic modulation of fluid flow parameters such as velocity, pressure and temperature to enhance heat transfer. By introducing controlled fluctuations in the flow, this technique can disrupt the thermal boundary layer and improve the existing heat transfer rate. It is used in heat exchangers, refrigeration systems and microfluidic devices.

8.5 EMERGING COOLING TECHNIQUES

The emerging cooling methods can be classified into various types and some of them are as shown in the Figure 8.5 below.

Most of the emerging techniques such as heat pipes [35], free cooling [36], micro-channel and micro-pumping [37], thermoelectric coolers (TECs) [38], PCM-based cooling [39–41] have shown great potential in handling the high-power density modern electronic devices.

8.5.1 Heat pipes

Heat pipes are a very promising means of cooling electronic devices such as computer, laptop, satellite modules, etc. as shown as Figure 8.6 [42]. It works based on the phase change of working fluid inside the pipes. Due to their

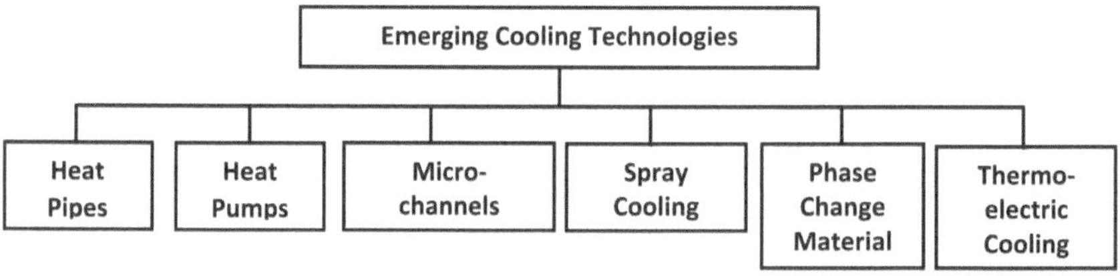

Figure 8.5 Classification of the emerging cooling technologies for electronic devices.

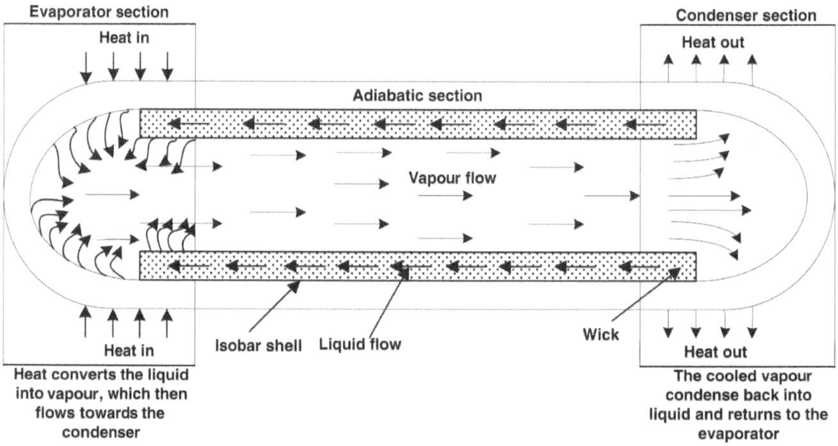

Figure 8.6 Working principle of heat pipe.

extremely high effective thermal conductivity [43] and very low effective thermal resistance (typically ranging from 0.05 to 0.4°C/W [44]) heat pipes are one of the most viable means of cooling high heat flux generating electronic devices such as the modern day CPUs. One of the highest volume applications of heat pipes is in the cooling of Pentium processors in laptops [44]. Very recently, Chen *et al.* [45] reported a comprehensive review on design, fabrication and performance analysis of small heat pipes for cooling small electronics devices and systems.

A heat pipe has three sections, viz., evaporator section, adiabatic section and condenser section. The wick provides the capillary pumping effect, which drives the fluid from the condenser section to the evaporator even against gravity. This allows the heat pipe to be operated in any orientation. When the evaporator section is heated from an external heat source (e.g., an electronic device or chip), the liquid evaporates there. The vapour raises the pressure and results in having pressure difference along the axial direction. This difference of pressure drives this vapour from evaporator to the condenser section, where it condenses while releasing the latent heat of vaporization to the heat sink. In the process, the vapour turns back to liquid phase. Meanwhile depletion of liquid by evaporation at the evaporator section causes the liquid/vapour interface to enter into the wick surface, and thus a capillary pressure develops there. This capillary pressure drives the condensed liquid back to the evaporator section and completes the cycle. Based on total capillary pressure drop, the maximum heat transport capacity of a heat pipe can be determined from the following equation:

$$Q_{max} = \frac{KA_w h_{fg}\rho}{\mu L}\left(\frac{2\sigma}{r} - \rho g L sin\varphi\right) \tag{8.3}$$

where σ is the liquid surface tension, μ is the liquid viscosity, h_{fg} is the latent heat of liquid, r is the effective pore radius of the wick structure, L is the effective length of heat pipe, K is the permeability, A_w is the cross-sectional area of the wick, ρ is liquid density, g is gravitational constant, and φ is orientation angle with respect to the horizontal plane.

8.5.2 Heat pumps

Heat pumps transfer heat from a low temperature fluid to a high temperature fluid using refrigerants as coolants. This cooling mode offers economic and environment friendly techniques of providing process heat for use in various industrial applications. Heat pumps are usually used for moderate cooling needs and are thus still not applied to electronics cooling systems. Kim *et al.* [46] performed numerical simulation using water/LiBr pair in a heat pump of dual micro-channel array evaporator. They showed that miniature absorption refrigeration system may be applicable for electronics cooling applications in the future.

8.5.3 Micro-channels

Micro-scale cooling systems are capable of cooling high heat generating electronic devices. The heat transfer performance of a micro-channel heat sink (MCHS) is much higher than that of any traditional heat exchangers. Thus, forced convective liquid cooling through MCHSs is one of the promising and high performance cooling technologies for high heat generating electronic devices. Moreover, this technology not only facilitates a significant reduction of the package size, but is also amenable to on-chip integration [47].

Micro-channels were first proposed as an advanced cooling solution [48] in 1981 when a very high heat flux of 790 W/cm^2 was achieved from an integral water-cooled micro-channel (50-μm wide and 300-μm deep) with a temperature rise of 71°C. Subsequently, micro-channel heat exchangers (MCHX) and MCHSs have improved electronics cooling by many folds. It has now become a key technology for cooling many devices and systems particularly electronics with high power densities such as VLSI and ultra-large-scale integrations (ULSI) [49, 50]. For the purpose of cooling very-high power chips such as microprocessors, a single-phase silicone micro-channel cooler was designed and implemented by Colgan *et al.* [51]. It was demonstrated that such a micro-channel cooler can cool chips with average power densities of 400 W/cm^2.

Dix *et al.* [52] optimized the performance of water-cooled MCHX for electronic cooling applications through the change of channel geometries. They suggested that for maximum heat transfer performance, it is necessary to consider the right balance between the surface area and pressure drop. Based on a numerical study on two-layer single phase flow micro-channel Abdoli *et al.* [53] showed that such a micro-channel configuration can manage heat from very high heat flux chips with hot spots. In a ground-breaking work, Atienza [54] considered inter-tier liquid as potential means to bring down the high temperatures in multi-processor system-on-chip (MPSoC) with hotspots. This new cooling technology involves liquid (water) injection through micro-channels between the tiers of a 3D stack. A view of 3D stacked MPSoC structure with interlayer micro-channel cooling is presented in the Figure 8.7 [54].

Qu and Mudawar [55] reported rapid increase in heat flux with small increments of wall temperature until the wall heat flux reached to a critical heat flux (CHF). Through the implementation of silicon micro-channels with nanostructured wall, significant enhancements in pool boiling heat flux was reported in a study by Kandlikar *et al.* [56].

8.5.4 Free cooling

Free cooling is one of the most economic cooling approaches. It uses the low external natural air temperature to cool data centres or server systems [57]. Water is also used in this cooling technique. If the ambient air temperature is much lower than that of the data centre, heat can naturally flow out from the data centre. Free cooling is usually classified into two groups—direct cooling

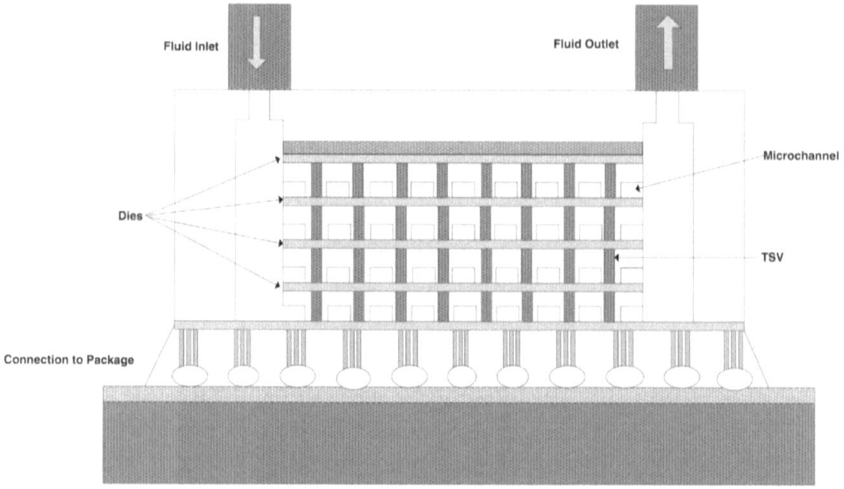

Figure 8.7 Micro-channel-based cooling.

and indirect cooling. Moreover, it is also divided into three categories based on the cooling flow direction and system used, viz., air-side free cooling, water-side free cooling, and heat pipe free cooling.

8.5.5 Phase change materials (PCMs)

As an emerging passive method PCM-based cooling has received great attention in recent years. It is a very promising cooling means as it can not only take the heat from the devices and appliances, but also store the heat for use in other purposes such as heating homes or offices. PCMs show numerous advantageous characteristics as heat transfer mediums including high latent heat of fusion, high specific heat capacity, temperature stability, and small volume change during phase change. If the heat output from the electronic device occurs during the phase transition, PCMs can absorb a very large quantity of heat without raising its own temperature. A novel design of a heat sink with PCM for cooling of microprocessors was proposed by Jaworski and Domanski [58]. They showed that even a small quantity of PCM in the heat sink can significantly increase its capability to stabilize microprocessor temperatures indicating high potential of PCM-based heat sink for electronics cooling. Although some studies have been reported [59], more extensive studies need to be conducted to establish the utility of PCMs in modern electronics.

8.5.6 Thermoelectric cooling

A TEC works on the principle of Peltier effect (thermoelectric effect). It involves the generation of a heat flux between the junctions of two different types of semiconductors. It transfers heat from one side of the device to the other with the consumption of electrical energy. TECs are compact and have no moving parts, thus offering a great potential to work as cooling systems for electronics [60]. TEC can also be integrated into electronic packages for cooling their hotspots [61]. It has been found that the coefficient of performance of a TEC is lower than that of an equivalent vapour-compression refrigeration system.

Saengchandr and Afzulpurkar [62] reported a numerical analysis on the combined thermoelectric module and heat pipe-based approach for cooling microprocessors and other computer chips. They claimed that their proposed cooling systems was of sufficient capacity to manage a heat dissipation of 200 W while maintaining temperatures that are significantly lower than that of other cooling systems. The cooling concept used in a TEC is shown in the Figure 8.8 below. The main component of a TEC is an array of two dissimilar conductors namely p- and n-type semiconductors.

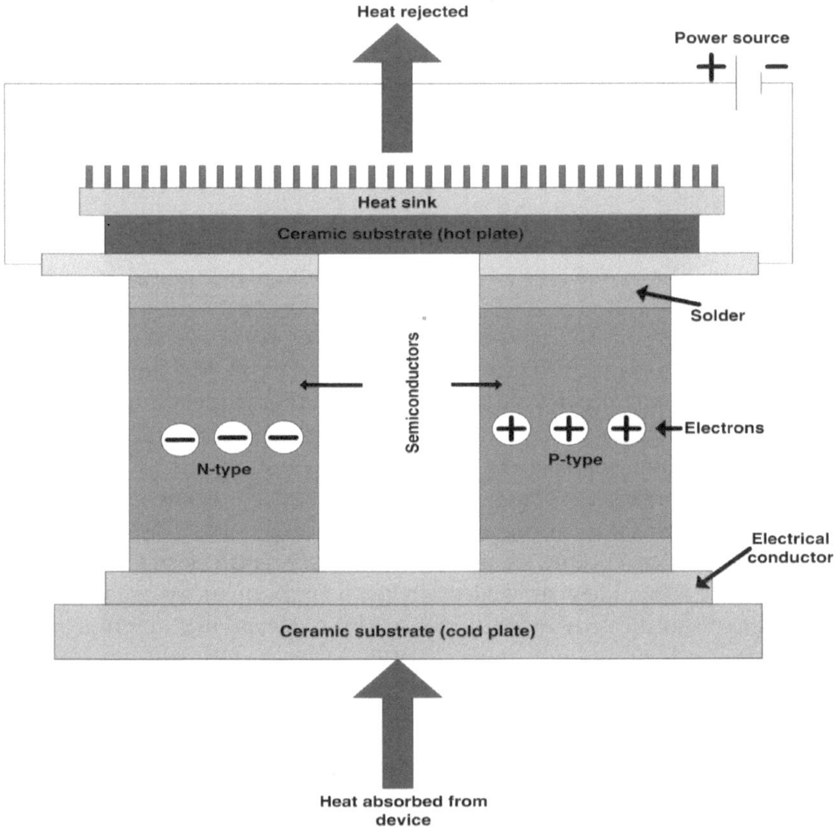

Figure 8.8 Working concept of thermoelectric cooling technology.

8.6 EVOLUTION OF COOLANTS (WORKING FLUIDS) IN THE COOLING TECHNOLOGIES

There are numerous traditional aqueous and non-aqueous coolants, which are used in various electronics cooling systems as shown in Figure 8.9. Properties and selection criteria of these coolants are discussed in this section. As a new and emerging type of coolants, nanofluids showed great potential as superior coolants for electronics applications compared to traditional coolants. Heat transfer properties including thermal conductivity, convective and boiling heat transfer performance of nanofluids have also been briefly reviewed here.

8.6.1 Conventional coolants

There are numbers of cooling fluids widely used in various cooling applications in the market. The choice of coolant in a particular application is governed primarily by the cooling technique that can be employed as well as the particular electronics device. It is established [63] that the liquid coolants for electronics must be non-flammable, nontoxic, and relatively inexpensive with excellent thermophysical properties (like high thermal conductivity, specific heat, and heat transfer coefficient together with a low viscosity).

The commonly used conventional coolants for electronics cooling are mainly classified into two groups: (i) dielectric and (ii) non-dielectric fluids.

8.6.2 Dielectric coolants

There are several types of fluids in this group of coolants, which are aromatics (benzene derivatives), aliphatics, silicones, and fluorocarbon-based fluids. Aromatics-based liquids are the most common mainly due to their cheaper costing and better performance. Examples of some aromatics coolants are diethyl benzene (DEB), toluene, xylene, etc. Aliphatic hydrocarbons of paraffinic and iso-paraffinic type (including mineral oils) are used in a variety of

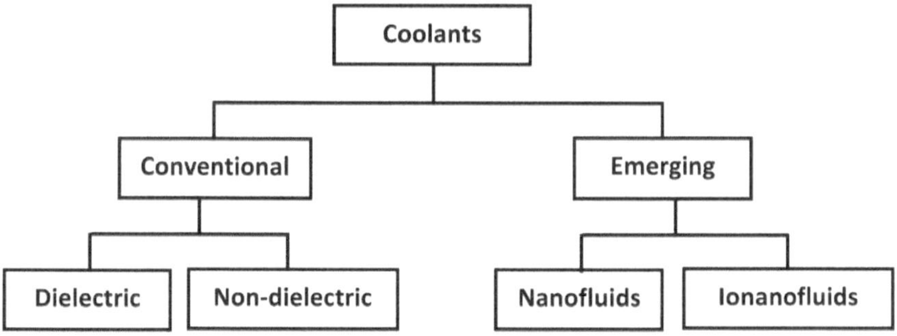

Figure 8.9 Classification of the coolants used in cooling systems.

direct cooling of electronics. Aliphatic polyalphaolefins (PAO) based fluids are also used in some electronics cooling applications. The silicon base liquids are another popular type of coolant widely known as silicone oils. The main advantage of this class of coolants is that their thermophysical properties such as viscosity and freezing point can be controlled by changing the chain length. The fluorocarbon-based liquids such as FC-40, FC-72, FC-77 and FC-87 are widely accepted in electronics industries. These liquids are relatively inert, stable and non-flammable. FC-72 and FC-77 are most commonly used for cooling applications in the electronics industry.

8.6.3 Non-dielectric coolants

These are often used for electronics cooling in preference to their counterparts because of their better thermal properties. They are normally aqueous solutions that exhibit high thermal conductivity and heat capacity as well as relatively low viscosity. Water, EG, and mixture of these two liquids are very popular and extensively used as coolants for many electronic devices. Some other popular non-dielectric coolants include propylene glycol (PG), water-methanol, water-ethanol, NaCl solution, potassium formate (KFO) solution, and liquid metals (e.g., Gallium). Mohapatra and Loikits [64] evaluated that among various non-dielectric coolants, KFO solution possesses the highest overall efficiency.

8.6.4 Emerging coolants

It is clear that the cooling challenges of modern day high-power density electronic devices cannot be completely fulfilled by the use of the conventional coolants due to their inherently poor thermal properties. However, the thermal properties of these coolants may be improved tremendously by dispersing particles of favourable thermal properties. The issue with this technique is that the introduction of even micro-sized metallic or non-metallic particles in the base liquids makes them unsuitable for application in small-sized electronic devices and cooling systems because they can be clogged and there can be sedimentation issues. With none of the aforementioned limitations and a superior set of thermal properties, nanofluids can suitably be used to meet the cooling demands of modern day small electronic devices [65].

8.7 APPLICATION OF NANOFLUIDS IN ELECTRONICS COOLING

Owing to the compact nature and excellent cooling performance of micro-channels in heat transfer, MCHS-based cooling systems have emerged as the top contender for solving the cooling problems of the future [66, 67]. Since the convective heat transfer coefficient is inversely proportional to the hydraulic diameter of the channel, micro-channels provide very high heat transfer performance. However, the use of conventional coolants imposes a limitation on the cooling performance of MCHS. Hence Nanofluids, due to their superior heat transfer capabilities, were expected to be highly effective in maximizing the efficiency of MCHSs.

Number of articles may be found in the literature where nanofluids were directly employed in cooling systems of commercial electronic or computing devices to evaluate their cooling performance. We mention a few of those articles only for the sake of brevity. Nguyen et al. [68] numerically investigated heat transfer performance of two nanofluids (γAl_2O_3-water and γAl_2O_3-EG) for cooling microprocessors of a high heat generation rate. The convective heat transfer coefficient of these nanofluids was reported to increase substantially with nanoparticle concentration and the flow rate (Re), which led to a considerable decrease in the maximum junction temperature of the microprocessors (a desirable outcome). Later, the same group of researchers [69] conducted another similar study evaluating the cooling performance of nanofluids (Al_2O_3-water) in commercial water block systems used for CPU cooling. The heat transfer performance of Al_2O_3-water nanofluids in a commercially available electronics cooling system (in a water block used to cool a CPU) was investigated by Roberts and Walker [70]. They reported an increase of almost 20% in the convective heat transfer coefficient in the straight heated tube as compared to that obtained by using water instead of nanofluid. Escher [71] performed a series of experiments and simulations to study the effectiveness of aqueous silica nanofluids in MCHSs for electronics cooling applications. It was reported that for a constant flow rate, the average heat transfer increases with increasing nanoparticle concentration.

8.8 CONCLUSION

The thermal management of high-performance electronic devices is and will be an important parameter governing the advancement and miniaturization of future products. Since the conventional cooling technologies and the traditional coolants are gradually falling short in the modern era of super-computing, innovative techniques and coolants are continuously evolving to meet the market demands. Studies reveal that techniques involving the implementation of heat pipes and MCHSs hold the greatest potential for future cooling systems. Apart from this, a new class of coolants nanofluids are continuously evolving that exhibit significantly better thermal properties such as thermal conductivity, convective and boiling heat transfer performance as compared to their base conventional fluids. In addition, results from the studies on application of nanofluids also confirmed that this emerging fluid performs better as coolants for electronic devices compared to conventional coolants. Nevertheless, continuous research is important to develop cooling systems for the future. The emerging cooling techniques like MCHSs together with these novel fluids can substantially increase their heat removal performance and can meet the cooling needs of those high-heat generating electronic devices. Thus, nanofluids based microchannel can be a possible next-generation cooling technology in the electronics industry. However, more exploration and experimentation is necessary for their realization and commercial scale applications in electronics industries.

REFERENCES

1. Eiamsa-ard S., Wongcharee K. Single-phase heat transfer of CuO/water nanofluids in micro-fin tube equipped with dual twisted-tapes. Int Commun Heat Mass Transf. 2012; 39: 1453–9.
2. Dey D, Sahu DS. A review on the application of the nanofluids. Heat Transf. 2021; 50: 1113–55. https://doi.org/10.1002/htj.21920.
3. Teng T-P, Hsu Y-C, Wang W-P, Fang Y-B. Performance assessment of an air-cooled heat exchanger for multiwalled carbon nanotubes-water nanofluids. Appl Therm Eng. 2015; 89: 346–55.
4. Rahmoune I, Bougoul S. Effect of magnetic field and magnetic nanofluid on heat transmission improvement in a curved minichannel. J Therm Anal Calorim. 2024; 149(2), 729–44.
5. Tao Q, Fu S, Yuan Y, Liu X, He D, Chen Y. Experimental study on enhanced nanofluid heat transfer limit and wetting predicted model in the microgroove under electric field. Int J Therm Sci. 2024; 197: 108829.
6. Harish S, Ishikawa K, Einarsson E, Aikawa S, Chiashi S, Shiomi J, et al. Enhanced thermal conductivity of ethylene glycol with single-walled carbon nanotube inclusions. Int J Heat Mass Transf. 2012; 55: 3885–90.
7. Cengel YA, Heat Transfer: A Practical Approach. Chapter 15, New York: McGraw–Hill; 2003.
8. Moore GE. Cramming more components onto integrated circuits. Proc IEEE. 1998; 86 (1): 82–5.
9. Pedram M, Nazarian S. Thermal modeling, analysis, and management in VLSI circuits: Principles and methods. Proc IEEE. 2006; 94: 1487–518.
10. Black JR. Electromigration—A brief survey and some recent results. IEEE Trans Elect Devi. 1969; ED-16: 338–47.
11. Fallah F, Pedram M. Standby and active leakage current control and minimization in CMOS VLSI circuits. IEICE Trans Elect (Special Section on Low-Power LSI and Low Power IP). 2005: E88-C(4): 509–19.
12. Leslie SG. Cooling options and challenges of high power semiconductor modules. Elect Cooling. 2006: 12(4), 20.
13. Lasance C, Simons R. Advances in high performance cooling for electronics. Elect Cooling. 2005; 11(4): 22–39.
14. Murshed SMS, Leong KC, Yang C. Thermophysical and electrokinetic properties of nanofluids—A critical review. Appl Therm Eng. 2008; 28: 2109–25
15. Choi SUS, Zhang ZG, Keblinski P. Nanofluids. In Encyclopedia of Nanoscience and Nanotechnology, Ed., Nalwa HS, 757–73. Los Angeles: American Scientific Publishers; 2004.
16. Wong KV, De Leon O. Applications of nanofluids: Current and future. Adv Mech Eng. 2010; 2010: 11.
17. Murshed SMS, Nieto de Castro CA. Nanofluids as advanced coolants. In Green Solvents I: Properties and Applications in Chemistry, Ed., Mohammad A and Inamuddin, 397–415. London: Springer; 2012.

18. Nieto de Castro CA, Murshed SMS, Lourenço MJV, Santos FJV, Lopes MLM, França JMP, Enhanced thermal conductivity and heat capacity of carbon nanotubes-ionanofluids. Int J Therm Sci. 2012; 62: 34–9.
19. França JMP, Vieira SIC, Lourenço MJV, Murshed SMS, Nieto de Castro CA. Thermal conductivity of ionanofluids of [C4mim][NTf2] and [C2mim][EtSO4] with carbon nanotubes. Experiment and theory. J Chem Eng Data. 2013; 58: 467–76.
20. Nieto de Castro CA, Murshed SMS, Lourenço MJV, Santos FJV, Lopes MLM, França JMP. Ionanofluids–new heat transfer fluids for green process development. In Green Solvents I: Properties and Applications in Chemistry. Eds. Ali M and Inamuddin, 233–249. London: Springer; 2012
21. Scott WA. Cooling of Electronic Equipment, New York: John Wiley and Sons; 1974
22. Mongia R, Masahiro K, DiStefano E, Barry J, Chen W, Izenson M, et al. Small scale refrigeration system for electronics cooling within a notebook computer. In: Proceedings of the 10th Intersociety Conference on Thermal and Thermomechanical Phenomena in Electronics Systems. USA. 2006, pp. 751–758.
23. Tiwari S, Swaminathan M, Harender, Singh DB. Performance enhancement of the photovoltaic system with different cooling methods. Environ Sci Pollut Res. 2022; 29(30): 45107–130. https://link.springer.com/article/10.1007/s11356-022-20330-x
24. Anandan SS, Ramalingam V. Thermal management of electronics: A review of literature. Therm Sci. 2008; 12(2): 5–26.
25. Outokesh M, Ajarostaghi SSM, Bozorgzadeh A, Sedighi K. Numerical evaluation of the effect of utilizing twisted tape with curved profile as a turbulator on heat transfer enhancement in a pipe. J Therm Anal Calorim. 2020; 140: 1537–53.
26. Manglik, RM and Bergles AE. Heat Transfer Enhancement of in Tube Flows in Process Heat Exchangers by Means of Twisted-Tape Inserts. 1991.
27. SyamSundar L. Heat transfer, friction factor and exergy efficiency analysis of nanodiamond-Fe3O4/water hybrid nanofluids in a tube with twisted tape inserts. Ain Shams Eng J. 2023; 14: 102068.
28. Manglik RM, Bergles AE. Heat transfer and pressure drop correlations for twisted-tape inserts in isothermal tubes: Part II—Transition and Turbulent Flows. J Heat Transfer. 1993; 115: 890–6.
29. Durga Prasad PV, Gupta AVSSKS. Experimental investigation on enhancement of heat transfer using Al_2O_3/water nanofluid in a u-tube with twisted tape inserts. IntCommun Heat Mass Transf. 2016; 75:154–61.
30. Maddah H, Alizadeh M, Ghasemi N, Wan Alwi SR. Experimental study of Al2O3/water nanofluid turbulent heat transfer enhancement in the horizontal double pipes fitted with modified twisted tapes. Int J Heat Mass Transf. 2014; 78: 1042–54.
31. Luo J, Alghamdi A, Aldawi F, Moria H, Mouldi A, Loukil H, Deifall AF, Ghoushchi SP. Thermal-frictional modelling of new special shape twisted tape and helical coiled wire turbulators in engine heat exchangers system. Case Stud Therm Eng. 2024; 53: 103877.
32. Webb RL. Performance evaluation criteria for use of enhanced heat transfer surfaces in heat exchanger design. Int J Heat Mass Transf. 1981; 24: 715–26.
33. Shajahan MI, Stephen C, Michael JJ, Arulprakasajothi M, Rathnakumar P, Parthasarathy M. Heat transfer investigations of in-line conical strip inserts using MWCNT/water nanofluid under laminar flow condition. Int J Therm Sci. 2023; 183: 107844.
34. Nakhchi ME, Esfahani JA. Numerical investigation of turbulent CuO–water nanofluid inside heat exchanger enhanced with double V-cut twisted tapes. J Therm Anal Calorim. 2021; 145: 2535–45.

35. Maydanik YF, Vershinin SV, Korukov MA, Ochterbeck JM. Miniature loop heat pipes-a promising means for electronics cooling. IEEE Trans Compon Pack Technol. 2005; 28(2): 290–6.
36. Zhang H, Shao S, Xu H, Zou H, Tian C. Free cooling of data centers: A review. Renew SustEnergRevi. 2014; 35: 171–82.
37. Garimella SV, Singhal V and Liu D. on-chip thermal management with microchannel heat sinks and integrated micropumps. Proc IEEE. 2006; 94(8): 1534–48.
38. Simons RE. Application of thermoelectric coolers for module cooling enhancement. Elect Cooling. 2000. https://www.electronics-cooling.com/2000/05/application-of-thermoelectric-coolers-for-module-cooling-enhancement/.
39. Tan FL, Tso CP. Cooling of mobile electronic devices using phase change materials. ApplThermEng. 2004; 24(2–3): 159–69.
40. Kandasamy R, Wang XQ, Mujumdar AS. Application of phase change materials in thermal management of electronics. Appl Therm Eng. 2007; 24 (17–18): 2822–32
41. Kandasamy R, Wang XQ and Mujumdar AS. Transient cooling of electronics using phase change material (PCM)-based heat sinks. Appl Therm Eng. 2008; 28: 1047–57.
42. Faghri A. Review and advances in heat pipe science and technology. J Heat Transf. 2012; 134(12): 123001.
43. El-Nasr AA, El-Haggar SM. Effective thermal conductivity of heat pipes. Heat Mass Transf. 1996; 32(1): 97–101.
44. Garner SD. Heat pipes for electronics cooling applications. Elect Cooling 1996; 2(3).
45. Chen X, Ye H, Fan X, Ren T, Zhang G, A review of small heat pipes for electronics. ApplThermEng. 2016; 96: 1–17.
46. Kim YJ, Joshi YK, Fedorov AG. An absorption based miniature heat pump system for electronics cooling. Int J Refrig. 2008; 31: 23–33.
47. Wei Y and Joshi YK. Stacked microchannel heat sinks for liquid cooling of microelectronic components. J Elect Pack. 2004; 126: 60–6.
48. Tuckerman D, Pease R. High-performance heat sinking for VLSI. IEEE Electron Dev Lett. 1981: 2(5): 126–29.
49. Kandlikar SG. High heat flux removal with microchannels—A roadmap of challenges and opportunities. Heat Transf Eng. 2005; 26(8): 5–14.
50. Kandlikar SG, Colin S, Peles Y, Garimella S, Pease RF, Brandner JJ, Tuckerman DB. Heat transfer in microchannels—2012 status and research needs. J Heat Transf. 2013; 135:091001.
51. Colgan EG, Furman B, Gaynes M, Graham WS, LaBianca NC, Magerlein JH, ... Schmidt RR. A practical implementation of silicon microchannel coolers for high power chips. IEEE Trans Compon Pack Technol. 2007; 30(2): 218–25.
52. Dix J, Jokar A, Martinsen R. A microchannel heat exchanger for electronics cooling applications. In: Proceedings of 6th International Conference on Nanochannels, Microchannels and Minichannels. Germany. 2008.
53. Abdoli A, Dulikravich GS, Vasquez G, Rastkar S. Thermo-fluid-stress-deformation analysis of two-layer microchannels for cooling chips with hot spots. J Electron Packag. 2015; 137(3): 031003.
54. Atienza D. Thermal-aware design for 3D multi-processors. Flash Inform. 2010; 10: 34–37.
55. Qu W, Mudawar I. Measurement and correlation of critical heat flux in two-phase microchannel heat sinks. Int J Heat Mass Transf. 2004; 47: 2045–59.
56. Yao S, Lu YW, Kandlikar SG. Pool boiling heat transfer enhancement through nanostructures on silicon microchannels. ASME J Nanotechnol Eng Med. 2012; 3: 031002.
57. Ebrahimi K, Jones GF, Fleischer AS. A review of data center cooling technology, operating conditions and the corresponding low-grade waste heat recovery opportunities. Renew Sustain Energy Rev. 2014; 31: 622–38.

58. Jaworski M, Domanski R. A novel design for heat sink with PCM for electronic cooling. In: Proceedings of the 10th International Conference on Thermal Energy Storage, Stockton, 31 May–2 June 2006.
59. Ling Z, Zhang Z, Shi G, Fang X, Wang L, Gao X, Fang Y, Xu T, Wang S, Liu X. Review on thermal management systems using phase change materials for electronic components, Li-ion batteries and photovoltaic modules. Renew Sustain Energy Rev. 2014; 31: 427–438.
60. Simons RE, Chu RC. Application of thermoelectric cooling to electronic equipment: A review and analysis. In: Proceedings of the 16th Annual IEE Semiconductor Thermal Measurement and Management Symposium. USA. 2000. pp. 1–9.
61. Seifert W, Pluschke V, Hinsche N F. Thermoelectric cooler concepts and the limit for maximum cooling. J Phys Condens Matter. 2014; 26: 255803.
62. Saengchandr B, and Afzulpurkar NV. A novel approach for cooling electronics using a combined heat pipe and thermoelectric module. Am J Eng Appl Sci. 2009; 2 (4): 603–10.
63. Mohapatra SC. An overview of liquid coolants for electronics cooling. Electron Cooling. 2006. https://www.electronics-cooling.com/2006/05/an-overview-of-liquid-coolants-for-electronics-cooling/.
64. Mohapatra S, Loikits D. Advances in liquid coolant technologies for electronics cooling. In: Proceedings of the 21st IEEE Semiconductor Thermal Measurement and Management Symposium. USA. 2005, pp. 354–360.
65. Murshed SMS, Nieto de Castro CA. Superior thermal features of carbon nanotubes based nanofluids- a review. Renew Sustain Energy Rev. 2014; 37: 155–67.
66. Mohammed HA, Bhaskaran G, Shuaib NH, Saidur R. Heat transfer and fluid flow characteristics in microchannels heat exchanger using nanofluids: A review. Renew Sustain Energy Rev. 2011; 15: 1502–12.
67. Hassan I, Phutthavong P, Abdelgawad M. Microchannel heat sinks: An overview of the state-of-the-art. Microsc Thermophys Eng. 2004; 8: 183–205.
68. Nguyen CT, Roy G, Lajoie PR, Maïga SEB. Nanofluids heat transfer performance for cooling of high heat output microprocessor. In: Proceedings of the 3rd IASME/WSEAS International Conference on Heat Transfer, Thermal Engineering and Environment. Greece. 2005, pp. 160–5.
69. Nguyen CT, Roy G, Gauthier C, Galanis N. Heat transfer enhancement using Al_2O_3– water nanofluid for an electronic liquid cooling system. Appl Therm Eng. 2007; 27: 1501–6.
70. Roberts NA, Walker DG. Convective performance of nanofluids in commercial electronics cooling systems. Appl Therm Eng. 2010; 30(16): 2499–2504.
71. Escher W, Brunschwiler T, Shalkevich N, Shalkevich A, Burgi T, Michel B, Poulikakos D. On the cooling of electronics with nanofluids. J Heat Transf. 2011; 133: 051401.

Chapter 9

Economic and environmental aspects of heat exchanger applications

Neha Singh Raghuvanshi, Arpita Johri, Mudit Saxena, Ashwani Kumar, and Yatika Gori

9.1 INTRODUCTION

9.1.1 Overview

Heat exchangers are specialized devices designed to transfer heat between two or more fluids, which may be liquids, gases, or a combination of both. Their main purpose is to efficiently transfer thermal energy from one fluid to another without allowing them to mix. This heat transfer typically occurs through a solid barrier or surface, with the fluids flowing on either side. Key considerations in the design of heat exchangers include heat transfer efficiency, pressure drop, and thermal resistance. They are essential for maintaining optimal temperatures across a broad range of applications, from cooling systems to heating processes. Heat exchangers are crucial in many industries due to their effectiveness in thermal management. In the energy sector, they are used in power plants to transfer heat between steam and water, thereby improving energy production and operational efficiency. In the HVAC (heating, ventilation, and air conditioning) industry, they are vital for controlling indoor temperatures and enhancing energy conservation. They also play a key role in chemical processing by ensuring precise temperature control during chemical reactions, and in the automotive industry, they help manage engine temperatures and boost performance. By facilitating efficient heat transfer, heat exchangers help reduce energy consumption, lower operational costs, and minimize environmental impact [1, 2].

9.1.2 Economic and environmental significance

This chapter highlights the economic and environmental significance of heat exchangers. Economically, they contribute to cost savings by enhancing energy efficiency and reducing operational expenses across various systems. By optimizing energy use and minimizing waste, heat exchangers reduce the need for additional fuel. Environmentally, their efficiency leads to lower greenhouse gas emissions and reduced heat loss, supporting sustainable energy practices. Their role in improving renewable energy systems, such as solar and geothermal, further underscores their contribution to environmental sustainability

DOI: 10.1201/9781003534785-9

[3]. The chapter aims to provide an in-depth overview of heat exchangers, covering their types, applications, and advancements. It will examine different heat exchanger designs and their functions, and how they benefit various industries. The chapter will also explore recent technological innovations, including advancements in materials, design, and smart control systems, and their impact on performance and efficiency. Additionally, it will address future research opportunities and emerging trends in heat exchanger technology, offering insights into potential advancements and their benefits [4, 5].

9.2 ECONOMIC ASPECTS OF HEAT EXCHANGER APPLICATIONS

Heat exchangers enhance economic efficiency by optimizing energy use and reducing operational costs across various industries. They enable significant cost savings through improved thermal management, reduced fuel consumption, and lower maintenance requirements. Their role in energy conservation and process efficiency contributes to both direct financial benefits and broader economic sustainability [6].

9.2.1 Cost analysis

For heat exchangers involve a comprehensive evaluation of all associated expenses throughout their lifecycle, from acquisition to disposal. This begins with capital costs, which include purchasing the heat exchanger and any additional equipment required for installation, such as piping, pumps, and control systems. Installation costs also account for labor and site modifications. Operational expenses, a crucial component of the analysis, include energy consumption, fuel costs, and potential downtime due to inefficiencies. Maintenance costs, including regular inspections, cleaning, and part replacements, contribute to the long-term investment. In performing a lifecycle cost analysis (LCCA), it's essential to consider depreciation, potential tax benefits, and savings from energy-efficient designs, which may offset higher initial expenses. This holistic approach allows businesses to make informed decisions that align with both their budget and operational goals (Figure 9.1) [7].

9.2.1.1 *Initial capital costs*

9.2.1.1.1 Purchase and installation costs

The initial capital costs for heat exchangers include both the purchase price and the associated installation expenses. These costs can vary significantly depending on the type, size, and complexity of the heat exchanger. For instance, shell-and-tube heat exchangers are generally more affordable due to their simpler design and lower material costs compared to more advanced options like microchannel or plate heat exchangers. Installation costs also depend on factors such as how easily the unit integrates into existing systems,

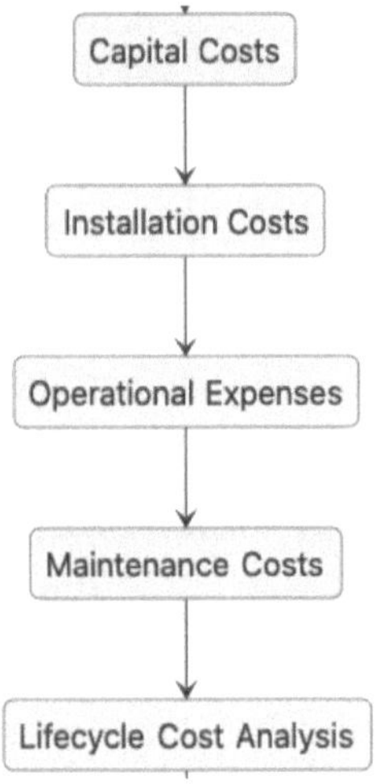

Figure 9.1 Cost analysis for heat exchangers.

the need for site preparation, and labor expenses. Proper installation is essential as it affects both the operational efficiency and longevity of the heat exchanger. Specialized or custom-designed heat exchangers might require specific installation techniques, which can further increase initial costs [8].

9.2.1.1.2 Comparison of different types of heat exchangers

The initial costs of various heat exchanger types can differ widely. Shell-and-tube heat exchangers, known for their durability and ease of maintenance, are often less costly but may offer lower heat transfer efficiency compared to plate or microchannel heat exchangers. Plate heat exchangers, although potentially more expensive upfront, provide enhanced heat transfer efficiency and compactness, potentially leading to lower long-term operational costs (Table 9.1) [9–11].

9.2.1.2 Operating costs

9.2.1.2.1 Energy consumption and operational efficiency

Operating costs of heat exchangers are heavily influenced by their energy consumption and overall efficiency. More efficient heat exchanger models, such as microchannel or plate heat exchangers, excel at transferring heat with greater effectiveness, thereby reducing the energy needed for temperature regulation. These advanced designs offer higher thermal performance by increasing surface area and improving fluid dynamics, which allows for faster heat transfer with less energy. As a result, less power is required to achieve desired temperature changes, significantly lowering operating expenses over

Table 9.1 Comparison of different heat exchangers

Category	*Shell-and-tube heat exchanger*	*Plate heat exchanger*	*Microchannel heat exchanger*
Initial cost	Lower initial cost due to simpler design and widespread use. Suitable for budget-conscious applications but may lack high-efficiency performance.	Moderate to high initial cost. More expensive than shell-and-tube, but offers compactness and improved heat transfer efficiency.	Higher initial cost due to advanced design and manufacturing requirements. Offers high efficiency but requires significant upfront investment.
Heat transfer efficiency	Moderate efficiency. Typically used in applications where space is less of a constraint and high performance is not a priority.	High efficiency due to increased surface area for heat transfer, making it ideal for compact systems with high thermal demands.	Superior efficiency due to the small channels that enhance heat transfer, making it suitable for space-saving designs and applications requiring precise temperature control.
Space requirement	Large, requiring more space for installation due to its bulky design.	Compact and efficient, requires less space than shell-and-tube models, making it ideal for confined areas.	Highly compact design with a small footprint, which is excellent for space-constrained applications.
Energy consumption	Higher energy consumption compared to more efficient models. Less effective heat transfer requires more energy to reach desired temperatures.	Lower energy consumption due to higher heat transfer efficiency. Reduces energy costs over time by quickly reaching desired temperature with less energy.	Very low energy consumption due to enhanced heat transfer efficiency and design, leading to long-term savings.

time. Furthermore, efficient heat exchangers reduce the burden on auxiliary systems, such as pumps, compressors, or cooling units, further minimizing energy consumption and related costs. However, the more efficient the system, the higher the initial purchase price tends to be. This creates a trade-off where the up-front investment in energy-efficient models must be weighed against the long-term savings in operational costs [12–14].

9.2.1.2.2 Maintenance and repair costs

Maintenance and repair expenses are key components of the total operating costs for heat exchangers. Regular maintenance is crucial for maintaining optimal performance and extending equipment lifespan. Shell-and-tube heat exchangers are easier to clean but may require periodic gasket and tube replacements. Plate heat exchangers often face more frequent maintenance due to issues like fouling and scaling. Microchannel heat exchangers, with their complex designs, may need specialized maintenance procedures. Implementing effective maintenance practices and using durable materials can help reduce repair costs and prevent costly downtime [15, 16].

9.2.1.3 Lifecycle costs

Lifecycle costs for a heat exchanger encompass all expenses from acquisition to disposal, providing a holistic view of the total cost of ownership. These costs begin with capital expenses, which include the initial purchase, installation, and any necessary modifications to integrate the exchanger into existing systems. Operating expenses, such as energy consumption and auxiliary system requirements, add to the long-term financial impact. Maintenance costs, covering routine inspections, part replacements, and repairs must also be factored in to ensure continuous performance and prevent costly breakdowns. Over time, the system may require upgrades or retrofitting to enhance efficiency or comply with new regulations, adding to the overall lifecycle cost. Eventually, decommissioning or replacement brings additional expenses in terms of safe disposal and system downtime. By considering all these factors, a comprehensive understanding of the heat exchanger's financial impact can be achieved [17–19].

9.2.1.3.1 Lifecycle cost analysis (LCCA)

LCCA is a technique used to assess the total financial impact of a heat exchanger by considering all associated costs throughout its operational life. LCCA involves calculating and comparing the costs of different heat exchanger options, including initial purchase price, operating costs, and maintenance expenses. This method helps in making informed decisions by identifying the most cost-effective solution that balances performance, efficiency, and long-term expenses. By employing LCCA, organizations can select heat exchangers that offer the best overall value, ensuring optimal allocation of financial resources [20, 21].

9.2.2 Return on investment (ROI)

Return on investment measures the financial return from investing in heat exchangers by evaluating energy savings, efficiency improvements, and overall system performance. It includes calculating the payback period to assess how quickly the investment will be recouped through cost savings and operational benefits.

9.2.2.1 Economic benefits

Investing in advanced heat exchangers, such as microchannel and plate models, can significantly reduce energy consumption due to their superior heat transfer efficiency. These modern designs maximize surface area and fluid flow, enabling quicker and more effective temperature regulation with less energy input. As a result, they require less power for heating or cooling processes, leading to lower utility bills and ongoing operational costs. Though advanced heat exchangers often come with a higher initial price tag, the substantial energy savings they provide typically offset these costs over time. By reducing energy demand and optimizing system performance, these efficient models deliver a favorable return on investment (ROI), making them a smart choice for businesses seeking long-term operational cost reductions. This balance between upfront costs and future savings highlights the financial and environmental benefits of upgrading to more efficient heat exchanger technologies [5, 15].

9.2.2.2 *Payback period*

9.2.2.2.1 Calculation of payback period

The payback period measures the time required for the savings from a heat exchanger to cover its initial cost. To calculate this, one must determine the total initial investment, including purchase and installation costs, and estimate the annual savings from energy and operational cost reductions. By dividing the initial investment by the annual savings, the payback period is obtained. For instance, if a heat exchanger costs $50,000 and saves $10,000 annually, the payback period would be 5 years. This metric helps assess the financial feasibility of the investment [6, 16].

9.2.2.2.2 Examples of ROI in different applications

The ROI for heat exchangers varies by application. In residential HVAC systems, high-efficiency heat exchangers might offer a payback period of 3–5 years through reduced energy bills and enhanced comfort. In industrial settings, where heat exchangers handle significant thermal loads, ROI can be more substantial. For instance, in a manufacturing facility, investing in a high-performance heat exchanger could lead to considerable energy savings and operational efficiencies, resulting in a payback period of 2–4 years. These examples demonstrate that the ROI for heat exchangers is often favorable, with long-term savings and performance improvements outweighing the initial investment [7].

9.2.3 Financial incentives and support

Financial incentives and support mechanisms such as government subsidies, tax credits, grants, and funding opportunities significantly enhance the feasibility and attractiveness of heat exchanger projects. These incentives help offset the often-high initial costs associated with acquiring advanced, energy-efficient technologies like plate or microchannel heat exchangers (Figure 9.2). By reducing upfront capital expenditure, businesses can adopt more sustainable

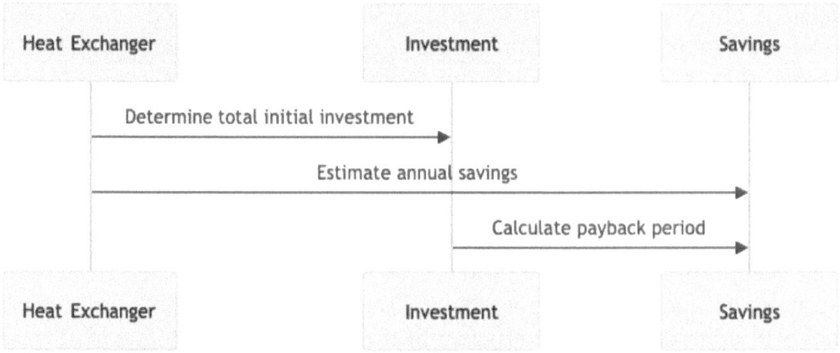

Figure 9.2 Calculation of payback period.

systems without compromising on efficiency or performance. This financial assistance not only improves the ROI but also encourages the adoption of eco-friendly technologies that comply with environmental regulations [22–24].

9.2.3.1 Government subsidies and tax credits

9.2.3.1.1 Overview of available incentives

Government subsidies and tax credits play a crucial role in encouraging the adoption of energy-efficient technologies, including advanced heat exchangers. These financial incentives aim to reduce the cost barrier for businesses and individuals looking to invest in high-efficiency systems. Subsidies typically cover a portion of the initial capital outlay, making it more affordable to purchase and install advanced heat exchangers. Tax credits, on the other hand, provide direct reductions in the amount of taxes owed, based on the value of the energy-efficient investment. This dual approach helps alleviate the upfront costs, which can be higher for advanced models like microchannel or plate heat exchangers. Many governments have introduced these programs as part of broader efforts to reduce energy consumption, lower carbon emissions, and promote the use of renewable energy sources. For instance, in the United States, the federal government offers tax credits through the Energy Policy Act, and similar initiatives exist in the European Union (EU) and other regions. These incentives may cover not only the purchase but also the installation of heat exchangers, making the financial investment more manageable [25, 26].

9.2.3.2 Grants and funding opportunities

9.2.3.2.1 Sources of funding for heat exchanger projects

Grants and funding opportunities offer additional financial support for heat exchanger projects, particularly those focused on research, development, and

Table 9.2 Sources of funding for heat exchanger projects [21–24]

Aspect	*Description*
Sources of funding	**Government agencies:** Departments such as the DOE in the United States offer grants for energy efficiency and environmental impact reduction. **Private foundations:** Non-profits and environmental organizations provide grants focused on sustainability and innovation. **Industry associations:** Specific to sectors, these organizations may offer funding to advance technology within their field. **International organizations:** Entities like the UN or World Bank offer grants for global initiatives including energy efficiency and climate impact.
Impact on project feasibility	**Financial relief:** Grants and funding reduce the initial cost burden of heat exchanger projects, making advanced technologies more accessible. **Supports R&D:** These resources aid in covering research and development expenses, fostering innovation in heat exchanger technologies. **Enhances ambition:** With financial backing, organizations can undertake more ambitious projects, such as implementing cutting-edge or high-efficiency systems. **Promotes adoption:** Financial support accelerates the adoption of innovative solutions, contributing to broader sustainability and efficiency goals.

the implementation of innovative technologies. Funding sources include government agencies, private foundations, industry associations, and international organizations. For example, government bodies like the Department of Energy (DOE) in the United States or similar agencies abroad often provide grants aimed at enhancing energy efficiency and reducing environmental impacts (Table 9.2).

9.3 ENVIRONMENTAL ASPECTS OF HEAT EXCHANGER APPLICATIONS

Heat exchangers play a critical role in reducing environmental impact across various industries by improving energy efficiency and supporting sustainable operations. By transferring heat more effectively, heat exchangers minimize the amount of energy needed for heating and cooling processes, which directly reduces fossil fuel consumption and greenhouse gas emissions. In HVAC systems, for example, efficient heat exchangers contribute to lower energy demand and, consequently, a smaller carbon footprint for buildings. In industrial applications, they enhance thermal management, reducing waste heat and optimizing resource use, which further lessens environmental strain.

9.3.1 Environmental impact assessment

An environmental impact assessment (EIA) for heat exchanger applications evaluates the direct and indirect environmental effects of implementing these systems. This involves assessing the reduction in energy consumption and emissions due to the enhanced efficiency of modern heat exchangers. Efficient heat exchangers help mitigate air and water pollution by lowering fuel combustion and reducing the need for cooling water in industrial processes. The EIA also considers the lifecycle impacts, including material sourcing, manufacturing, and disposal of heat exchangers. Eco-friendly designs that use sustainable materials and allow for recyclability further reduce environmental harm. Therefore, an EIA highlights how heat exchangers contribute to environmental sustainability, while identifying areas for further improvement, such as adopting even greener technologies and materials in future designs [27–29].

9.3.1.1 Resource utilization

9.3.1.1.1 Raw materials and energy consumption

The environmental impact of heat exchangers starts with the raw materials used and the energy consumed during their manufacturing. Heat exchangers are typically made from materials like metals (stainless steel, copper, aluminum) or composites, each with varying degrees of environmental impact. For instance, the production of stainless steel involves significant energy consumption and the emission of greenhouse gases. According to the World Steel Association, producing one ton of stainless steel generates approximately 1.8 tons of CO_2 [30,31].

Energy consumption is another critical factor. Manufacturing processes for heat exchangers often require high temperatures, which consume substantial amounts of energy (Figure 9.3). For example, producing a standard shell-and-tube heat exchanger may involve energy-intensive processes like welding and machining. The energy used in these processes contributes to the overall environmental footprint of the heat exchanger. Reducing energy consumption in manufacturing and sourcing materials with lower environmental impacts are crucial steps toward minimizing the environmental footprint.

9.3.1.1.2 Environmental footprint of manufacturing

The environmental footprint of manufacturing heat exchangers extends beyond direct emissions from production processes to encompass the entire

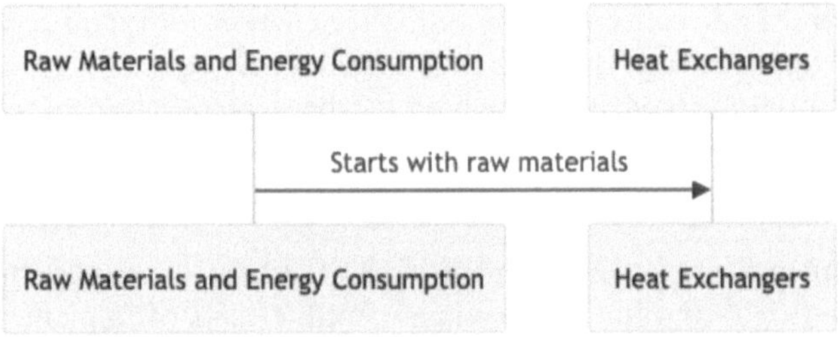

Figure 9.3 Raw materials and energy consumption.

supply chain, including the transportation of raw materials and finished products. Transporting materials from suppliers to manufacturing facilities, and later delivering the completed heat exchangers to end-users, results in additional carbon emissions. These logistical activities can significantly increase the overall environmental impact, particularly when raw materials are sourced from distant locations. To mitigate these effects, manufacturers can implement several sustainable practices. First, adopting energy-efficient production technologies can lower emissions during the manufacturing phase. Second, reducing waste during production, by optimizing material use and recycling excess materials, can further decrease the environmental footprint. Lastly, sourcing raw materials locally or regionally can significantly reduce the emissions associated with transportation, as shorter distances require less fuel and energy. By integrating these strategies, manufacturers can minimize the carbon footprint of heat exchanger production, contributing to a more sustainable lifecycle for these essential systems while aligning with environmental regulations and consumer expectations for eco-friendly products [32, 33].

9.3.1.2 *Waste and emissions*

9.3.1.2.1 Waste management and disposal

Waste generated during the manufacturing of heat exchangers can include offcuts of metal, used lubricants, and other by-products. Proper waste management practices are essential to minimize the environmental impact. Recyclable materials, such as metal scraps, should be collected and processed to reduce landfill waste. For example, according to the Environmental Protection Agency (EPA), recycling metal can save up to 95% of the energy required to produce new metal from ore. Implementing recycling programs and ensuring safe disposal of hazardous materials can significantly reduce the environmental footprint [34, 35].

9.3.1.2.2 Emissions during operation and their mitigation

During operation, heat exchangers can contribute to emissions indirectly by influencing the efficiency of the systems they are part of. For instance, inefficient heat exchangers may cause higher energy consumption in HVAC systems or industrial processes, leading to increased emissions from power plants or other energy sources. Mitigating these emissions involves improving the efficiency of heat exchangers through advanced designs and materials. High-efficiency heat exchangers, such as those with enhanced heat transfer surfaces or advanced microchannel configurations, can reduce energy consumption and, consequently, emissions from associated energy sources. Additionally, incorporating sustainable practices in the operation and maintenance of heat exchangers, such as regular cleaning and optimizing performance, can further reduce their environmental impact. By focusing on resource utilization, waste management, and emissions, the environmental impact of heat exchangers can be managed more effectively, leading to more sustainable and eco-friendly practices in their lifecycle [36–38].

9.3.2 Energy efficiency and conservation

Heat exchangers play a pivotal role in enhancing energy efficiency and conservation across various applications by optimizing the transfer and use of thermal energy. By effectively capturing and reusing heat that would otherwise be wasted, these systems significantly reduce the overall energy demand of heating and cooling processes [39–41].

9.3.2.1 Contribution to energy efficiency

The contribution of heat exchangers to energy efficiency is substantial as they facilitate the transfer of heat between fluids without mixing them. This process allows for the recovery and reuse of thermal energy from exhaust or waste streams, which would otherwise dissipate into the environment. For example, in industrial settings, heat exchangers can capture waste heat from high-temperature processes and transfer it to incoming cold fluids, reducing the need for additional energy to heat these fluids. In HVAC systems, efficient heat exchangers improve the performance of heating and cooling units by ensuring that less energy is required to reach the desired temperatures. This not only lowers energy consumption but also reduces the operational costs associated with heating and cooling. Additionally, by decreasing the energy needed for temperature control, heat exchangers contribute to a reduction in greenhouse gas emissions and other pollutants, aligning with broader sustainability goals. Thus, their ability to enhance energy efficiency and promote conservation makes heat exchangers essential for both economic and environmental benefits, supporting the transition toward more sustainable and cost-effective energy practices [42, 43].

9.3.2.1.1 Role of heat exchangers in reducing energy consumption

Heat exchangers play a crucial role in enhancing energy efficiency by transferring heat between fluids with minimal energy loss. Their primary function is to recover and reuse thermal energy that would otherwise be wasted. In HVAC systems, for example, heat exchangers can recover heat from exhaust air to preheat incoming fresh air, thereby reducing the energy required for heating. This process lowers the load on heating systems and improves overall efficiency. Similarly, in industrial processes, heat exchangers can capture waste heat from high-temperature streams and use it to heat other fluids or generate power, leading to significant energy savings (Figure 9.4).

Several case studies highlight the energy-saving potential of heat exchangers. In a study conducted by the American Council for an Energy-Efficient Economy (ACEEE), the installation of a high-efficiency heat exchanger in

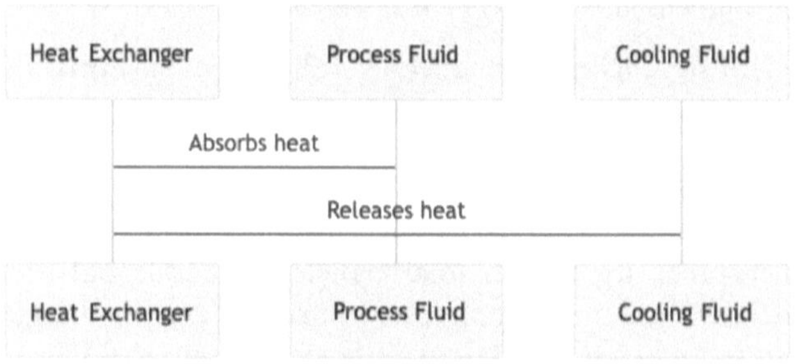

Figure 9.4 Role of heat exchangers in reducing energy consumption.

a commercial building resulted in a 20% reduction in annual energy consumption. Another example is a manufacturing facility that integrated a heat recovery heat exchanger into its process. The facility reported a 30% reduction in energy use and a payback period of just 18 months due to the efficient recovery of waste heat. These examples demonstrate the significant impact that advanced heat exchanger technologies can have on reducing energy consumption and operational costs [44–46].

9.3.2.2 Environmental benefits

9.3.2.2.1 Reduction in greenhouse gas emissions

Heat exchangers play a vital role in environmental sustainability by significantly reducing greenhouse gas emissions through enhanced energy efficiency. Their primary function is to facilitate the transfer of heat between two or more fluids, optimizing energy use and minimizing waste. By improving the efficiency of heat transfer processes, heat exchangers decrease the overall demand for energy, which often relies on fossil fuels like coal, oil, and natural gas. The reduction in energy consumption directly translates to lower carbon dioxide (CO_2) emissions, a major greenhouse gas contributing to global warming and climate change. For example, according to the US EPA, implementing energy-efficient technologies that cut energy consumption by just 10% can result in a similar percentage decrease in CO_2 emissions. Heat exchangers contribute to this reduction by enhancing heat recovery systems. They allow for the reuse of heat that would otherwise be wasted, reducing the need for additional energy inputs from less sustainable sources. This energy efficiency not only lowers operational costs for businesses and facilities but also diminishes their environmental footprint. The more effective heat recovery and reduced energy demands achieved through modern heat exchangers contribute to a more sustainable energy cycle, ultimately supporting broader climate goals. By integrating advanced heat exchangers into systems, industries can make substantial progress toward reducing greenhouse gas emissions, improving air quality, and combating climate change [47–50].

9.3.2.2.2 Improvement in overall system sustainability

Heat exchangers significantly enhance overall system sustainability by optimizing energy use and minimizing waste, particularly in renewable energy systems. In solar thermal or geothermal systems, for instance, heat exchangers play a crucial role in improving the efficiency of energy transfer and storage. By efficiently transferring heat between different fluids, they ensure that renewable energy captured from solar panels or geothermal sources is utilized more effectively. For example, incorporating heat exchangers in solar thermal systems can boost the system's efficiency by up to 25%, allowing for greater heat absorption and reduced energy loss. This improved efficiency translates into more effective use of renewable resources, making energy systems more sustainable and reducing reliance on fossil fuels. By minimizing energy waste and enhancing the performance of renewable energy technologies, heat exchangers support broader environmental sustainability goals and promote eco-friendly practices. Their role in advancing energy efficiency and resource utilization underscores their importance in developing sustainable energy solutions [12].

9.3.3 Life Cycle Assessment (LCA)

Life Cycle Assessment (LCA) is a method for evaluating the environmental impacts of a product or system throughout its entire life cycle—from raw material extraction through production, use, and disposal. It provides a comprehensive view of environmental performance, helping identify opportunities for improvements and more sustainable practices.

9.3.3.1 Methodology

LCA is a comprehensive method for evaluating the environmental impacts associated with a product or system throughout its entire lifecycle, from initial concept to final disposal. The LCA process is structured into four primary stages: Goal and Scope Definition, Inventory Analysis, Impact Assessment, and Interpretation.

9.3.3.1.1 Goal and scope definition

The first stage of LCA, Goal and Scope Definition, establishes the framework and objectives for the assessment. This phase outlines the purpose of the LCA, such as determining the environmental benefits of a new heat exchanger design or comparing different models. It involves defining the scope of the study, which includes specifying the system boundaries—such as which life cycle stages will be included (e.g., production, usage, end-of-life)—and the functional unit, which quantifies the performance of the heat exchanger to be assessed (e.g., heat transfer capacity per unit). The results' intended use is also clarified, such as guiding product improvements or informing stakeholders. Clear goal and scope definition ensures that the LCA is focused and relevant to the questions at hand.

9.3.3.1.2 Inventory analysis (Life Cycle Inventory, LCI)

Inventory analysis, or Life Cycle Inventory (LCI), is the second stage and involves a detailed data collection process. This phase gathers information on all inputs and outputs throughout the heat exchanger's life cycle. Inputs include raw materials, energy consumption, and other resources needed during production, while outputs encompass emissions, waste, and by-products generated at each stage. The data collection covers several phases: material extraction, manufacturing, transportation, usage, and end-of-life disposal. Accurate and comprehensive inventory data are critical as they form the basis for assessing environmental impacts in subsequent stages.

9.3.3.1.3 Impact assessment (Life Cycle Impact Assessment, LCIA)

The third stage, Impact Assessment (or Life Cycle Impact Assessment, LCIA), involves analyzing the data collected during the inventory phase to evaluate potential environmental impacts. This stage categorizes and quantifies the environmental effects associated with the heat exchanger's life cycle. Key indicators assessed include greenhouse gas emissions, energy consumption, water usage, and resource depletion. For example, the carbon footprint measures the total greenhouse gas emissions, while energy use assesses the overall energy consumed throughout the product's life. This stage helps in understanding the relative significance of various impacts, such as identifying whether emissions during the usage phase are more critical than those during production.

9.3.3.1.4 Interpretation

The final stage, Interpretation, involves analyzing the results from the impact assessment to provide actionable insights. This phase focuses on identifying significant environmental impacts, understanding their causes and evaluating the trade-offs between different impacts. Interpretation also includes assessing the reliability and completeness of the data and results. Recommendations for improving the heat exchanger's environmental performance may be developed, such as suggesting design modifications or alternative materials. The interpretation phase ensures that the findings are used effectively to inform decision-making, drive improvements, and communicate results to stakeholders. In summary, LCA is a structured approach that provides a detailed understanding of the environmental impacts associated with a heat exchanger throughout its lifecycle. By systematically defining goals, analyzing inventory data, assessing impacts, and interpreting results, LCA helps in making informed decisions that enhance environmental sustainability and support more responsible product development (Table 9.3).

Table 9.3 *Life cycle assessment (LCA) stages*

Stage	*Description*	*Key activities*
Goal and scope definition	Establishes the framework and objectives for the assessment, outlining the purpose, scope, and functional unit. Defines the system boundaries and the intended use of the results.	• Define assessment objectives (e.g., environmental benefits, model comparison). • Specify system boundaries (e.g., production, usage). • Quantify functional unit (e.g., heat transfer capacity). • Clarify intended use of results (e.g., product improvements).
Inventory analysis (LCI)	Collects detailed data on all inputs and outputs throughout the heat exchanger's life cycle, including raw materials, energy consumption, emissions, waste, and by-products.	• Gather data on raw materials and energy. • Record emissions and waste. • Document all life cycle stages (production, usage, disposal).
Impact assessment (LCIA)	Analyzes inventory data to evaluate potential environmental impacts, categorizing and quantifying effects such as greenhouse gas emissions, energy consumption, water usage, and resource depletion.	• Assess key indicators (e.g., carbon footprint, energy use). • Categorize and quantify environmental impacts. • Compare impacts across different life cycle stages.
Interpretation	Analyzes results from the impact assessment to identify significant environmental impacts, understand their causes, and evaluate trade-offs. Provides recommendations for improving environmental performance and ensures data reliability.	• Identify and analyze significant impacts. • Evaluate data reliability and completeness. • Develop recommendations (e.g., design modifications). • Communicate results to stakeholders.

9.3.3.2 Case studies

9.3.3.2.1 Examples of LCA for different heat exchanger technologies

Several case studies have used LCA to evaluate the environmental impacts of various heat exchanger technologies. For example, research comparing shell-and-tube heat exchangers with plate heat exchangers indicates that plate heat exchangers typically have a lower environmental impact due to their superior heat transfer efficiency and reduced material requirements. Another study focused on microchannel heat exchangers found that, despite their higher energy consumption during manufacturing, their overall environmental benefit is substantial due to their high efficiency and lower operational energy needs.

These case studies highlight several key lessons and best practices. A significant finding is the importance of including both operational and end-of-life stages in the LCA to obtain an accurate EIA. Effective data collection and precise inventory analysis are crucial for reliable results. Incorporating design improvements, such as using recyclable materials or enhancing energy efficiency, can substantially reduce the environmental footprint. Overall, LCA proves valuable in guiding decisions toward more sustainable heat exchanger technologies and practices, enabling organizations to minimize environmental impacts and make more informed choices in design and selection.

9.4 COMPARATIVE ANALYSIS

9.4.1 Comparison of heat exchanger types

Comparing heat exchanger types reveals distinct differences in economic and environmental performance. Traditional models, such as shell-and-tube, contrast with advanced technologies like plate and microchannel heat exchangers in terms of cost, efficiency, and environmental impact. Understanding these differences helps in selecting the most suitable heat exchanger for specific applications.

9.4.1.1 Traditional vs. Advanced Technologies

Heat exchangers come in various designs, each offering unique economic and environmental performance features. This comparison examines traditional heat exchangers, such as shell-and-tube models, alongside advanced technologies like plate and microchannel heat exchangers.

9.4.1.1.1 Shell-and-tube heat exchangers

Shell-and-tube heat exchangers are a conventional design favored for their durability and moderate initial investment. They are widely used in applications such as power generation and chemical processing. Although they are economically feasible with moderate initial costs and low maintenance requirements, their efficiency is generally lower compared to advanced technologies. This lower efficiency leads to higher energy consumption and a larger environmental footprint due to their greater size and material usage.

9.4.1.1.2 Plate heat exchangers

Plate heat exchangers offer superior energy efficiency relative to shell-and-tube designs, thanks to their compact and enhanced heat transfer surfaces. This improved efficiency results in reduced energy consumption and operational costs. Despite their higher initial cost, plate heat exchangers provide long-term cost benefits through lower maintenance needs and reduced environmental impact. Their more efficient design results in less material use and better performance, leading to a smaller environmental footprint compared to traditional models.

9.4.1.1.3 Microchannel heat exchangers

Microchannel heat exchangers represent the cutting edge of heat exchanger technology, renowned for their exceptional heat transfer efficiency and compact size.

Table 9.4 Comparison of different heat exchangers

Heat exchanger type	*Initial capital cost*	*Energy efficiency*	*Maintenance cost*	*Environmental impact*
Shell-and-tube	Moderate	Moderate	Low	Higher material use; less efficient heat transfer
Plate	Higher	High	Moderate	Reduced material use; higher efficiency
Microchannel	Highest	Very high	High	Very efficient; compact; high material and manufacturing costs

They require the highest initial investment but deliver significant energy savings and operational efficiency. While their manufacturing process involves higher material and production costs, their overall environmental impact is minimized due to lower energy consumption. Microchannel heat exchangers are ideal for applications with strict efficiency requirements and space constraints, such as in HVAC systems and advanced industrial processes (Table 9.4) [38–41].

9.4.1.2 Industry-specific applications

Different types of heat exchangers are optimized for specific industry needs based on their performance attributes.

9.4.1.2.1 Power generation

Shell-and-tube heat exchangers are widely utilized in power plants because of their durability and capability to handle high pressures and temperatures. Their robust design allows them to withstand the demanding conditions typical of power generation processes while ensuring reliable operation. For example, in steam power plants, these heat exchangers facilitate effective heat transfer between steam and water, which is essential for maintaining operational efficiency and safety.

9.4.1.2.2 HVAC systems

Plate heat exchangers are favored in HVAC systems due to their compact design and high heat transfer efficiency. Their space-efficient structure makes them particularly suitable for installations with limited space, such as in residential and commercial buildings. The enhanced thermal performance of plate heat exchangers improves energy conservation by optimizing heating and cooling processes, which leads to reduced energy costs and better control of indoor climate.

9.4.1.2.3 Automotive

Microchannel heat exchangers are increasingly popular in the automotive industry because of their compact size and superior heat transfer capabilities.

These exchangers are employed in vehicle cooling systems to manage engine temperatures more effectively. Their high efficiency results in improved engine performance and reduced fuel consumption, thereby enhancing vehicle reliability and decreasing environmental impact. For instance, in high-performance vehicles, microchannel heat exchangers handle higher heat loads efficiently, contributing to optimal engine operation.

9.4.1.3 Industry-specific applications

9.4.1.3.1 Comparative analysis across sectors

9.4.1.3.1.1 HVAC SYSTEMS

In HVAC systems, heat exchangers are pivotal for optimizing energy efficiency and space utilization. Plate heat exchangers are often chosen for their superior thermal performance and compact footprint. For example, according to a study by the US DOE, plate heat exchangers can achieve up to 90% efficiency in heat recovery applications, surpassing the approximately 70% efficiency of traditional shell-and-tube models. Their compact design is particularly advantageous for residential and commercial settings with limited space. This efficiency translates into notable energy savings; residential HVAC systems using high-efficiency plate heat exchangers can see a reduction in annual energy bills of about 10–15%, based on data from the American Society of Heating, Refrigerating, and Air-Conditioning Engineers (ASHRAE) [13].

9.4.1.3.1.2 CHEMICAL PROCESSING

In the chemical processing sector, heat exchangers are essential for efficient thermal management. Shell-and-tube heat exchangers are favored for their durability and capacity to withstand corrosive substances and high pressures. For example, in sulfuric acid production facilities, shell-and-tube heat exchangers are utilized to handle the severe operational conditions while ensuring consistent performance. However, advancements in heat exchanger technology, such as plate heat exchangers, are gaining attention due to their higher heat transfer efficiency and smaller footprint. A report from the *Chemical Engineering Journal* notes that plate heat exchangers can enhance thermal efficiency by up to 30% compared to traditional designs, though they may come with higher initial costs [14].

9.4.1.3.1.3 POWER GENERATION

In power generation, shell-and-tube heat exchangers are commonly used due to their robustness and ability to manage high-pressure steam effectively. For instance, in coal-fired power plants, these heat exchangers facilitate the transfer of heat between steam and water, essential for maintaining the efficiency of the power generation process. Given their ability to handle high pressure and temperature, shell-and-tube heat exchangers are well-suited for these demanding environments. According to the *International Journal of Thermal Sciences*, these heat exchangers typically achieve around 85% thermal efficiency, making them reliable for power generation applications [15].

9.4.2 Benchmarking against alternatives

9.4.2.1 Alternative technologies

Benchmarking heat exchangers involves comparing them with alternative heat transfer technologies, such as air-cooled and water-cooled systems. Air-cooled heat exchangers, for example, avoid the use of water, which can reduce operational costs and minimize water consumption. However, they are generally less efficient compared to liquid-cooled systems. Water-cooled heat exchangers, including both shell-and-tube and plate designs, offer higher heat transfer efficiency but require ongoing water management and treatment. Industry data from Heat Transfer Research, Inc., indicates that water-cooled systems can achieve efficiency levels up to 95%, whereas air-cooled systems typically range from 70% to 80% efficiency [16].

9.4.2.2 Cost and environmental performance

Benchmarking heat exchanger involves assessing various metrics such as initial cost, energy efficiency, maintenance needs, and environmental impact. For instance, microchannel heat exchangers, while having the highest initial capital costs, offer significant long-term savings due to their exceptional efficiency. The International Energy Agency reports that microchannel heat exchangers can reduce energy consumption by up to 20% compared to conventional models. Environmental performance is evaluated based on factors like carbon footprint and resource consumption. High-efficiency heat exchangers contribute to lower greenhouse gas emissions and reduced operational costs. Benchmarking standards from organizations like the American Society of Mechanical Engineers (ASME) and the International Organization for Standardization (ISO) provide criteria for evaluating these metrics, ensuring a fair comparison across different heat exchanger technologies [17].

9.5 INNOVATIONS AND TRENDS

9.5.1 Technological advances

The landscape of heat exchangers is rapidly evolving with significant technological advances. Innovations in design, materials, and efficiency are driving improvements in performance and sustainability. Cutting-edge technologies, such as enhanced heat transfer surfaces and advanced manufacturing techniques, are setting new standards for energy efficiency and operational

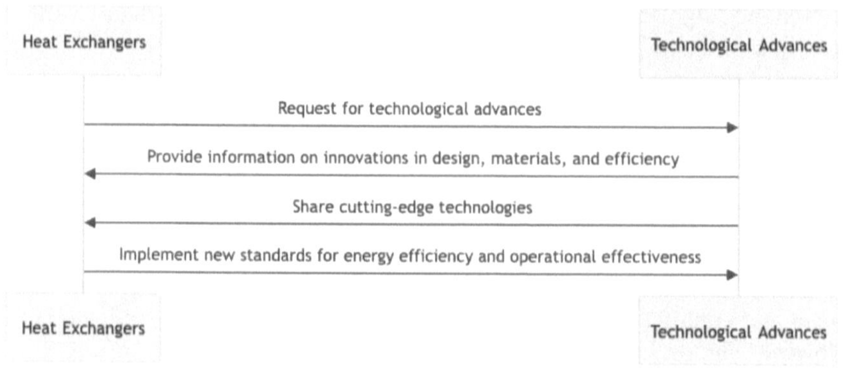

Figure 9.5 Innovations and trends.

effectiveness, paving the way for more effective and eco-friendly heat exchange solutions (Figure 9.5).

9.5.1.1 Emerging materials and designs

Recent advancements in materials and design have significantly improved the performance of heat exchangers. Innovations in materials science have led to the development of more efficient, durable, and environmentally friendly heat exchangers. For instance, the introduction of advanced materials like titanium and high-performance alloys has enhanced the resistance of heat exchangers to corrosion and high temperatures. Titanium heat exchangers, in particular, are highly resistant to corrosive environments, making them ideal for use in chemical processing and seawater cooling systems [18].

9.5.1.2 Smart technologies and IoT

The integration of smart technologies and the Internet of Things (IoT) is revolutionizing heat exchanger systems, driving enhanced efficiency and performance. Smart technologies facilitate real-time monitoring and control, leading to optimized performance and predictive maintenance. For example, IoT-enabled sensors continuously monitor parameters like temperature, pressure, and flow rates. This data is transmitted to central systems where it is analyzed to detect inefficiencies, predict potential failures, and adjust operations for optimal performance. A notable application is in HVAC systems, where smart controllers use data from IoT sensors to dynamically adjust heat exchanger operations based on real-time conditions and environmental changes. This results in improved energy efficiency and reduced operational costs. Advanced data analytics further support long-term performance assessments, aiding in maintenance and upgrade decisions. In industrial settings, smart heat exchangers equipped with IoT capabilities offer significant improvements in operational efficiency and reliability. For instance, predictive maintenance algorithms analyze both historical and real-time data to anticipate equipment failures, thereby minimizing downtime and extending the lifespan of the heat exchangers [42, 43].

9.5.2 Future trends

9.5.2.1 Sustainability and green technologies

The heat exchanger industry is progressively shifting toward sustainability and green technologies, driven by the imperative to reduce environmental impact and enhance energy efficiency. Key trends in this evolution include the following (Figure 9.6).

9.5.2.1.1 Use of sustainable materials

The growing emphasis on sustainability has prompted significant changes in heat exchanger manufacturing, particularly with the adoption of eco-friendly and recyclable materials. Manufacturers are increasingly incorporating recycled metals, such as aluminum and steel, which offer the same durability and efficiency as newly mined materials but with a much lower environmental footprint. Additionally, bio-based polymers, derived from renewable resources like plants, are being used for specific components within heat exchangers. These materials not only help reduce the consumption of non-renewable resources but also cut down on waste generated during production.

9.5.2.1.2 Advances in energy efficiency

Energy efficiency is a cornerstone of innovation in heat exchanger technology. Modern advancements focus on maximizing heat transfer while minimizing

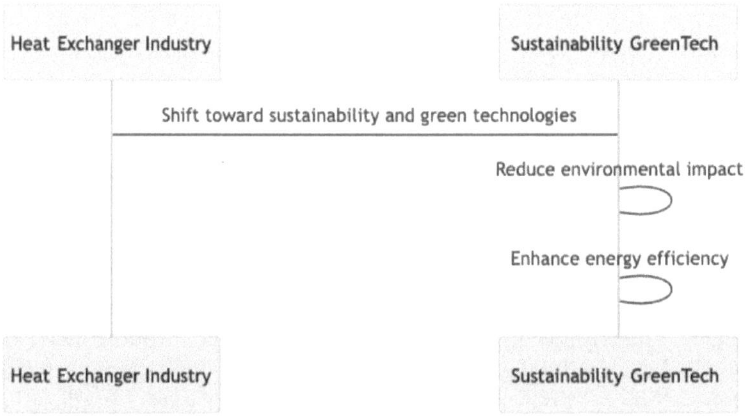

Figure 9.6 Sustainability and green technologies.

energy consumption, which is essential for both cost savings and environmental benefits. One of the key trends is the development of high-performance heat transfer surfaces, which are designed to increase the rate of heat exchange without requiring more energy. Additionally, the use of optimized flow designs, particularly through 3D printing, allows manufacturers to create intricate internal structures that enhance fluid dynamics, resulting in more efficient heat transfer. For instance, these high-performance heat exchangers can reduce energy consumption by up to 20%, providing significant savings on operational costs for industries and households alike. This reduction in energy use also aligns with global efforts to reduce carbon emissions and combat climate change, making these innovations not only economically beneficial but also environmentally critical [44, 45].

9.5.2.1.3 Integration with renewable energy systems

The integration of heat exchangers with renewable energy systems is a growing trend that enhances the overall efficiency of sustainable energy solutions. In solar thermal systems, for example, heat exchangers are used to transfer heat collected from solar panels to either storage tanks or directly to heating systems in buildings. This efficient transfer of heat is crucial for optimizing the performance of solar thermal systems, ensuring that the energy captured from the sun is utilized effectively. Similarly, in geothermal energy systems, heat exchangers play a pivotal role in transferring heat from the earth to residential or industrial heating systems.

9.5.2.1.4 Improved recyclability

As the concept of a circular economy gains traction, the design of heat exchangers is increasingly focused on recyclability and reducing waste. Modern heat exchangers are being designed with modular components that can be easily disassembled at the end of their lifecycle, allowing for the recovery and reuse of valuable materials. For instance, metals like aluminum and steel, commonly used in heat exchangers, are highly recyclable and can be reprocessed with minimal loss of quality. By facilitating easier recycling, manufacturers are able to reduce the amount of waste that ends up in landfills, contributing to a more sustainable production cycle.

9.5.2.2 Policy and regulatory changes

Changes in policies and regulations are significantly impacting the heat exchanger industry, influencing both current practices and future developments.

9.5.2.3 Policy and regulatory changes

Changes in policies and regulations are shaping the future of the heat exchanger industry, driving both technological innovation and the adoption

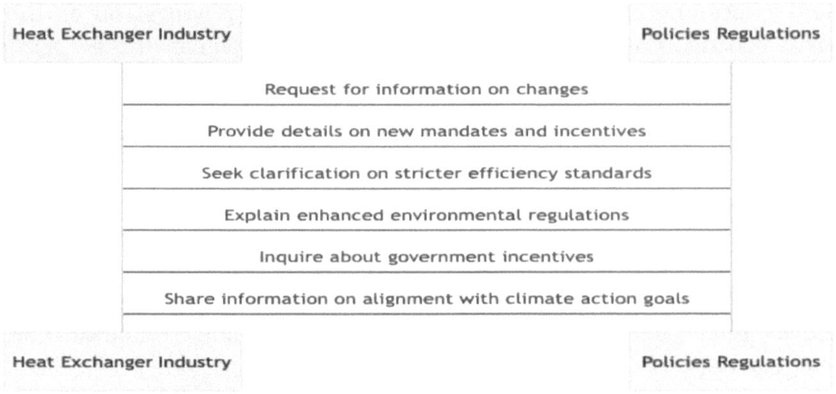

Figure 9.7 Policy and regulatory changes.

of sustainable practices. The industry is being transformed by new mandates and incentives designed to improve energy efficiency and reduce environmental impact. Key areas of regulatory change include stricter efficiency standards, enhanced environmental regulations, government incentives, and alignment with climate action goals (Figure 9.7).

9.5.2.3.1 Stricter efficiency standards

One of the most significant regulatory changes impacting the heat exchanger industry is the implementation of stricter energy efficiency standards. Governments and regulatory bodies, such as the EU with its Ecodesign Directive and the US DOE, have introduced higher minimum efficiency requirements for heat exchangers. These standards are designed to reduce energy consumption and associated environmental impacts. Manufacturers are required to develop more advanced and efficient heat exchangers to meet these regulatory demands, leading to innovations in heat transfer technology, material science, and product design.

9.5.2.3.2 Enhanced environmental regulations

Environmental regulations are becoming increasingly stringent, particularly concerning greenhouse gas emissions and the use of high global warming potential (GWP) refrigerants in heat exchangers. For example, the EU's F-Gas regulation and similar global initiatives are phasing out the use of refrigerants with high GWP due to their significant contribution to climate change. In response, manufacturers are transitioning to low-GWP refrigerants and developing heat exchangers that comply with these new environmental requirements. Additionally, industries using heat exchangers must now meet stricter carbon emission limits, which has accelerated the demand for energy-efficient designs. This shift toward environmentally friendly refrigerants and higher-efficiency systems helps reduce the overall carbon footprint of heating, ventilation, and cooling technologies, making them more sustainable in the long term [19].

9.5.2.3.3 Incentives and support programs

Governments around the world are introducing financial incentives and support programs to encourage the adoption of sustainable heat exchanger technologies. These programs include tax credits, grants, and subsidies that help offset the initial costs of implementing advanced heat exchangers, particularly those designed for energy conservation and low environmental impact. For instance, in the United States, businesses that invest in high-efficiency heat exchangers can benefit from various tax incentives provided under energy efficiency programs, while similar schemes are in place across Europe and other regions. These incentives not only make the transition to greener technologies more affordable for businesses but also accelerate the adoption of innovative heat exchangers that contribute to energy savings and environmental sustainability. Support from governments ensures that companies have the financial backing to invest in these technologies, ultimately fostering industry-wide change.

9.5.2.3.4 Climate action goals

National and international climate action goals, such as those set out in the Paris Agreement, are exerting additional pressure on industries to reduce their carbon footprints. Countries across the globe have committed to achieving net-zero emissions by mid-century, and industries are expected to play a crucial role in meeting these targets. Heat exchangers, as essential components of energy systems, are increasingly being designed with climate action goals in mind. This involves creating systems that reduce energy consumption and emissions, contributing to a cleaner, more sustainable future. The heat exchanger industry is responding by developing products that align with these global objectives, ensuring that heat transfer technologies contribute to carbon reduction efforts. Industries that adopt these advanced heat exchangers are not only reducing their own environmental impact but also helping nations meet their climate commitments.

9.6 CASE STUDIES AND PRACTICAL INSIGHTS

9.6.1 Industry case studies

In exploring case studies and practical insights, this section delves into real-world applications of heat exchangers, examining successful implementations and lessons learned. It highlights how different industries leverage innovative technologies to enhance efficiency, reduce costs, and address challenges. By analyzing these practical examples, valuable insights into best practices and effective solutions for optimizing heat exchanger performance and sustainability are gained.

9.6.1.1 Successful implementations

9.6.1.1.1 Industrial manufacturing

A notable example of successful heat exchanger implementation comes from a major industrial manufacturing facility in Germany. The plant, which previously relied on traditional shell-and-tube heat exchangers, faced several operational challenges, including high energy costs and frequent maintenance issues. To address these problems, the facility transitioned to plate heat exchangers, known for their compact design and enhanced heat transfer efficiency. This switch resulted in a 30% reduction in energy consumption and a 20% decrease in operational costs, significantly improving the plant's energy footprint and overall operational effectiveness. The compact nature of plate heat exchangers also allowed the plant to save valuable space, further optimizing their systems. The German government supported this transition through energy efficiency incentives, helping the facility offset the initial investment costs, making the upgrade both environmentally and economically beneficial [20].

9.6.1.1.2 Commercial HVAC systems

In the commercial sector, a large office building in New York City serves as an example of successful heat exchanger integration within a HVAC system. To reduce energy consumption and enhance climate control, the building opted to incorporate microchannel heat exchangers into its system. These exchangers, known for their high thermal performance and compact design, were ideal for the limited space available in the building's infrastructure. The integration of microchannel heat exchangers resulted in a 25% boost in energy efficiency, which led to substantial annual energy savings. Despite the higher upfront costs of microchannel technology, an ROI analysis demonstrated that the energy savings allowed the investment to pay off within three years. The building's HVAC system not only became more energy-efficient but also contributed to lower operational costs and a reduced environmental impact, making it a key success story in commercial energy optimization [21].

9.6.1.2 Lessons learned

9.6.1.2.1 Managing transition costs

One of the primary challenges in upgrading to advanced heat exchanger technologies is the significant initial investment. Both the industrial manufacturing plant in Germany and the commercial office building in New York City faced substantial financial outlays to replace outdated systems with more efficient ones. These upfront costs included not only the purchase of cutting-edge heat exchanger units but also the associated installation expenses, infrastructure modifications, and labor costs. Managing these transition costs required strategic financial planning. In both cases, leveraging financial incentives and support programs was crucial in making the upgrades feasible. For example, the manufacturing plant in Germany benefited from government-backed energy efficiency incentives, which helped mitigate the high initial costs. Similarly, in New York, the office building took advantage of local government tax credits and energy-saving grants designed to encourage sustainable building practices. Conducting thorough cost-benefit analyses was essential for both projects to ensure that the long-term energy savings would justify the initial expenditure.

9.6.1.2.2 Integration and installation challenges

The integration of new heat exchanger technologies into existing systems posed significant challenges for both the manufacturing plant and the office building. At the manufacturing facility, replacing the older shell-and-tube heat exchangers with plate heat exchangers required modifications to the existing infrastructure. This process included rearranging pipelines, retrofitting existing connections, and ensuring that the new exchangers could function within the plant's operational framework. These changes resulted in temporary disruptions to operations as sections of the plant needed to be shut down during the installation. To manage this, the facility adopted a phased implementation approach, carefully planning the installation in stages to minimize downtime and avoid halting production completely. Each phase was coordinated to coincide with scheduled maintenance periods, reducing the overall impact on productivity. Similarly, in the office building, the installation of microchannel heat exchangers required precise coordination with HVAC engineers.

9.6.1.2.3 Ongoing maintenance and monitoring

Once the new heat exchanger systems were in place, maintaining and monitoring them became critical to ensure long-term performance and efficiency. In the manufacturing plant, the introduction of plate heat exchangers necessitated a shift in the maintenance regime. Unlike the older shell-and-tube models, plate heat exchangers have more intricate designs and require periodic cleaning to prevent fouling and ensure optimal heat transfer efficiency. To address this, the plant implemented a rigorous maintenance schedule, training its staff to perform regular inspections, cleanings, and minor repairs.

9.6.2 Best Practices

9.6.2.1 Economic optimization

To maximize economic benefits in heat exchanger systems, several strategies can be implemented to reduce costs and enhance ROI.

9.6.2.1.1 Lifecycle cost analysis

A comprehensive LCCA is a vital tool for assessing the total cost of ownership for heat exchangers. This analysis takes into account all aspects of a heat exchanger's financial impact, including the initial capital expenditure, operational costs, maintenance, and eventual end-of-life disposal expenses. By evaluating these factors, organizations can make informed decisions about which heat exchanger technologies offer the best long-term financial returns. For instance, microchannel heat exchangers, despite having a higher upfront cost compared to traditional models, can lead to substantial savings in energy consumption over time. This efficiency results in lower operating costs and improved ROI. Additionally, a lifecycle analysis allows for comparisons between different heat exchanger configurations, helping organizations choose technologies that best align with their operational needs and budget constraints. The LCCA also provides insights into potential costs related to system retrofits, upgrades, and end-of-life management, ensuring that organizations account for long-term costs rather than focusing solely on initial outlays.

9.6.2.1.2 Energy efficiency enhancements

Investing in high-efficiency heat exchangers is one of the most effective ways to improve ROI, as reducing energy consumption directly impacts operational costs. Advanced heat exchanger technologies, such as plate heat exchangers with enhanced heat transfer surfaces, can significantly boost energy efficiency by optimizing heat exchange between fluids. The integration of cutting-edge designs, including optimized flow channels and innovative materials, can reduce energy use by improving heat transfer rates and minimizing heat loss. For example, heat exchangers designed with improved heat transfer surfaces can achieve higher efficiency without the need for larger or more complex systems, thereby lowering both energy usage and operational expenses. Regular performance evaluations are also critical to maintaining optimal efficiency. Organizations can further enhance energy savings by adjusting system parameters, such as flow rates and temperature differentials, to match real-time operational needs.

9.6.2.1.3 Proactive maintenance and upgrades

Adopting a proactive maintenance strategy is essential for ensuring that heat exchangers operate efficiently over their full lifespan, reducing both downtime and repair costs. Regularly scheduled maintenance, such as cleaning and inspecting components, helps to prevent performance degradation caused by fouling, corrosion, or mechanical wear. By addressing these issues before they lead to system failures, organizations can minimize the likelihood of costly breakdowns and extend the lifespan of their heat exchangers. Timely upgrades and retrofits, such as replacing aging components or incorporating newer, more efficient technologies, also ensure that the heat exchanger continues to meet the evolving demands of the facility. Proactive maintenance is further enhanced by implementing advanced monitoring systems, which provide real-time data on key performance indicators such as pressure, temperature, and flow rates.

9.6.2.1.4 Utilizing financial incentives

Taking advantage of financial incentives and support programs is another crucial factor in improving ROI when investing in heat exchanger upgrades. Many governments and regulatory bodies offer rebates, tax credits, and grants to encourage the adoption of energy-efficient technologies. These incentives can significantly reduce the financial burden associated with upgrading or replacing outdated heat exchangers, making the initial investment more manageable. For example, in regions that prioritize sustainability and energy efficiency, businesses may qualify for financial assistance if they implement heat exchangers that meet certain efficiency standards. In the EU, for instance, companies can benefit from the Ecodesign Directive, which offers support for energy-saving technologies. Similarly, in the United States, the DOE provides tax credits for businesses that invest in energy-efficient systems, helping to offset the upfront costs of adopting advanced heat exchangers. These financial incentives not only reduce the capital investment required but also accelerate the payback period by lowering operational costs and increasing long-term savings.

9.6.2.2 Environmental sustainability

Reducing the environmental impact of heat exchangers involves adopting best practices that promote sustainability:

9.6.2.2.1 Eco-friendly materials

The use of sustainable and recyclable materials in heat exchanger construction has become a key strategy for reducing environmental impact. Traditional materials, while effective, often contribute to higher-energy consumption and carbon emissions during production and disposal. By shifting to eco-friendly alternatives like recycled metals and bio-based polymers, manufacturers can significantly reduce the carbon footprint of heat exchangers throughout their lifecycle. Recycled aluminum is a prime example, as it boasts excellent thermal conductivity and is highly recyclable. Aluminum's ability to maintain its properties after recycling makes it an ideal material for sustainable heat exchanger design, cutting down on the need for new raw materials. Additionally, bio-based polymers, derived from renewable resources, offer another avenue for reducing reliance on fossil fuels in manufacturing.

9.6.2.2.2 Design for recyclability

Designing heat exchangers for recyclability is an essential practice that supports the circular economy by reducing waste and promoting the recovery of valuable materials. At the end of their lifecycle, many conventional heat exchangers are difficult to disassemble, which complicates the recycling process and often results in large amounts of waste. Modern heat exchanger designs, however, emphasize ease of disassembly and the use of recyclable materials. For example, modular designs allow individual components to be removed and recycled separately, ensuring that metals and other materials are properly recovered and reused. This approach not only reduces the volume of waste sent to landfills but also ensures that valuable resources are conserved. The use of easily separable materials, such as metals that can be melted down and repurposed, ensures that heat exchangers can be recycled efficiently.

9.6.2.2.3 Compliance with environmental standards

Adhering to environmental regulations and standards is crucial for minimizing the environmental impact of heat exchangers and ensuring compliance with laws that govern emissions and resource use. Governments and regulatory bodies are increasingly mandating that manufacturers use low-GWP refrigerants and design heat exchangers that meet stricter emissions reduction requirements. High-GWP refrigerants have been a significant contributor to greenhouse gas emissions, and regulations now require the adoption of alternatives with a lower impact on the environment. Heat exchanger manufacturers are also tasked with ensuring their products meet standards such as the EU's Ecodesign Directive or the US DOE's energy efficiency guidelines. Staying informed about regulatory changes and incorporating compliant technologies not only helps organizations avoid penalties but also ensures they are contributing to broader environmental goals. For example, switching to low-GWP refrigerants and reducing emissions through more efficient designs can help companies meet their environmental targets while also reducing their operating costs.

9.7 CONCLUSION

This chapter on the Economic and Environmental Aspects of Heat Exchanger Applications highlighted the crucial interplay between cost-effectiveness and environmental impact in heat exchanger systems. We reviewed how technological advancements, including high-efficiency and eco-friendly designs, can significantly improve both economic performance and environmental sustainability. By comparing traditional and modern heat exchanger technologies, we outlined their respective economic and environmental impacts across different industries, such as power generation and HVAC systems. Additionally, best practices for optimizing economic performance and minimizing environmental impact were discussed, focusing on LCCA, proactive maintenance, and the use of sustainable materials and compliance with regulations. Ultimately, adopting innovative technologies and practices ensures that heat exchangers enhance operational efficiency while supporting broader sustainability objectives.

REFERENCES

1. Kakac, S., & Liu, H. (2002). *Heat Exchangers: Selection, Rating, and Thermal Design*. CRC Press.
2. Incropera, F.P., & DeWitt, D.P. (2002). *Fundamentals of Heat and Mass Transfer*. Wiley.
3. Chen, G., & Hwang, Y (2009). *Energy Efficiency and Sustainable Development: Heat Exchangers in Renewable Energy Systems*. Springer. Kays, W.M., & London, A.L. (1998). *Compact Heat Exchangers*. McGraw-Hill.
4. Khosrow-Pour, M. (Ed.). (2018). *Encyclopedia of Information Science and Technology* (4th ed., Vol. 3, pp. 4383–4391). IGI Global. https://doi.org/10.4018/978-1-5225-2255-3.ch382
5. American Society of Heating, Refrigerating, and Air-Conditioning Engineers. (2017). *HVAC Systems and Equipment*. 4th ed. ASHRAE.
6. Pugh, S. (2009). *Total Cost Management: The Integrated Approach to Project Control*. CRC Press.
7. Smith, P., & Jones, A. (2018). *Energy Efficiency and Return on Investment: A Guide to Evaluating Industrial and Residential Systems*. Wiley.
8. U.S. Department of Energy. (2021). *Tax credits, rebates, and incentives for energy efficiency and renewable energy*. https://www.energy.gov/
9. World Steel Association. (2020). *Sustainability indicators: Greenhouse gas emissions*. https://worldsteel.org/
10. U.S. Environmental Protection Agency. (2021). *Recycling and waste reduction: The benefits of recycling metal*. https://www.epa.gov/
11. American Council for an Energy-Efficient Economy. (2020). *Case studies on energy savings from high-efficiency heat exchangers*. https://www.aceee.org/
12. U.S. Environmental Protection Agency. (2021). *Greenhouse gas emissions and energy efficiency*. https://www.epa.gov/
13. U.S. Department of Energy. (2023). *Heat recovery and ventilation systems*. https://www.energy.gov/
14. Chemical Engineering Journal. (2022). Advancements in Heat Exchanger Technologies. *Chemical Engineering Journal*, 451, 1–12. https://doi.org/10.1016/j.cej.2022.138468
15. International Journal of Thermal Sciences (2022). Performance Evaluation of Shell-and-Tube Heat Exchangers in Power Generation. *International Journal of Thermal Sciences*, 165, 105618.
16. Chakrabarty, D., Patil, P., & Sharma, K. (2017). Experimental Study on Fluid Flow and Heat Transfer from a Rectangular Prism Approaching the Wall of a Wind Tunnel. *Heat Transfer—Asian Research*, 46(7), 793–814. https://doi.org/10.1002/htj.21244.
17. Tiwari, S., Swaminathan, M., Harender, & Singh, D. B. (2022). Performance enhancement of the photovoltaic system with different cooling methods. *Environmental Science and Pollution Research*, 29(30), 45107–45130. https://link.springer.com/article/10.1007/s11356-022-20330-x.
18. Wang, J., Zhang, Y., & Liu, H. (2021). Additive Manufacturing of Heat Exchangers: Advances and Applications. *Journal of Thermal Science and Engineering Applications*, 13(4), 045003. https://doi.org/10.1115/1.4050984

19. U.S. Environmental Protection Agency (EPA). (2022). *Regulations for Greenhouse Gas Emissions from Industrial Facilities.* https://www.epa.gov/
20. Sharma, S. K., Mallick, A., Singh, D. B., & Tiwari, G. N. (2022). Experimental Study of Solar Energy–Based Water Purifier of Single-Slope Type by Incorporating a Number of Similar Evacuated Tubular Collectors. *Environmental Science and Pollution Research*, 29(5), 6837–6856. http://dx.doi.org/10.1007/s11356-021-16123-3
21. Kanouni, B., Badoud, A. E., Mekhilef, S., Bajaj, M., & Zaitsev, I. (2024). Advanced Efficient Energy Management Strategy Based on State Machine Control for Multi-Sources PV-PEMFC Batteries System. *Scientific Reports*, 14(1), 7996. http://dx.doi.org/10.1038/s41598-024-58785-2
22. Awasthi, M.K, Editors (2024). *Computational Fluid Flow and Heat Transfer: Advances, Design, Control & Applications.* CRC Press. https://doi.org/10.1201/9781003465171.
23. Kumar, Ashwani, Dutt, Nitesh, & Kumar Awasthi, Mukesh (2024). *Heat Transfer Enhancement Techniques: Thermal Performance, Optimization and Applications.* Publisher Wiley Scrivener.
24. Kundu, Arijit, & Kumar, Ashwani (2024). An Explorative Study of Earth-Air Heat Exchanger for Sustainable Building Applications. *International Journal of Energy for a Clean Environment.* https://doi.org/10.1615/InterJEnerCleanEnv.2024052513.
25. Dutt, Nitesh (2024). Effect of Duct Length Variation on Solar Air Heater Performance for Smooth and D-Shaped Roughened Absorber Plate. *Heat Transfer.* https://doi.org/10.1002/htj.23115.
26. Dutt, Nitesh (2024). Thermo-Hydraulic Performance Investigation of Solar Air Heater Duct Having Staggered D-Shaped Ribs: Numerical Approach. *Heat Transfer*, 53(3), 1501–1531. https://onlinelibrary.wiley.com/doi/10.1002/htj.22998.
27. Awasthi, MK (2024). Electrohydrodynamic Capillary Instability of Rivlin–Ericksen Viscoelastic Fluid Film With Mass and Heat Transfer. *Heat Transfer*, 53(1), 115–133. https://doi.org/10.1002/htj.22944.
28. Verma, Vikas (2023). Ground and Solar Assisted Heat Pump Systems for Space Heating and Cooling Applications in the Northern Region of India – A Study on Energy and CO_2 Saving Potential. *SETA*, 59, 103405. ISSN 2213-1388. https://doi.org/10.1016/j.seta.2023.103405.
29. Mokhtari, M., Jamei, S., Razaviyan, Z., Heidari, M., Thangavel, S., & Kumar, A. (2024). Numerical and Hydrodynamic Analysis of a Container Vessel due to the Change of Collision Wave Angle at Different Speeds in Regular Wave. *Journal of Mechanical Engineering and Sciences*, 18(4), 10247–10260. https://doi.org/10.15282/jmes.18.4.2024.3.0809.
30. Dutt, Nitesh (2023). Thermo-Hydraulic Performance of Solar Air Heater Having Discrete D-Shaped Ribs as Artificial Roughness. *ESPR.* https://doi.org/10.1007/s11356-023-28247-9.
31. Reddy, P. N. (2023). CFD Simulation and Thermal Performance Optimization of Flow in a Channel With Multiple Baffles. *Journal of Heat and Mass Transfer Research*, 10(2), 257–268. https://doi.org/10.22075/JHMTR.2023.31108.1458.
32. Meena, C. S., Prajapati, A. N., Kumar, A., & Kumar, M. (2022). Utilization of Solar Energy for Water Heating Application to Improve Building Energy Efficiency: An Experimental Study. *Buildings*, 12(12), 2166.
33. Meena, C. S., Kumar, A., Roy, S., Cannavale, A., & Ghosh, A. (2022). Review on Boiling Heat Transfer Enhancement Techniques. *Energies*, 15(15), 5759.
34. Awasthi, M.K. (2024). *Introduction to Mathematical and Computational Methods* (Chapter 1, pp. 01–20). CRC Press. https://doi.org/10.1201/9781003465171-1.
35. Awasthi, M. K. (2024). *Modeling Rayleigh-Taylor instability in Nanofluid Layers* (Chapter 13, pp. 249–262). CRC Press. https://doi.org/10.1201/9781003465171-13.

36. Saxena, Anjali (2023). Water Consumption Optimization of Hybrid Heat Pump Water Heating System. *Lecture Notes in Mechanical Engineering.* Springer. https://doi.org/10.1007/978-981-99-1894-2_61.
37. Kundu, A., Kumar, A., Awasthi, M. K., Dwivedi, G., & Shukla, A. K. (2025). An Explorative Study of Earth-Air Heat Exchanger for Sustainable Building Applications. *International Journal of Energy for a Clean Environment*, 26(3):1–22. https://doi.org/10.1615/InterJEnerCleanEnv.2024052513.
38. Kumar, A. (2024). *Safety and Environmental Considerations in Nanofluid-Based Systems* (Chapter 6, pp. 99–122). CRC Press. https://doi.org/10.1201/9781003494454-6.
39. Manji, A.N.S.H.A. (2024). *A Study on Waste Management Practices in Oman and a Way Forward to a Circular Economy* (Chapter 6, pp. 129–149). CRC Press. https://doi.org/10.1201/9781003502012-6.
40. Razaviyn, Z., Heidari, M., Thangavel, S., Verma, V., Kumar, A., & Yadav, A. K. (2025). Marine Energy Harvesting from Fluid Flow via Vortex Induced Vibrations. *International Journal of Thermofluids*, 25, 101015. https://doi.org/10.1016/j.ijft.2024.101015.
41. Thangavel, S. (2024). *Strategic and Technical Analysis of Sustainable Smart Cities Development in Oman* (Chapter 8, pp. 156–184). CRC Press. https://doi.org/10.1201/9781003496656-8.
42. Chitt, M. (2024). *Building Performance Prediction and Optimization.* CRC Press (Chapter 11, pp. 241–257). https://doi.org/10.1201/9781003496656-11.
43. Kundu, A. (2023). *Modelling and Simulation of Thermal Energy System for Design Optimization* (Chapter 7, pp. 103–140). CRC Press http://dx.doi.org/10.1201/9781003395768-7.
44. Pimoli, D. S., Tomar, P., Kumar, A., Villagómez-Galindo, M., Valencia, M., Dwivedi, G., & Shukla, A. (2025). Experimental Investigation of Friction Factor and Heat Transfer Enhancement Using Circular Perforated Twisted Tape Inserts in Heat Exchangers. *International Journal of Energy for a Clean Environment.* https://doi.org/10.1615/InterJEnerCleanEnv.2024053616.
45. Tomar, Pritosh (2024). Numerical Investigation of Thermal Performance Enhancement of Solar Reservoir Using Flash Cycle. *Journal of Advanced Research in Fluid Mechanics and Thermal Sciences*, 18(4). *123*(1), 197–221. https://doi.org/10.37934/arfmts.123.1.197221
46. Singh, V. P., Jain, S., Karn, A., Dwivedi, G., Kumar, A., Mishra, S., & Kamel, S. (2022). Heat Transfer and Friction Factor Correlations Development for Double Pass Solar Air Heater Artificially Roughened With Perforated Multi-V Ribs. *Case Studies in Thermal Engineering*, 39, 102461. https://doi.org/10.1016/j.csite.2022.102461.
47. Sharma, K. K., Kumar, R., & Jha, P. K. (2024). Physical Modeling and Numerical Investigation of Fluid Flow and Solidification Behavior in a Slab Caster Mold Using Hexa-Furcated Nozzle. *Heat Transfer*, 53(3), 1050–1072. https://doi.org/10.1002/htj.22980.
48. Saxena, A., Cuce, E., Singh, D. B., Sethi, M., Cuce, P. M., Sagade, A. A., & Kumar, A. (2023). Experimental Studies of Latent Heat Storage Based Solar Air Heater for Space Heating: A Comparative Analysis. *Journal of Building Engineering*, 69, 106282. http://dx.doi.org/10.1016/j.jobe.2023.106282.
49. Saxena, A., Joshi, S. K., Gupta, P., Tirth, V., Suryavanshi, A., Singh, D. B., & Sethi, M. (2023). An Experimental Comparative Analysis of the Appropriateness of Different Sensible Heat Storage Materials for Solar Cooking. *Journal of Energy Storage*, 61, 106761. http://dx.doi.org/10.1016/j.est.2023.106761.
50. Bhandari, P., Padalia, D., Ranakoti, L., Khargotra, R., András, K., & Singh, T. (2023). Thermo-Hydraulic Investigation of Open Micro Prism Pin Fin Heat Sink Having Varying Prism Sides. *Alexandria Engineering Journal*, 69, 457–468. https://doi.org/10.1016/j.aej.2023.02.016

Chapter 10

MHD viscous fluid flow with thermal radiation and chemical reactions in a converging channel

Sunmoni Mudoi

10.1 INTRODUCTION

This chapter investigates the flow of a viscous magnetohydrodynamic (MHD) fluid along a converging channel. It focuses on the flow's effects on thermal radiation, heat source, and chemical reaction. The flow behaviour of the MHD fluid is intriguing for many industrial and engineering applications, including electromagnetic pumping, chemical processing, and liquid metal cooling. Understanding the individual and combined impacts of thermal radiation, heat sources, and chemical reactions on the fluid's profiles is crucial since they complicate the flow dynamics. By investigating these phenomena, this study aims to advance the creation of more precise models and effective designs for MHD fluid flow systems.

Thermal radiation describes how energy is transferred using electromagnetic waves that a body emits as a result of its temperature. Thermal radiation is important in many disciplines, including heat transmission and thermodynamics, radiant cooling and heating, astrophysics and cosmology, materials science, and engineering. Thermal radiation plays a crucial role in MHD flows, especially in high-temperature applications like fusion reactors, astrophysical flows, and rocket propulsion (Das 2011; Turkyilmazoglu 2011; Misra and Sinha 2013; Krishna and Kumar 2024). A study by Sudarmozhi et al. (2023) looked into MHD Maxwell Fluid: Double Diffusion with Thermal Radiation, and Chemical Reactions in Porous Medium. Mishra et al. (2024) studied how thermal radiation, chemical reactions, and induced magnetic fields affect MHD flow across permeable media. Ometan et al. (2024) examined the combined impacts of mass and heat transfer for binary chemical processes and thermal radiation in unstable MHD flows. Aliakbar et al. (2009) looked at the impacts of heat radiation on Maxwellian fluid MHD flow. The impact of thermal radiation on an MHD nanofluid flow that is unstable on an infinite vertical flat plate whose temperature increases with heat input were comprehensively investigated by Reddy and Goud (2023).

Heat source is an entity or system that emits thermal energy and raises the temperature of its surroundings. It transmits energy to its surroundings through conduction, convection, or thermal radiation (Reddy et al. 2015;

 DOI: 10.1201/9781003534785-10

Aghalya and Tamizharasi 2024). A heat source is any source of thermal energy that the fluid absorbs or releases to change its temperature and dynamics in the context of MHD fluids (Dessie and Kishan 2014; Raghunatha et al. 2023; Abel and Nandeppanavar 2009). A mathematical study of MHD natural convection in a polar fluid inside two coaxial vertical cylinders was done by Mishra et al. (2024), accounting for heat generation and Hall current effects. Abel and Mahesha (2008) looked at the effects of radiation, a non-uniform heat source, and varying thermal conductivity on heat transmission in MHD viscoelastic fluid across a stretching sheet. An analysis of MHD Casson fluid flow, including heat generation, chemical reactions, and generalised Yang-Abdel-Cattani derivatives, was carried out by Sehra et al. (2023). Ibrahim et al. (2017) investigated effects of a heat source and chemical reaction on a nonlinear permeable stretching sheet during the dissipative MHD mixed convection flow of a Casson nanofluid. An understanding of the heat sources in MHD fluids is essential for a number of applications, including astrophysics and space plasmas; geodynamo and magnetosphere investigations; industrial operations, including chemical processing and metallurgy; and medical applications, including drug administration and cancer treatment.

Chemical species interact with the magnetic and velocity fields of the fluid during chemical reactions in MHD fluids. These reactions can strongly impact the temperature, content, and dynamics of the fluid. Chemical processes in MHD fluids can alter the fluid's composition and properties, change its flow and velocity, affect the dynamics of the magnetic field, change temperature and heat transfer, and affect the fluid's overall behaviour and stability (Li et al. 2023; Oyelami et al. 2023; Afify 2004). The effects of thermal radiation and chemical reactions on the MHD boundary layer flow of a Maxwell fluid across a stretched sheet containing nanoparticles were studied by Kumar and Reddy (2023). Vasa et al. (2024) looked into how heat and mass transfer affect MHD flow across a porous plate that is vertically stacked. They did this while chemical reactions and heat production happened. The impact of thermal radiation on convective heat transfer in an MHD boundary layer of a carrier fluid with chemical processes was examined by Shah et al. (2023). Khan (2014) studied the two-dimensional chemical reaction fluid boundary layer flow across a shaking sheet using injection and suction. Chemical reactions in MHD fluids have several applications, such as environmental clean-up, chemical processing and manufacturing, and fusion energy research.

A fluid that resists deformation and flow due to its resistance to shear stress is referred to as viscous. A fluid's viscosity determines its thickness or flow resistance. Compared to non-viscous fluids like water or air, viscous fluids have a higher viscosity. In MHD, the term "viscous fluid" refers to the behaviour of viscous electrically conducting fluids in the presence of magnetic fields. Viscosity is a crucial component of MHD because it affects the flow of the fluid, the dynamics of the magnetic field, and its overall behaviour (Sajid et al. 2007; Krishna 2020). The finite difference model of time-varying MHD viscous flow was examined by Ibrahim et al. (2023). The conjugate heat transfer of MHD through a vertical permeable plate in a viscous fluid was investigated by Bhargavi et al. (2023). Akbar et al.'s (2021) study used a ciliated tube with entropy generation to investigate the heat transport of a viscous MHD fluid. The effects of deformation-induced work and viscous dissipation on the MHD flow and heat transfer of a viscoelastic fluid across a stretching sheet were investigated by Cortell (2006). In the radiative MHD flow of viscous fluid via a spinning cone, Khan et al. (2020) investigated the key elements of the Dufour and Soret effects, with entropy production implicated. Applications for viscous fluid MHD include space plasmas; industrial operations, such as chemical and metallurgical processing; and biomedical applications.

This research looks at how MHD forces change the flow and heat transfer in a channel that is converging because of chemical reactions, heat sources, and thermal radiation all happening at the same time. Due to its significance for numerous technical applications, including nozzles, diffusers, and mixing devices, the converging channel shape is of considerable importance.

Finding the critical variables that govern flow behaviour and heat transfer properties, as well as gaining a thorough grasp of how these physical phenomena interact, are our two main objectives. The results obtained from this research will have a substantial impact on the understanding of natural phenomena like astrophysical flows and geothermal systems, as well as on the design and optimisation of a variety of engineering systems, including MHD generators, pumps, and flow control devices.

10.2 PROBLEM FORMULATION

Examine a constant, 2-D MHD, viscous, incompressible fluid flow via a convergent channel under the impacts of thermal radiation when a heat source and chemical reaction are present. We take it that the y-axis is normal to the wall and the x-axis is parallel to the converging channel's wall. We apply a magnetic field in reverse direction of the channel. When we supply a magnetic field $B(x)$, we assume a very tiny magnetic Reynolds, which results in a minimal induced magnetic field.

By the assumption, general governing equations of motions are:

$$\frac{\partial u}{\partial x}+\frac{\partial u}{\partial y}=0 \tag{10.1}$$

$$u\frac{\partial u}{\partial x}+v\frac{\partial u}{\partial y}=U\frac{\partial u}{\partial x}+v\frac{\partial^2 u}{\partial y^2}+\frac{\sigma B^2}{\rho}(U-u)+g\beta(T-T_\infty)+g\beta^*(C-C_\infty) \tag{10.2}$$

$$u\frac{\partial T}{\partial x}+v\frac{\partial T}{\partial y}=\frac{k}{\rho C_P}\frac{\partial^2 T}{\partial y^2}+\frac{Q_0}{\rho C_P}(T-T_\infty)-\frac{1}{\rho C_P}\frac{\partial q_r}{\partial y} \tag{10.3}$$

$$u\frac{\partial C}{\partial x}+v\frac{\partial C}{\partial y}=D\frac{\partial^2 C}{\partial y^2}-K_c\left(C-C_\infty\right)+D_T\frac{\partial^2 T}{\partial y^2} \tag{10.4}$$

The boundary conditions:

$$y=0;u=0;v=0;T=T_w;C=C_w \tag{10.5}$$

$$y\to\infty;u\to U\left(x\right);T\to T_\infty,\ C\to C_\infty \tag{10.6}$$

Consider $U(x)=-\frac{a}{x}$, where $a>0$ is the strength of the sink placed at the origin, $B(x)=\frac{B_0}{x}$, here B_0 constant, dimensionless variable $\eta=y\sqrt{-\frac{U(x)}{xv}}=\frac{y}{x}\sqrt{\frac{a}{v}}$, stream function $\psi(x,y)=-\sqrt{va}f(\eta)$ and $u=\frac{\partial\psi}{\partial y}\,and\,v=-\frac{\partial\psi}{\partial x}$.

We present the subsequent parameters and non-dimensional variables:

$$\theta=\frac{T-T_\infty}{T_w-T_\infty},\varphi=\frac{C-C_\infty}{C_w-C_\infty},Gr=\frac{g\beta\left(T_w-T_\infty\right)x^3}{a^2},Gm=\frac{g\beta^*\left(C_w-C_\infty\right)x^3}{a^2}$$

$$M=\frac{\sigma B_0^2}{\rho a},Pr=\frac{\mu C_p}{k},S=\frac{Qx}{\rho C_pU},\gamma=\frac{K_cx}{U},E=\frac{U^2}{C_p\left(T_w-T_\infty\right)},\ Sc=\frac{v}{D}$$

The amended versions of Equations (10.2–10.4) are as follows:

$$f'''+\left(f'\right)^2+Mf'=\left(M+l\right)-Gr\theta-Gm\phi \tag{10.7}$$

$$\left(1+R\right)\theta''+Prf\theta'+PrS\theta=0 \tag{10.8}$$

$$\phi''+Sc\gamma\phi+Sr\theta''=0 \tag{10.9}$$

The dimensionless boundary conditions:

$$\eta=0;f'=0,\ \theta=1,\ \phi=1 \tag{10.10}$$

$$\eta\to\infty;f'\to 1,\theta\to 0,\phi\to 0 \tag{10.11}$$

10.3 METHODS OF SOLUTION

The dimensionless quantities transform Equations (10.1–10.4), with boundary conditions in Equations (10.5–10.6), into non-dimensional Equations (10.7–10.9), with boundary conditions in Equations (10.10–10.11). The bvp4c method in MATLAB can then solve these equations. We can also use this method to find numerical values for the fluid's temperature, velocity, and species concentration. We also analyse the various skin friction, Nusselt and Sherwood numbers.

10.4 RESULTS

An analysis has been conducted on the characteristics of the boundary layer flow via a converging conduit of an electrically conducting MHD viscous fluid when there is simultaneous heat and mass transfer, chemical reaction, viscous dissipation, thermophoresis, and heat source. Figures provide a graphic representation of the pertinent flow properties. We look at the impact of the study's prominent non-dimensional parameters on this problem. We find the solution for parameters with varying values.

Figures 10.1–10.2 depict how f' changes in relation to *Gr and Gm.* When *Gr* and *Gm* increase, it indicates a stronger buoyancy force, potentially leading to an increased fluid velocity, enhanced heat transfer, as well as mass transfer.

Figures 10.3–10.8 display the fluid profile variations for the parameters M and Pr. M and Pr increase the fluid's velocity, as shown in Figures 10.3 and 10.8. The electromagnetic force acting on the fluid grows with M; this may result in increased fluid acceleration, causing the boundary layer to thin and the fluid to move more quickly in the vicinity of the boundaries. The increase in fluid velocity with Pr is due to enhanced electromagnetic forces and momentum transfer, respectively, which can lead to accelerated fluid motion and increased velocity gradients. The fluid's temperature rises as M decreases, but fall as Pr increases. An increase in the magnetic parameter typically increases

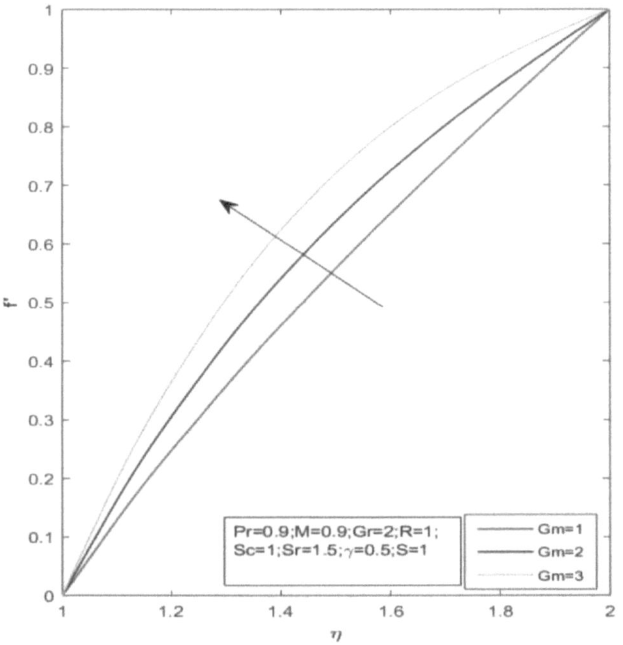

Figure 10.1 Changes of f' with Gm.

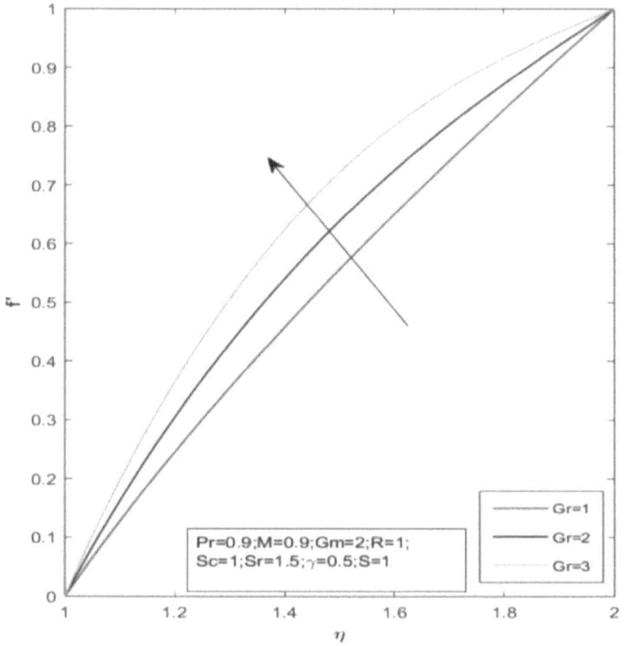

Figure 10.2 Changes of f′ with Gr.

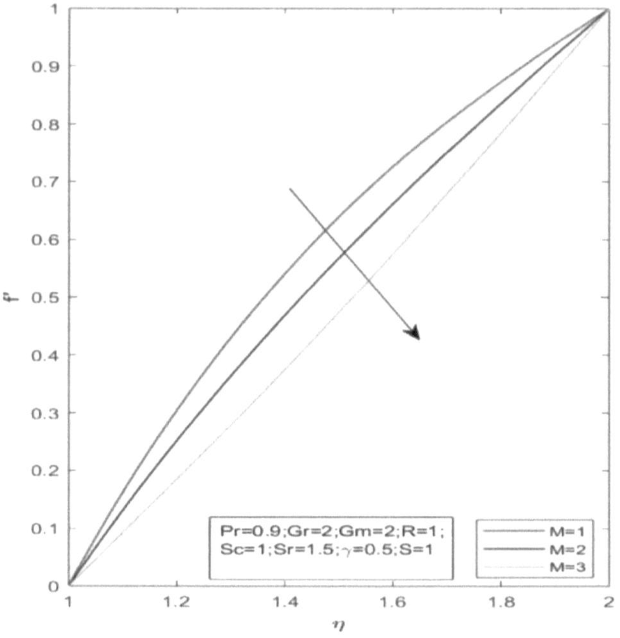

Figure 10.3 Changes of f′ with M.

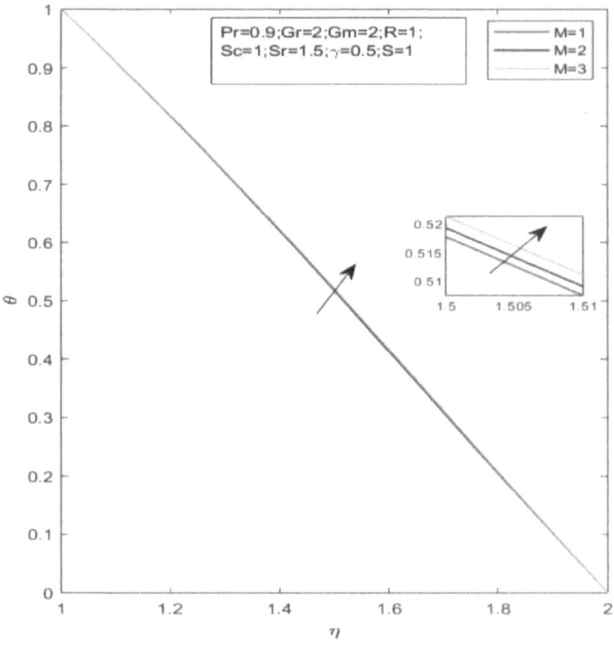

Figure 10.4 Changes of θ with M.

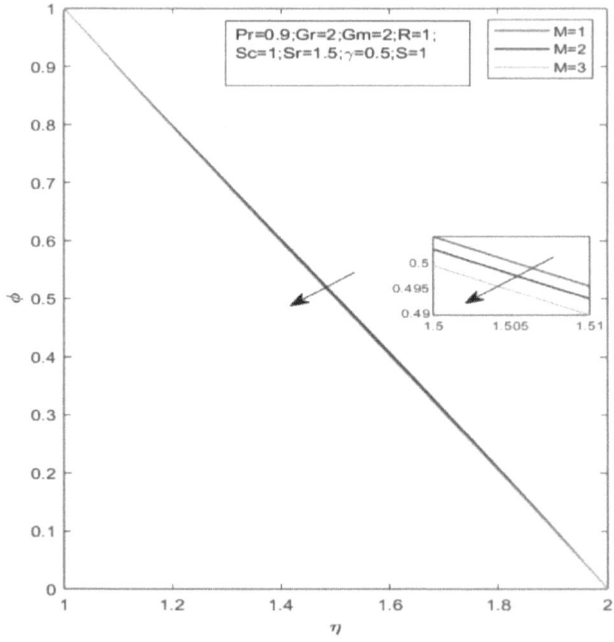

Figure 10.5 Changes of ϕ with M.

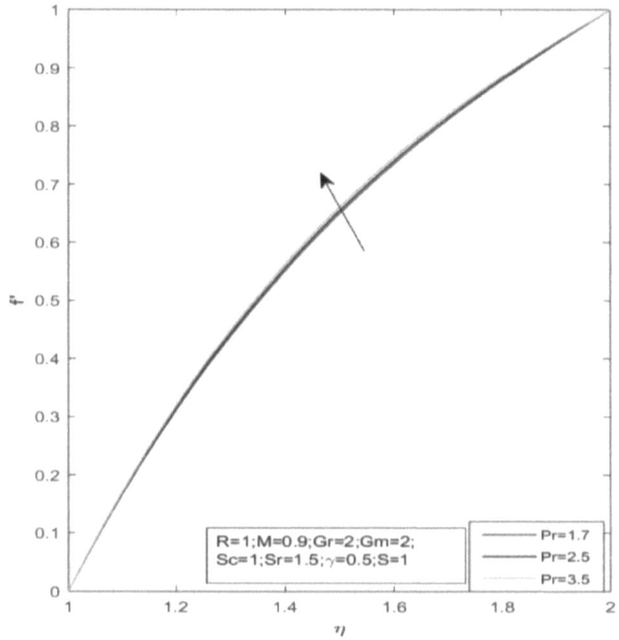

Figure 10.6 Changes of f′ with Pr.

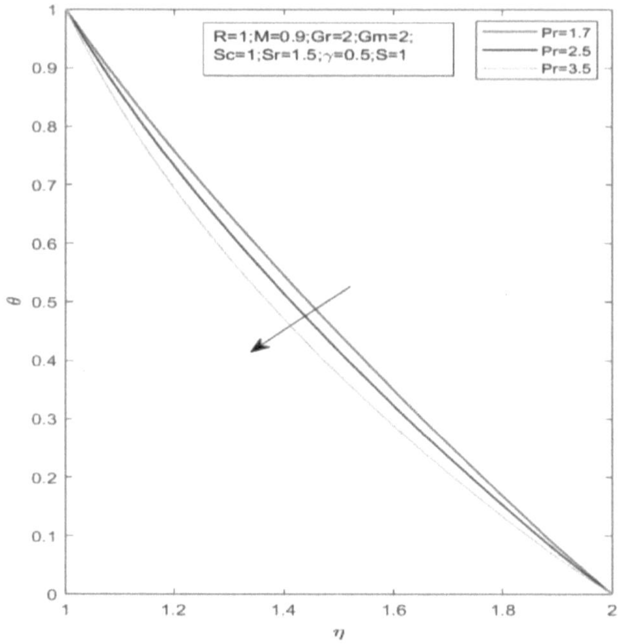

Figure 10.7 Changes of θ with Pr.

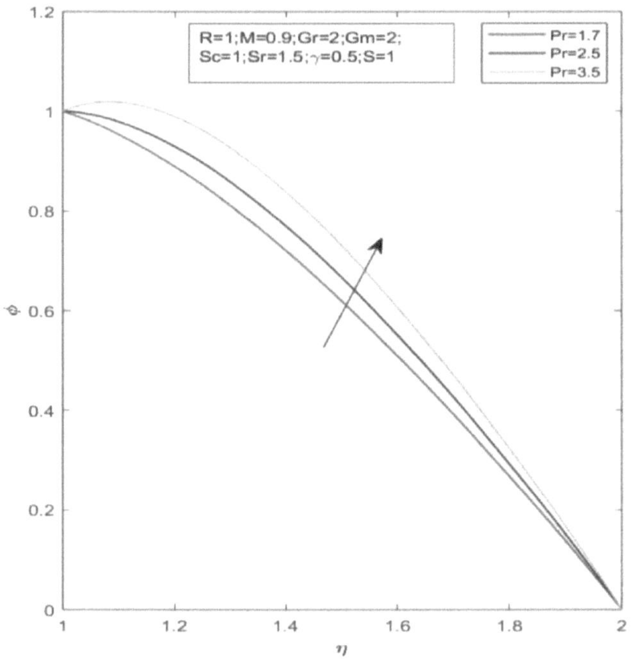

Figure 10.8 Changes of ϕ with Pr.

the temperature of the fluid because the magnetic field induces a Lorentz force, which opposes the fluid motion and generates heat due to viscous dissipation, thereby raising the temperature. A higher Prandtl number means the fluid has a lower thermal diffusivity, which implies that heat diffuses more slowly. As a result, increasing the Prandtl number generally decreases the temperature of the fluid because the thermal boundary layer becomes thinner, allowing less heat to be retained in the fluid. Figures 10.5 and 10.8 illustrate how the fluid's concentration rose with Pr and dropped with M. A higher Pr indicates better heat retention in the fluid since it correlates with lower thermal diffusivity. As a result of the thermal boundary layer's influence on the solutal boundary layer, which raises concentration, this may improve convective flow and raise solute concentration in the fluid. Because of the Lorentz force that M produces, the fluid's particle motion may be dampened, which lowers the solute's diffusion and, ultimately, its concentration.

The elevated concentration profiles for Sc and Sr are shown in Figures 10.9 and 10.10. A decrease in mass diffusivity relative to momentum diffusivity is indicated by an increase in the Schmidt number. Since reduced mass diffusivity restricts the solute particles' ability to spread, larger Schmidt numbers typically result in bigger concentration gradients near the surface and an increase in concentration profiles. Since a greater Soret number denotes a stronger influence of temperature gradients on mass diffusion, increases in

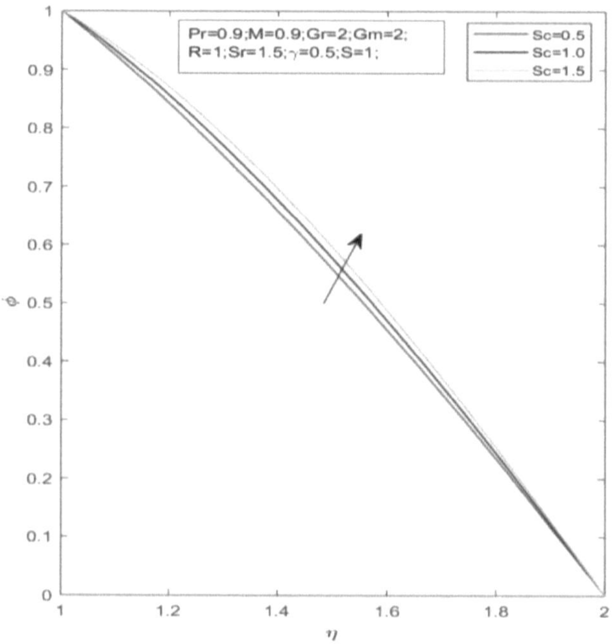

Figure 10.9 Changes of ϕ with Sc.

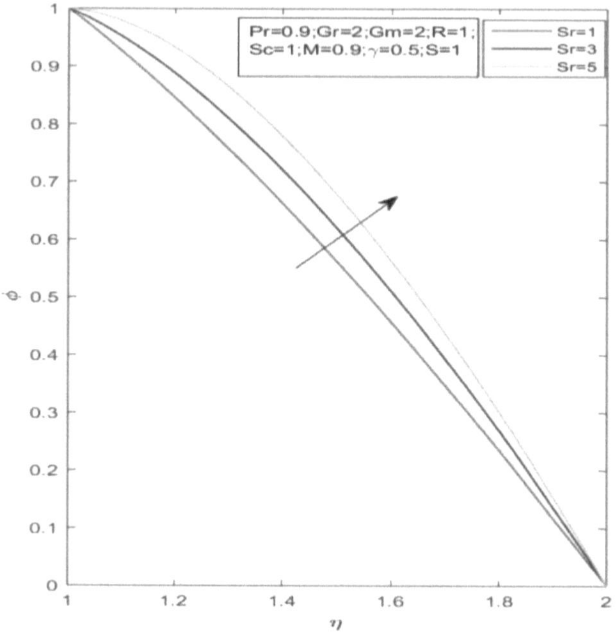

Figure 10.10 Variation of ϕ with Sr.

Table 10.1 Variation of physical parameters with different parameters

Pr	*S*	*R*	*M*	*Sc*	*Sr*	$f''(0)$	$-\theta'(0)$	$-\phi'(0)$
1.7						1.7560	1.2979	0.3654
2.5	2	2	2	1	3	1.7719	1.5044	0.4467
3.5						1.7943	1.8409	0.4706
	1					1.7419	1.1370	0.6144
	2					1.7267	0.9644	0.8819
	3					1.7099	0.7793	1.1688
		1				1.7419	1.1370	0.6144
		2				1.7371	1.0867	0.6926
		3				1.7348	1.0633	0.7289
			1			1.7094	1.1349	0.6176
			2			1.3520	1.1109	0.6537
			3			0.9193	1.0798	0.7005
				0.5		1.7262	1.1354	0.7085
				1.0		1.7419	1.1370	0.6144
				1.5		1.7585	1.1387	0.5168
					1	1.7294	1.1357	0.6868
					3	1.7798	1.1409	0.3895
					5	1.8310	1.1462	0.0712

the Soret number generally improve the concentration profiles. As a result, concentration gradients become more noticeable.

Table 10.1 illustrates how the numerical values of several physical parameters vary across different parameters. According to the table, the coefficient of skin friction rises for *Pr, Sc, and Sr* but falls for *S, R, and M*. There are a number of variables that can affect skin friction when viscous fluids and MHD are involved. Increased skin friction is the result of higher velocities leading to greater shear rates. A magnetic field in MHD can make a fluid more resistant to flowing, which raises skin friction. Skin friction is reduced at lower velocities because shear rates are reduced. Skin friction can be reduced by a weaker pressure gradient since it will exert less force on the fluid as it approaches the boundary. With *Pr, Sc, and Sr*, the Nusselt number value rises, but falls with *S, R, and M*. A greater Nusselt number indicates increased convective heat transfer, which can lead to faster heat transfer rates, improved heat exchanger performance, and more efficient cooling or heating. A lower Nusselt number indicates reduced convective heat transfer, which can lead to slower heat transfer rates, worse heat exchanger performance, and less efficient cooling or heating. For *Pr, S, R, and M*, the coefficient of Sherwood number increases; for *Sc and Sr*, it decreases. A greater Sherwood number indicates greater convective mass transfer, which leads to faster mass transfer rates and improved chemical reactions. A greater Sherwood number may indicate a thinner boundary layer, leading to a lower mass transfer resistance and an enhanced mass transfer coefficient. A lower Sherwood number indicates slower mass transfer rates due to reduced convective mass transfer.

10.5 CONCLUSIONS

It appears that you are interested in the properties of viscous fluid flow along a converging channel while taking chemical reactions, heat sources, and thermal radiation into account. Future research on this topic might look into the best channel geometry and flow conditions to achieve particular performance metrics, like minimum pressure drop or maximum heat transfer; how multiple immiscible fluids or fluid-particle mixtures behave in converging channels, with implications for chemical processing and biomedical engineering; how converging channels affect the yield and kinetics of chemical reactions, with implications for process intensification and miniaturisation; and how to construct more precise and effective numerical schemes to simulate these intricate flows, incorporating sophisticated turbulence models, radiation transport, and chemical kinetics, among other things.

The principal findings of this study are:

1. Velocity profiles of the flow increase with Gr, Gm, and Pr, but reduces with M, R, and S.
2. Temperature of the fluid rises by M, R, and S, but falls down with Pr.
3. Concentration of the fluid grows with Pr, Sc, and Sr, but reduces with M, R, and S.
4. The coefficient of skin friction rises for Pr, Sc, and Sr, but falls for S, R, and M.
5. With Pr, Sc, and Sr, the Nusselt number value rises, but falls with S, R, and M.
6. For Pr, S, R, and M, the coefficient of Sherwood number increases; for $Sc\,and\,Sr$, it decreases.

NOMENCLATURE

$U(x)$	the velocity of potential flow
Cp	specific heat
σ	is electric conductivity
$(u\ v)$	velocity components along the $x\,and\,y\,axes$
K_C	is the rate of chemical reaction
$T_w\,and\,T_\infty$	temperatures
$C_w\,and\,C_\infty$	concentrations
C	dimensional species concentration
T	dimensional species temperature
S	Heat source parameter

ρ	is the fluid density
R	radiation parameter
M	Hartmann Number
Sc	Schimdt number
ν	is the kinematic viscosity
Gm	Grashof number for mass transfer
D	coefficient of chemical molecular diffusivity
κ	is the thermal conductivity
Gr	Grashof number for heat transfer
Q	heat generation coefficient
Sr	Soret Number

BIBLIOGRAPHY

M. S. Abel and N. Mahesha, "Heat transfer in MHD viscoelastic fluid flow over a stretching sheet with variable thermal conductivity, non-uniform heat source and radiation," *Applied Mathematical Modelling*, vol. 32, no. 10, pp. 1965–1983, Oct. 2008, https://doi.org/10.1016/j.apm.2007.06.038.

M. S. Abel and M. M. Nandeppanavar, "Heat transfer in MHD viscoelastic boundary layer flow over a stretching sheet with non-uniform heat source/sink," *Communications in Nonlinear Science and Numerical Simulation*, vol. 14, no. 5, pp. 2120–2131, May 2009, https://doi.org/10.1016/j.cnsns.2008.06.004.

Ahmed A. Afify, "MHD free convective flow and mass transfer over a stretching sheet with chemical reaction," *Heat and Mass Transfer*, vol. 40, no. 6–7, Feb. 2004, https://doi.org/10.1007/s00231-003-0486-0.

T. Aghalya and R. Tamizharasi, "Effect of heat source/sink and thermal radiation on an unsteady MHD Casson flow past over an exponentially accelerated vertical plate," *Journal of Thermal Analysis and Calorimetry*, Mar. 2024, https://doi.org/10.1007/s10973-024-12950-x.

Noreen Sher Akbar, S. Akhtar, E. N. Maraj, A. E. Anqi and R. Z. Homod, "Heat transfer analysis of MHD viscous fluid in a ciliated tube with entropy generation," *Mathematical Methods in the Applied Sciences*, vol. 46, no. 10, pp. 11495–11508, Nov. 2021, https://doi.org/10.1002/mma.7906.

V. Aliakbar, A. Alizadeh-Pahlavan and K. Sadeghy, "The influence of thermal radiation on MHD flow of Maxwellian fluids above stretching sheets," *Communications in Nonlinear Science and Numerical Simulation*, vol. 14, no. 3, pp. 779–794, Mar. 2009, https://doi.org/10.1016/j.cnsns.2007.12.003.

N. Bhargavi, T. Poornima and Basma Souayeh, "Magnetohydrodynamic conjugate heat transfer analysis on a viscous fluid past a vertical permeable plate," *International Journal of Modern Physics B/International Journal of Modern Physics b*, Jun. 2023, https://doi.org/10.1142/s0217979224502114.

R. Cortell, "Effects of viscous dissipation and work done by deformation on the MHD flow and heat transfer of a viscoelastic fluid over a stretching sheet," *Physics Letters*, vol. 357, no. 4–5, pp. 298–305, Sep. 2006, https://doi.org/10.1016/j.physleta.2006.04.051.

K. Das, "Impact of thermal radiation on MHD slip flow over a flat plate with variable fluid properties," *Heat and Mass Transfer*, vol. 48, no. 5, pp. 767–778, Oct. 2011, https://doi.org/10.1007/s00231-011-0924-3.

H. Dessie and N. Kishan, "MHD effects on heat transfer over stretching sheet embedded in porous medium with variable viscosity, viscous dissipation and heat source/sink," *Ain Shams Engineering Journal*, vol. 5, no. 3, pp. 967–977, Sep. 2014, https://doi.org/10.1016/j.asej.2014.03.008.

M. Ibrahim, E. Al Sarairah, H. Hejazi and T. Saeed, "Finite difference scheme of time-dependent MHD viscous fluid flow," *International Journal of Modern Physics B*, Apr. 2023, https://doi.org/10.1142/s0217979224501376.

S. M. Ibrahim, G. Lorenzini, P. Vijaya Kumar and C. S. K. Raju, "Influence of chemical reaction and heat source on dissipative MHD mixed convection flow of a Casson nanofluid over a nonlinear permeable stretching sheet," *International Journal of Heat and Mass Transfer*, vol. 111, pp. 346–355, Aug. 2017, https://doi.org/10.1016/j.ijheatmasstransfer.2017.03.097.

S. A. Khan, T. Hayat, M. Ijaz Khan and A. Alsaedi, "Salient features of Dufour and Soret effect in radiative MHD flow of viscous fluid by a rotating cone with entropy generation," *International Journal of Hydrogen Energy*, vol. 45, no. 28, pp. 14552–14564, May 2020, https://doi.org/10.1016/j.ijhydene.2020.03.123.

Y. Khan, "Two-dimensional boundary layer flow of chemical reaction MHD fluid over a shrinking sheet with suction and injection," *Journal of Aerospace Engineering*, vol. 27, no. 5, Sep. 2014, https://doi.org/10.1061/(asce)as.1943-5525.0000274.

M. Veera Krishna, "Hall and ion slip effects on MHD laminar flow of an elastico-viscous (Walter's-B) fluid," *Heat Transfer*, vol. 49, no. 4, pp. 2311–2329, Mar. 2020, https://doi.org/10.1002/htj.21722.

M. Veera Krishna and A.G. Vijaya Kumar, "Chemical reaction, slip effects and non-linear thermal radiation on unsteady MHD Jeffreys nanofluid flow over a stretching sheet," *Case Studies in Thermal Engineering*, pp. 104129–104129, Feb. 2024, https://doi.org/10.1016/j.csite.2024.104129.

M. Anil Kumar and Y. Dharmendar Reddy, "Thermal radiation and chemical reaction influence on MHD boundary layer flow of a Maxwell fluid over a stretching sheet containing nanoparticles," *Journal of Thermal Analysis and Calorimetry*, vol. 148, no. 13, pp. 6301–6309, Apr. 2023, https://doi.org/10.1007/s10973-023-12097-1.

S. Li *et al.*, "Effects of activation energy and chemical reaction on unsteady MHD dissipative Darcy–Forchheimer squeezed flow of Casson fluid over horizontal channel," *Scientific Reports*, vol. 13, no. 1, p. 2666, Feb. 2023, https://doi.org/10.1038/s41598-023-29702-w.

M. Mishra, J. P Panda and S. S. Sahoo, "Investigations concerning the effects of thermal radiation, induced magnetic field, and chemical reaction on MHD flow through a permeable medium," *International Journal of Modern Physics B/International Journal of Modern Physics b*, vol. 38, no. 02, Mar. 2023, https://doi.org/10.1142/s0217979224500309.

M. Mishra, J. P. Panda and S. S. Sahoo, "An analytical investigation of MHD natural convective flow of a polar fluid in two vertical concentric cylinders with hall current and heat source," *International Journal of Applied and Computational Mathematics*, vol. 10, no. 2, Mar. 2024, https://doi.org/10.1007/s40819-024-01714-8.

J. C. Misra and A. Sinha, "Effect of thermal radiation on MHD flow of blood and heat transfer in a permeable capillary in stretching motion," *Heat and Mass Transfer*, vol. 49, no. 5, pp. 617–628, Mar. 2013, https://doi.org/10.1007/s00231-012-1107-6.

S. O. Ometan, M. O. Alabi and A. M. Okedoye, "Analyzing the impact of heat and mass transfer on unsteady MHD flow with thermal radiation and binary chemical reaction," *European Journal of Theoretical and Applied Sciences*, vol. 2, no. 3, pp. 267–280, May 2024, https://doi.org/10.59324/ejtas.2024.2(3).23.

H. Oyelami, None Bidemi Olumide Falodun, N. Ebenezer and None Olaide Yetunde Saka-Balogun, "Numerical study of chemical reaction and magnetic field effects on MHD boundary layer flow over a flat plate," *CFD Letters*, vol. 16, no. 3, pp. 55–68, Dec. 2023, https://doi.org/10.37934/cfdl.16.3.5568.

K. R. Raghunatha, Y. Vinod, Suma Nagendrappa Nagappanavar and None Sangamesh, "Unsteady Casson fluid flow on MHD with an internal heat source," *Journal of Taibah University for Science*, vol. 17, no. 1, Oct. 2023, https://doi.org/10.1080/16583655.2023.2271691.

M. Gnaneswara Reddy, P. Padma and B. Shankar, "Effects of viscous dissipation and heat source on unsteady MHD flow over a stretching sheet," *Ain Shams Engineering Journal*, vol. 6, no. 4, pp. 1195–1201, Dec. 2015, https://doi.org/10.1016/j.asej.2015.04.006.

Y. Dharmendar Reddy and B. Shankar Goud, "Comprehensive analysis of thermal radiation impact on an unsteady MHD nanofluid flow across an infinite vertical flat plate with ramped temperature with heat consumption," *Results in Engineering*, vol. 17, pp. 100796–100796, Mar. 2023,https://doi.org/10.1016/j.rineng.2022.100796.

M. Sajid, T. Javed and T. Hayat, "MHD rotating flow of a viscous fluid over a shrinking surface," *Nonlinear Dynamics*, vol. 51, no. 1–2, pp. 259–265, Feb. 2007, https://doi.org/10.1007/s11071-007-9208-3.

None Sehra, H. Sadia, Sami Ul Haq and I. Khan, "Time fractional Yang–Abdel–Cattani derivative in generalized MHD Casson fluid flow with heat source and chemical reaction," *Scientific Reports*, vol. 13, no. 1, Oct. 2023, https://doi.org/10.1038/s41598-023-43630-9.

A. Shah, A. Hassan, Hanen Karamti, Abdullah Alhushaybari, S. M. Eldin and A. M. Galal, "Effect of thermal radiation on convective heat transfer in MHD boundary layer Carreau fluid with chemical reaction," *Scientific Reports*, vol. 13, no. 1, Mar. 2023, https://doi.org/10.1038/s41598-023-31151-4.

K. Sudarmozhi, D. Iranian, I. Khan, J. Alzahrani, A. s Al-johani and S. M. Eldin, "Double diffusion in a porous medium of MHD Maxwell fluid with thermal radiation, heat generation and chemical reaction," *Case Studies in Thermal Engineering*, vol. 43, pp. 102700–102700, Mar. 2023, https://doi.org/10.1016/j.csite.2023.102700.

M. Turkyilmazoglu, "Thermal radiation effects on the time-dependent MHD permeable flow having variable viscosity," *International Journal of Thermal Sciences*, vol. 50, no. 1, pp. 88–96, Jan. 2011, https://doi.org/10.1016/j.ijthermalsci.2010.08.016.

N. Vasa, N. Raja and B. Shankar, "heat and mass transfer significance on MHD flow over a vertical porous plate in the presence of chemical reaction and heat generation," *CFD Letters*, vol. 16, no. 5, pp. 9–20, Jan. 2024, https://doi.org/10.37934/cfdl.16.5.920.

Chapter 11

Numerical analysis of magnetised forced convective micropolar Casson hybrid nanofluid flow over a vertical stretching surface

K. Gnanaprasanna, R. Akshaya, C. Nirmala, and Abhishek Kumar Singh

11.1 INTRODUCTION

Magnetohydrodynamics involves the theory that combines the study of fluid dynamics and electromagnetism. The theory of laminar flow of an electrically conducting fluid in a homogenous magnetic field was pioneered by Hartmann who proposed that fluid flowing across a magnetic field is capable of producing electric current that is capable of influencing the flow behaviour and also the temperature of the fluid. A fluid which possesses zero shear rate and infinite viscosity properties is termed as Casson fluid. This non-Newtonian fluid physically possesses visco elastic shear thinning property and due to this apparent viscous nature, it increases the rate of shear stress. The major applications include nuclear reactor, fire and combustion modelling electronic chip and semiconductor wafers.

Ahmed et al [1] investigated the thermally stratified single and multi-walled carbon nanotube of micropolar fluid over vertical plate and revealed that temperature variations are enhanced for hybrid nanofluid and the velocity profiles show an inverse trend for varying values of micropolar parameter and nanoparticle concentrations. Gorla et al [2] utilised fourth-order Runge-Kutta (RK) method and presented the boundary layered with micropolar fluid flow over a flat plate for wide range of Prandtl number ($10 \leq Pr \leq 1000$) and material parameter K ($0 \leq K \leq 10$), and found that the thermal boundary layer thickness has an opposite trend for Pr. Chen et al [3] investigated the numerical solutions of mixed convective flows over stretching sheet and formulated the heat transfer effects of exponential forms of velocity, temperature parameter, buoyancy parameter and Prandtl number using finite difference method. Spectral local linearisation method subject to Casson-Williamson fluid flows over Riga surface with generalised Fick's and Fourier law with thermal radiative and variable thermophysical properties were studied by Tijani et al [4]. Their overall study concluded that Willamson fluid has stronger magnetic influence and high thermal conductivity and diffusivity when compared to Casson fluid. The effects of buoyancy-induced MHD flow with non-linear thermal radiation over exponentially stretching sheet using efficient shooting method and fourth-order RK method have been studied by Rajput et al. [5]. The study revealed

DOI: 10.1201/9781003534785-11

that for increased values of permeability parameter, thermal radiation parameter and Eckert number, the temperature profile is raised. Stagnation point flows of MHD micropolar hybrid nanofluid in a flat sheet with thermal radiation, suction-injection and Joule heating effects were presented by Showkat et al [6]. Water-based magnetohydrodynamic Blasius flows over vertical surface is investigated by Farizza et al [7] and proved that the incremented values of magnetic parameter along with volume concentration of copper-alumina nanoparticles increased the skin friction and heat transfer coefficient values.

Gorla et al [8] revealed that the material property parameter as well as magnetic parameter increased the temperature profile for buoyancy-induced magnetic forces of micropolar fluid moving in a hot vertical plate. Hosam et al [9] modelled the comparative study of dusty micropolar ironoxide – silver hybrid nanofluid flows over stretching sheet and delivered that temperature and velocity variations are larger for magnetic parameter and Eckert number. The combined free and forced convective asymptotic micropolar fluid flow over semi-infinite flat plate were studied by Gorla et al [10]. The numerical results using fourth-order RK method revealed that both drag coefficient and surface heat transfer values are reduced for micropolar fluids. The effects of buoyancy and viscous dissipative mixed convective boundary layer flow using fifth-order RK Fehlberg integration method was solved numerically by Abd et al [11]. The study has its applications in polymer extrusions, hot rolling and casting metals, etc. The influences of comparative analysis of SWCNT and MWCNT of micropolar hybrid nanofluid flow over exponentially stretching Riga surface using bvp4c numerical method is analysed by Abbas et al [12]. The impacts of chemically radiative stretching/shrinking ternary micropolar fluid flow using incomplete gamma functions method analytically was discussed by Vanitha et al [13]. Waini et al [14] proposed the micropolar magnetite nanofluid flow of stagnation region in a stretching-shrinking sheet using bvp4c MATLAB software. Mahdy et al [15] investigated the influences of thermal radiations of hybrid micropolar nanofluid flow in a porous saturated interface by applying Chebyshev differential quadrature method and concluded that for hybrid nanofluid, the internal heat generation is enhanced. Maryam et al [16] presented the comparative study of copper-water (nanofluid) and copper-titanium oxide water (hybrid nanofluid) involving micropolar fluid flow in an exponentially stretching surface and revealed rate of heat transfer for hybrid nanofluid is larger when compared to nanofluid for varying values of micropolar, porosity and rotation parameters.

James et al [17] highlighted a detailed review on major thermophysical properties of various hybrid nanofluids and concluded that bio nanofluid possesses low viscosity and GNP-Ag hybrid nanofluid has high thermal conductivity. Gorla et al [18] developed the mixed convective micropolar fluid flow from a semi-infinite plate by considering isothermal boundary conditions with constant surface heat flux for varying values of material properties and Prandtl number numerically. Mohammadien et al [19] studied the heat transfer behaviour of temperature and wall heat flux across stretching surface of

Table 11.1 The thermal and physical properties of ethylene glycol and Cu – Al_2O_3 Casson hybrid nanofluid properties

Properties	*Ethylene glycol*	*Cu*	*Al_2O_3*
$C_p(Jkg^{-1}K^{-1})$	2400	385	765
$\rho(kgm^{-3})$	1110	8933	3970
$k(Wm^{-1}K^{-1})$	0.26	400	40
$\sigma(sm^{-1})$	1.07×10^{-4}	59.6×10^{6}	35×10^{6}

Table 11.2 The comparison table for varying values of micropolar parameter Δ of present results with available work in the literature for case of steady state with $Pr = 10, \beta = 0, \epsilon = 0, \phi = 0; \nu = 0, \lambda = 1.5, B_0 = 0, q_r = 0, Q_0 = 0$ for $G_\eta(0)$ at $\eta = 0$

Δ	*0.0*	*0.01*	*0.15*	*0.5*	*5.5*
Gorla (1983)	0.7085	0.88138	0.92380	0.99967	1.11470
Gorla (1990)	0.7171	0.81640	0.90271	0.96042	1.17800
Gorla (2000)	0.7214	0.85656	0.91552	0.97032	1.15410
Present results	0.7046	0.88262	0.90087	0.98971	1.20616

micropolar fluid flow and concluded that boundary layer thickness and angular velocity increases for increasing values of micro-rotation parameter. Khan et al [20] investigated the buoyancy effects of water-based silver-titanium oxide micropolar hybrid nanofluid flow over a permeable vertical sheet using bvp4c MATLAB software.

Motivated by the above literature, the present work majorly focusses on micropolar fluid flow over vertical stretching surface under the influence of external magnetic field and Joule heating effects of copper-alumina hybrid nanoparticles. Non-Newtonian fluid flow through which heat and mass transfer occur simultaneously are applied in drying, distribution of temperature, moisture of agriculture crops due to freezing, evaporation of surface of a water body, energy transfer of a wet cooling tower and flow in a desert cooler. The thermal and physical properties of hybrid nanoparticles are given in Table 11.1. A comparison table for fixed values of buoyancy parameter λ, Prandtl number and varying values of micropolar material parameter Δ with previous existing results found to be in good agreement and is given in Table 11.2.

11.2 MATHEMATICAL FORMULATION

Consider a steady laminar forced convective two-dimensional non-Newtonian micropolar hybrid nanofluid flow over a vertical stretching surface. The thermal characteristics of CuO-Al_2O_3 suspended in ethylene glycol (basefluid) are considered. The x-axis is taken towards the moving plate with no slip boundary condition and velocity assumed to be $v = v_w$ along the y-axis in Figure 11.1. The momentum equation is associated with the effects of viscous dissipation,

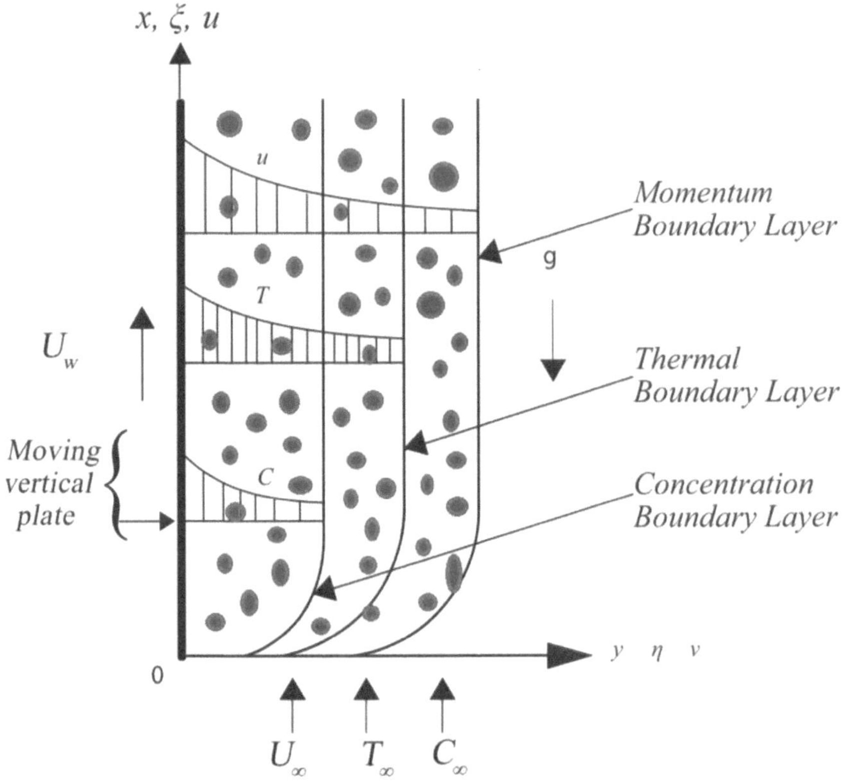

Figure 11.1 Physical description of the model.

micro-rotation, non-Newtonian parameter β and external magnetic field. The energy equation is associated with the effects of variable thermal conductivity, chemical reaction, radiative heat flux, Joule heating effect and viscous dissipation. The mentioned assumptions with Boussinesq's approximation the governing equations based on prior literature [2] are as follows:

The continuity equation:

$$\frac{\partial u}{\partial x}+\frac{\partial v}{\partial y}=0 \tag{11.1}$$

The momentum equation:

$$\left(u\frac{\partial u}{\partial x}+v\frac{\partial u}{\partial y}\right)=\left(u_e\frac{du_e}{dx}\right)+\frac{1}{\rho_{hnf}}\left(1+\frac{1}{\beta}\right)\frac{\partial}{\partial y}\left(\mu_{hnf}+S\right)\frac{\partial u}{\partial y}+\frac{S}{\rho_{hnf}}\frac{\partial N_k}{\partial y} \\ -\frac{\sigma_{hnf}}{\rho_{hnf}}B_0^2\left(u-u_e\right) \tag{11.2}$$

The micropolar equation for angular momentum:

$$!\left(u\frac{\partial N_k}{\partial x}+v\frac{\partial N_k}{\partial y}\right)=\frac{v_{hnf}}{\rho_{hnf}j}\frac{\partial^2 N_k}{\partial y^2}-\frac{S}{\rho_{hnf}j}\left(2N_k+\frac{\partial u}{\partial y}\right) \tag{11.3}$$

The energy equation:

$$\left(u\frac{\partial T}{\partial x}+v\frac{\partial T}{\partial y}\right)=\frac{1}{\left(\rho c_p\right)_{hnf}}\frac{\partial}{\partial y}\left(k_{hnf}\frac{\partial T}{\partial y}\right)+\frac{Q_0}{\left(\rho c_p\right)_{hnf}}(T-T_\infty)-\frac{1}{\left(\rho c_p\right)_{hnf}}\left(\frac{\partial q_r}{\partial y}\right)$$
$$+\frac{\sigma_{hnf}}{\left(\rho c_p\right)_{hnf}}B_0^2\left(u-u_e\right)^2+\frac{\mu_{hnf}}{\left(\rho c_p\right)_{hnf}}\left[\left(\frac{\partial u}{\partial y}\right)^2+\left(\frac{\partial w}{\partial y}\right)^2\right] \tag{11.4}$$

with boundary conditions

$$u=u_w\left(x\right);v=0;\ T=T_w\left(x\right);\ \text{at } y=0 \tag{11.5}$$

$$u\to u_\infty;v\to v_\infty;T\to T_\infty;\ \text{at } y\to y_\infty \tag{11.6}$$

The non-similar transformations:

$$\xi=\frac{x}{L};\eta=\sqrt{\frac{u_e}{v_\infty x}};\ N_k=\frac{u_e}{2v_\infty x}^{1/2}g(\eta);\ j=\frac{hv_\infty}{u_e};\ \lambda=\frac{v_{hnf}}{\rho_\infty j v_\infty}$$

$$\psi(x,y)=\left(u_e xv\right)^{1/2}f;\ u=\frac{\partial\psi}{\partial y};$$

$$v=-\frac{\partial\psi}{\partial x};\ f_\eta=F;\ \theta(\xi,\eta)=\frac{\left(T-T_\infty\right)}{\left(T_w-T_\infty\right)}$$

$$N_k=u_e\sqrt{\frac{u_e}{2v_\infty x}}g(\xi,\eta);\ \Delta=\frac{S}{\rho_\infty v_\infty}$$

$$u=u_e f_\eta;\ \epsilon=\frac{U_\infty(x)}{U_w(x)+U_\infty(x)};\ Re=\frac{u_\infty L}{v_\infty}$$

$$\beta=\mu_a\frac{\sqrt{2\pi}}{P_y};\ v=\left(-vu_e\right)^{1/2}x^{-1/2}\left[\frac{(s+1)}{2}f+\frac{(s+1)}{2}\xi f_\xi+\frac{(s-1)}{2}\eta f_\eta\right]$$

$$v_\infty=\frac{\mu_{bf}}{\rho_{bf}};\ Pr=\frac{\mu_\infty}{Cpk_\infty};\ Ec=\frac{u_e^2\left(T_w-T_\infty\right)}{\left(C_p\right)_{bf}};\ St=\frac{B_0^2L\sigma_{bf}}{\rho_{bf}u_\infty}R_1=\frac{16\sigma^* T^3}{3K^* K_\infty};$$

The conventional Casson base fluid and hybrid nanofluid for various physical parameters for dilute suspensions of copper and aluminium oxide nanoparticles whose base fluid is ethylene glycol are as follows:

$$\text{Viscosity: } (D_1) = \frac{\mu_{hnf}}{\mu_{bf}}$$

$$\frac{\mu_{hnf}}{\mu_{bf}} = \frac{\mu_\infty}{(1+a_1\theta)(1-\phi)^{2.5}}$$

$$\text{Density: } (D_2) = \frac{\rho_{hnf}}{\rho_{bf}}$$

$$\frac{\rho_{hnf}}{\rho_{bf}} = (1-\phi) + \frac{\rho_{Cu}\phi_{Cu}}{\rho_{bf}}$$

$$\text{Electrical conductivity: } (K_1) = \frac{\sigma_{hnf}}{\sigma_{bf}}$$

$$\frac{\sigma_{hnf}}{\sigma_{bf}} = 1 + \left[\frac{3\phi\left(\psi_1 - \phi\sigma_{bf}\right)}{\psi_2 - \left(\psi_1 - \phi\sigma_{bf}\right)\phi}\right]$$

where

$$\psi_1 = (\sigma\phi)_{Cu} + (\sigma\phi)_{Al_2O_3}; \quad \psi_2 = \psi_1 + 2\phi\sigma_{bf}$$

$$\psi_3 = (\phi k)_{Cu} + (\phi k)_{Al_2O_3}; \quad \psi_4 = 2\phi\sigma_{bf}$$

$$\text{Specific heat capacity: } (D_4) = \frac{(\rho C_p)_{hnf}}{(\rho C_p)_{bf}}$$

$$\frac{\left(\rho C_p\right)_{hnf}}{\left(\rho C_p\right)_{bf}} = (1-\phi) + \frac{\left(\rho C_p\right)_{Cu}\phi_{Cu}}{\left(\rho C_p\right)_f} + \frac{\left(\rho C_p\right)_{Al_2O_3}\phi_{Al_2O_3}}{\left(\rho C_p\right)_f}$$

$$\text{Thermal conductivity: } (D_5) = \frac{k_{hnf}}{k_{bf}}$$

$$\frac{k_{hnf}}{k_{bf}} = \frac{2k_{bf} + \frac{\psi_3}{\phi} - 2\left(\phi k_{bf} - \psi_3\right)}{2k_{bf} + \phi k_{bf} + \left(\frac{\psi_3}{\phi} - \psi_3\right)}$$

where

$$\psi_3 = \frac{\phi_{Cu}k_{Cu} + \phi_{Al_2O_3}k_{Al_2O_3}}{\phi_{Cu} + \phi_{Al_2O_3}}; \phi = \phi_{Cu} + \phi_{Al_2O_3}; k_{bf} = k_\infty\left(1 + b_1\theta\right)$$

$$\frac{k_{bnf}}{k_{bf}} = \frac{x_1\theta + x_2}{x_3\theta + x_4}$$

$$x_1 = 2\phi k_\infty b_1 - 2\phi^2 k_\infty b_1;\ x_2 = 2\phi k_\infty + \psi_3 - 2\phi^2 k_\infty + 2\phi\psi_3$$

$$x_3 = 2\phi k_\infty b_1 + \phi^2 k_\infty b_1;\ x_4 = 2\phi k_\infty + \phi^2 k_\infty + \psi_3 + \psi_3\phi$$

Applying the congenial non-similar transformations to Equations (11.2–11.4), the dimensionless form of the equations are as follows:

$$2sF^2 - 1 + 2\xi FF_\xi - (s+1)fF_\eta - 2\xi f_\xi F_\eta = 2s - 2\left[\frac{K_1}{D_1}\xi\frac{(F-1)St}{\left(1-\epsilon e^{\xi}\right)}\right]$$

$$+\left[\frac{1}{D_1 D_2}\left(1+\frac{1}{\beta}\right)\frac{F_{\eta\eta}}{(1+a_1\theta)}\right] + \frac{1}{D_1}\left(1+\frac{1}{\beta}\right)\Delta F_\eta + \frac{\Delta}{N_1}g_\eta$$

$$3mgF - gF - (s+1)fg_\eta - 2\xi f_\xi g_\eta = \frac{\lambda}{D_1}g_{\eta\eta} - \frac{2g\Delta}{D_1} - \frac{\Delta}{D_1}F_\eta$$

$$\xi F\theta_\xi - \xi f_\xi\theta_\eta - \left(\frac{s+1}{2}\right)f\theta_\eta = \frac{1}{D_4}\left[\frac{b_1\theta_\eta^2 x_5}{Pr}\right] + \frac{1}{D_4}\left[\frac{(1+b_1\theta)\theta_{\eta\eta}x_5}{Pr}\right] + \frac{R_1\theta_{\eta\eta}}{D_4 Pr}$$

$$+\frac{1}{D_4}\left[\frac{(1+b_1\theta)\theta_{\eta\eta}^2 x_6}{Pr}\right] + \frac{1}{D_4}\left[\frac{\theta\xi qRe}{1-\epsilon e^{\xi}}\right] + \frac{1}{D_2 D_4}\left[\frac{F_\eta^2 Ec}{1+a_1\theta}\right]$$

$$+\frac{K_1}{D_1}\left[F^2\xi\left(1-\epsilon e^{\xi}\right)StEc\right] - \frac{K_1}{D_1}\left[F\xi\left(1-\epsilon e^{\xi}\right)StEc\right]$$

$$+\frac{D_3}{D_1}\left[\xi\left(1-\epsilon e^{\xi}\right)StEc\right] + \frac{\left(1+\frac{1}{\beta}\right)F_\eta^2 Ec}{D_1 D_4(1+a_1\theta)}$$

The dimensionless stream function is defined as:

$$f = \int_0^\eta F d_\eta + f_w;\ f_w = 0$$

The boundary conditions constituted in the dimensionless form is given by: when $\eta = 0$

$$F(\xi,\eta) = 0, g(\xi,\eta) = 0 \text{ and } \theta(\xi,\eta)$$

when $\eta \to \infty$

$$F(\xi,\eta) = 1, g(\xi,\eta) = 1 \text{ and } \theta(\xi,\eta) \to 0$$

The mathematical computations for the variable physical quantities of skin friction and heat transfer coefficient for the current study are as follows: The skin friction coefficient is

$$C_{fx} = \frac{\mu_{hnf}\left(\frac{\partial u}{\partial y}\right)_{y=0}}{\frac{1}{2}\rho_{bf}u_e^2} \tag{11.7}$$

$$C_{fx} = 2F_\eta Re_x^{-1/2}\frac{D_1}{D_2} \tag{11.8}$$

The Nusselt number is given by:

$$N_{ux} = -\frac{xq_w}{k_f\left(T_w - T_\infty\right)}$$

$$N_{ux} = -\theta_\eta Re_x^{1/2}D_5$$

11.3 SOLUTION METHODOLOGY

The dimensionless governing equations are linearised utilising quasilinearisation approach that is followed by implicit finite difference scheme coupled with iterative Varga's algorithm. The numerical solutions go through stable convergence conditions, as shown below. The linearised form of the nonlinear partial differential equations and the boundary conditions are as follows:

$$X_1^i F_\eta^{i+1} + X_2^i F_\eta^{i+1} + X_3^i F^{i+1} + X_4^i F_{xi}^{i+1} + X_5^i \theta_\eta^{i+1} + X_6^i \theta^{i+1} + X_7 i g_\eta^{i+1} = X_8^i$$

$$Y_{1i}g_{\eta\eta i+1} + Y_{2i}g_{\eta i+1} + Y_{3i}g_{i+1} + Y_{4i}g_{ji+1} + Y_{5i}F_{\eta i+1} + Y_{6i}F_{i+1} = Y_{7i}$$

$$Z_{1i}\theta_{\eta\eta i+1} + Z_{2i}\theta_{\eta i+1} + Z_{3i}\theta_{\xi i+1} + Z_{4i}\theta_{i+1} + Z_{5i}F_{\eta i+1} + Z_{6i}F_{i+1} + Z_{7i}H_{\eta i+1} = Z_{8i} \text{ where}$$

$$X_1 = 2\xi f_\xi + (m+1)f + \frac{1}{N_1N_2}\left(1+\frac{1}{\beta}\right)\frac{1}{1+a_1\theta}$$

$$X_2 = \left(1+\frac{1}{\beta}\right)\frac{a_1\theta_\eta}{\left(1+a_1\theta\right)^2} + \xi F + \left(1+\frac{1}{\beta}\right)\frac{\Delta}{N_1}$$

$$X_3 = -\xi F$$

$$X_4 = -\xi F_\xi - 2mF - \frac{K_1}{N_1}St\left(\frac{\xi}{1-\epsilon e^\xi}\right)$$

$$X_5 = \frac{-1}{ss1}\left(\frac{1}{1+\beta}\right)\frac{a_1 F_\eta}{(1+a_1\theta)^2}$$

$$X_6 = \frac{-a_1 F_{\eta\eta}}{(1+a_1\theta)^2}\frac{1}{ss1}\left(\frac{1}{1+\beta}\right) + \frac{2a_1^2 F_\eta \theta_\eta}{(1+a_1\theta)^3}\frac{1}{ss1}\left(\frac{1}{1+\beta}\right) + \frac{2Gr\xi}{Re^2}$$

$$X_7 = \frac{\Delta}{N_1} X_8 = -sF^2 - \xi FF_\xi - \frac{1}{N_1 N_2}\left(1+\frac{1}{\beta}\right)\frac{a_1 F_\eta \theta_\eta}{(1+a_1\theta)^2}$$

$$-\frac{1}{N_1 N_2}\left(1+\frac{1}{\beta}\right) - \frac{a_1 F_{\eta\eta}\theta}{(1+a_1\theta)^2}$$

$$+\frac{1}{N_1 N_2}\left(1+\frac{1}{\beta}\right)\frac{2a_1^2 F_\eta \theta\theta_\eta}{(1+a_1\theta)^3} - s - \frac{K_1}{N_1}\frac{St\xi}{\left(1-\epsilon e^{\xi}\right)}$$

$$Y_1 = \frac{\lambda}{N_1}$$

$$Y_2 = f + mf + 2\xi f_\xi$$

$$Y_3 = -\xi F$$

$$Y_4 = -\frac{2\Delta}{N_1} - 3mF + F$$

$$Y_5 = \frac{-\Delta}{N_1}$$

$$Y_6 = -3mg + g - \xi g_\xi + 2\xi g_\eta$$

$$Y_7 = 2\xi F g_\eta - 3mFg + Fg - \xi F g_{xi}$$

$$Z_1 = \frac{(1+b_1\theta)x_5}{PrN_4} + \frac{R_1}{PrN_4}$$

$$Z_2 = \xi f_\xi + \left(\frac{s+1}{2}\right)f + \frac{2b_1\theta_\eta x_5}{N_4 Pr} + \frac{1+b_1\theta x_5}{N_4 Pr}$$

$$Z_3 = -\xi F$$

$$Z_4 = \frac{b_1 x_5 \theta_{\eta\eta}}{N_4 Pr} + \frac{b_1 \theta_\eta^2 x_6}{N_4 Pr} + \frac{\xi Q Re}{\left(1 - \epsilon e^{\xi}\right) N_4} - \frac{a_1 F_\eta^2 Ec}{\left(1 + a_1 \theta\right)^2 N_2 N_4}$$

$$Z_5 = \frac{2 F_\eta Ec}{\left(1 + a_1 \theta\right) N_2 N_4}$$

$$Z_6 = \frac{2 N_3 F \xi \mathrm{StEc}\left(1 - \epsilon e^{\xi}\right)}{N_1} - \frac{2 N_3 \xi \mathrm{StEc}\left(1 - \epsilon e^{\xi}\right)}{N_1} - \xi \theta_\xi$$

$$Z_7 = \theta_\eta$$

$$Z_8 = \frac{-N_3}{N_1} \xi \left(1 - \epsilon e^{\xi}\right) StEc + \frac{N_3}{N_1} F^2 \xi \left(1 - \epsilon e^{\xi}\right) StEc + \frac{F_\eta^2 Ec}{\left(1 + a_1 \theta\right) N_2 N_4} + \frac{\left(1 + b_1 \theta\right) \theta_\eta^2}{N_4 Pr} + \frac{b_1 \theta_\eta^2 x_5}{N_4 Pr} - \frac{a_1 \theta F_\eta^2 Ec}{\left(1 + a_1 \theta\right)^2 N_2 N_4} + \frac{b_1 \theta_\eta^2 \theta x_6}{N_4 Pr} + \frac{b_1 x_5 \theta_{\eta\eta} \theta}{N_4 Pr} - \xi F \theta_\xi$$

11.4 MATHEMATICAL COMPUTATION OF FINITE DIFFERENCE SCHEME

The quasi-linearised dimensionless PDEs are converted into system of linear equations using the implicit finite difference scheme in which the central difference and backward difference operators are applied in ξ, η directions, respectively. The emerged matrix is structured in a block tri-diagonal manner. As the step size of any experiment play a major role, all the important results should be chosen wisely. The author chooses 0.05 and 0.01 as the step size for $\Delta\xi$ and $\Delta\eta$ in the ξ and η directions, respectively. When the relative difference between the two successive values lies below 0.0001, the iterative process is terminated. The numerical scheme employed to achieve the stable criteria of convergence to be followed as:

$$\mathrm{Max}\left|\left(F_\eta\right)_w^{i+1} - \left(F_\eta\right)_w^{i+1}\right|, \left|\left(G_\eta\right)_w^{i+1} - \left(G_\eta\right)_w^{i+1}\right|, \left|\left(H_\eta\right)_w^{i+1} - \left(H_\eta\right)_w^{i+1}\right| \leq 10^{-4} \begin{bmatrix} F_{m,n+1} \\ g_{m,n+1} \\ \theta_{m,n+1} \end{bmatrix}$$

$$\times \begin{bmatrix} a_{11} & a_{12} & a_{13} \\ a_{21} & a_{22} & a_{23} \\ a_{31} & a_{32} & a_{33} \end{bmatrix} + \begin{bmatrix} F_{m,n} \\ g_{m,n} \\ \theta_{m,n} \end{bmatrix} \times \begin{bmatrix} b_{11} & b_{12} & b_{13} \\ b_{21} & b_{22} & b_{23} \\ b_{31} & b_{32} & b_{33} \end{bmatrix} + \begin{bmatrix} F_{m,n-1} \\ g_{m,n-1} \\ \theta_{m,n-1} \end{bmatrix} \times \begin{bmatrix} c_{11} & c_{12} & c_{13} \\ c_{21} & c_{22} & c_{23} \\ c_{31} & c_{32} & c_{33} \end{bmatrix} = \begin{bmatrix} d_1 \\ d_2 \\ d_3 \end{bmatrix}$$

11.5 RESULTS AND DISCUSSION

The present study investigated the numerical approach for micropolar Casson hybrid nanofluid flow over stretching surface with variable viscosity and thermal conductivity. Copper and alumina nanoparticles are suspended in the base Casson fluid, ethylene glycol, and variations of Reynold's number (Re), stream function variable ξ, buoyancy parameter λ, micropolar parameter Δ on velocity, temperature, micro-rotation, skin friction and heat transfer profiles are examined graphically. The physical model of the system is portrayed in Figure 11.1.

It is explored that for variations of the stream function ξ, velocity profile decreases. This is because increasing the reaction rate raises the chemical reaction process, which decreases the velocity profile as displayed in the Figure 11.2(a). Also for increasing values of ξ, parameter increases, the diffusivity of the chemical substance decreases, leading to a decline in the velocity profile. In particular, for $\xi = 2$, the profile is enhanced in the region $0 \leq \eta \leq 2$ due to buoyancy force resistance in the flow, decreasing the velocity profile.

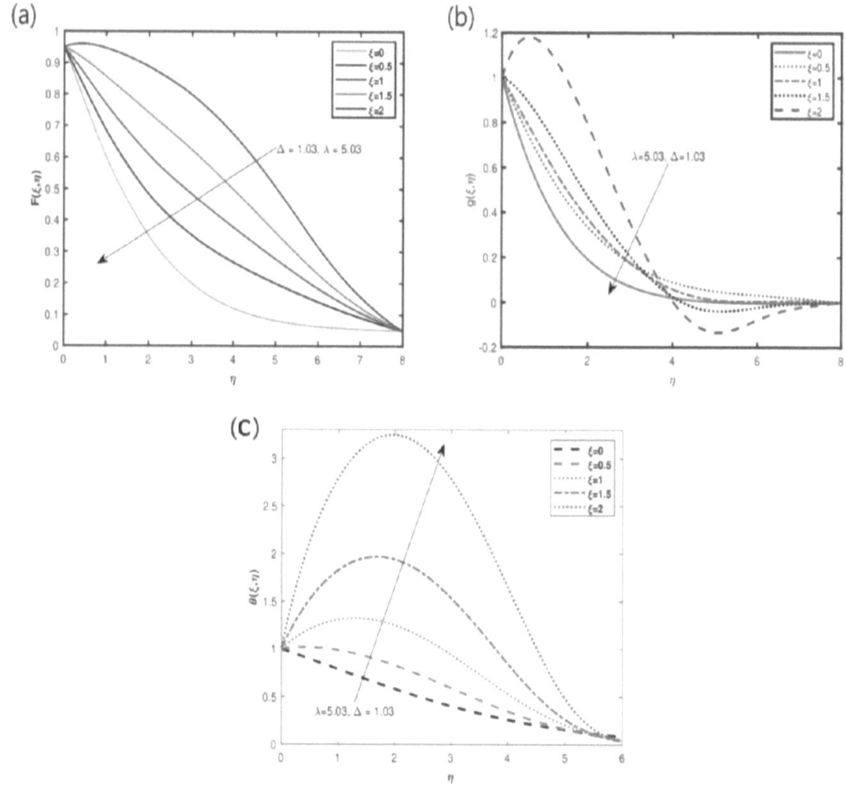

Figure 11.2 Effect of ξ variations on velocity profiles for $\beta = 0.5$, $\epsilon = 0.2$, St = 0.2, Rd = 0.5, q = 0.4, $\Delta = 1.03$, $\lambda = 5.03$ and $Pr = 0.7$.

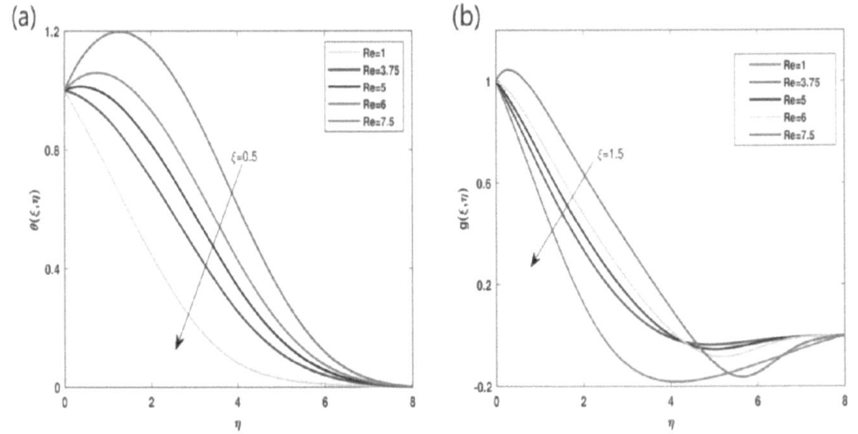

Figure 11.3 Effect of ξ variations on micro-rotational profiles for $\beta = 0.5$, $\epsilon = 0.2$, St = 0.2, Rd = 0.5, q = 0.4, $\Delta = 1.03$, $\lambda = 5.03$ and $Pr = 0.7$.

In Figure 11.2(b), we noticed that as the buoyancy parameter increases, the micro-rotation profile reduces towards the middle of the channel. It is discovered that at a particular point, the parameters have no effect on the micro-rotation distribution for the given boundary conditions. Figure 11.3 justified the variations of micropolar profile for varying values of stream function ξ in the range $0 \leq \xi \leq 2$. The physical reason suggested that due to the presence of magnetic parameter in the energy equation, the micro-rotational parameters increases. This parameter results in the augmentation of internal and external forces including electric and adhesive forces. These forces enhanced the micropolar fluid profile near the wall region and the profile reaches a steady state.

Figure 11.2(c) shows the temperature profile behaviour for buoyancy parameter ξ while keeping other physical parameters constant. When $0 \leq \lambda \leq 2$, increasing the buoyancy parameter results in a decrease in the temperature profile. The asymptotic behaviour of temperature profile for increased values of ξ is scrutinised in Figure 11.2(c). The presence of micromorphic structure with each element of the fluid is the reason behind the increased profile. A micropolar fluid is a unique form of fluid that shows distinct microscopic effects due to the constrained design and tiny movements of the fluid components.

Figure 11.3(a) portrayed Reynolds number reduces the thickness of the thermal boundary layer on their increment and so diminishes the temperature profile. It is observed that for fixed values of stream function, $\xi = 0.5$, and for increasing values of Reynold's number, the temperature profile is decreased for $\eta \geq 4$. The physical reason suggested that rotation and stretch allow the particles of the fluid to undergo intrinsic spins and homogenous deformations.

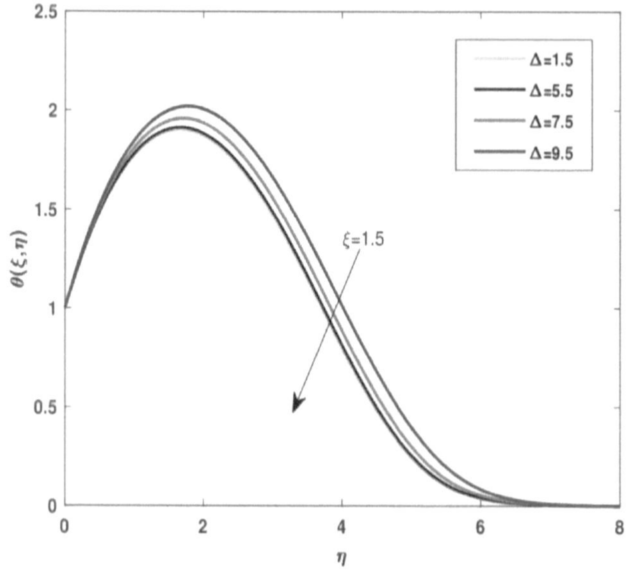

Figure 11.4 Effect of ξ variations on temperature profiles for $\beta = 0.5$, $\epsilon = 0.2$, St = 0.2, Rd = 0.5, q = 0.4, $\Delta = 1.03$, $\lambda = 5.03$ and $Pr = 0.7$.

In Figure 11.3(b), we observed that the micro-rotation profile decreases near the middle of the channel for the increasing values of the Reynolds number. It is observed that at a certain point, the parameters do not affect the micro-rotation distribution for the boundary constraints due to the dominating behaviour of inertial forces. Also, when Re is lowered, the micropolar parameter (Δ) also reduces the profile. The physical behaviour states that the viscous dissipation and Reynold's number causes the chemical reaction takes place in the flow and the flow is majorly due to the buoyancy effects arising due to the temperature difference.

Figure 11.4 depicts the impact of micro-rotation parameter on temperature profile by keeping other physical parameters fixed. This figure depicts that parameter boosts the fluid temperature to decrease its value within the boundary region. It is to be noted that for $1 \leq \eta \leq 3$, the temperature profile is increased. The physical interpretation suggested that the rise in temperature profile is due to the relationship between the buoyant force and fluid temperature is linear. The fluid increases in volume and loses density as the temperature rises.

Figure 11.5(a) points out that the micro-rotation profiles retards by enhancing the buoyancy parameter (λ). Figure 11.5(a) shows the decreasing micropolar profile for increasing values of buoyancy parameter λ in the range $5 \leq \eta \leq 6.5$. For constant values of stream function $\xi = 1$, $\epsilon = 0.2$, $\beta = 0.5$, $\lambda = 5.03$, $Pr = 0.7$, the micropolar profile is decreased at a steady rate. In case of micropolar fluid, the stress contains thermodynamic pressure tensor. When the pressure tensor arises from the dependence of the free energy on the micro-inertial,

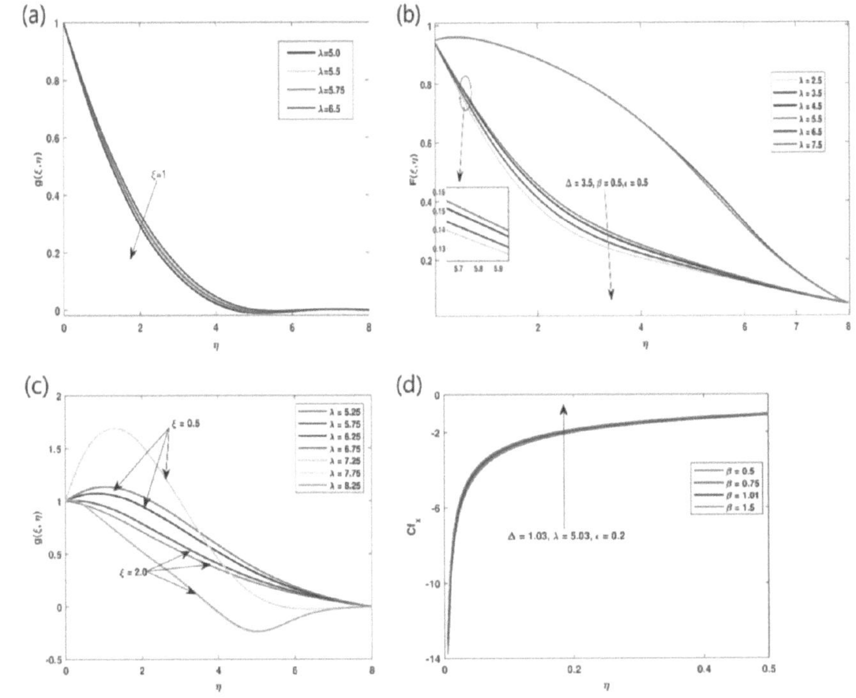

Figure 11.5 Effect of Reynold's number variations on temperature profiles for $\beta = 0.5$, $\epsilon = 0.2$, St = 0.2, Rd = 0.5, q = 0.4, $\Delta = 5.5$ and $Pr = 0.7$.

the thermodynamic pressure arises from the dependence of the stream function and due to this physical behaviour of micropolar fluid, the profile is reduced.

Figure 11.5(b) is diagrammed to examine the results of buoyancy parameter (λ) on velocity component. A decline in velocity field is established by growing the amounts of buoyancy parameter when $2.5 \leq \lambda \leq 7.5$. Physically, the stretching rate of sheet declines for greater rotational parameter and so the velocity profile reduces. It is to be noted that a two distinct velocity profile is characterised, particularly for λ values below 5.5, the profile is lowered and for λ values greater than 5.5, the profile is augmented near the wall region. Physically, when the magnetic parameter increases, a Lorentz force (opposing force) is created in the fluid, which retards the flow velocity.

Figure 11.5(c) shows the variations of buoyancy parameter (λ) on micropolar fluid component. A decline in micropolar fluid field is established by growing the amounts of buoyancy parameter (λ) when $5.25 \leq \lambda \leq 8.25$. Physically, the stretching rate of sheet declines for greater rotational parameter and so the micropolar fluid profile reduces. It is particularly noted that the micropolar profile is increased for $\lambda = 7.5$ in the region.

For $1 \leq \eta \leq 2$, the profile is lowered for values of $\lambda \geq 8.5$ in the region $4 \leq \eta \leq 6$. It is observed that the thickness of the thermal boundary layer grows as

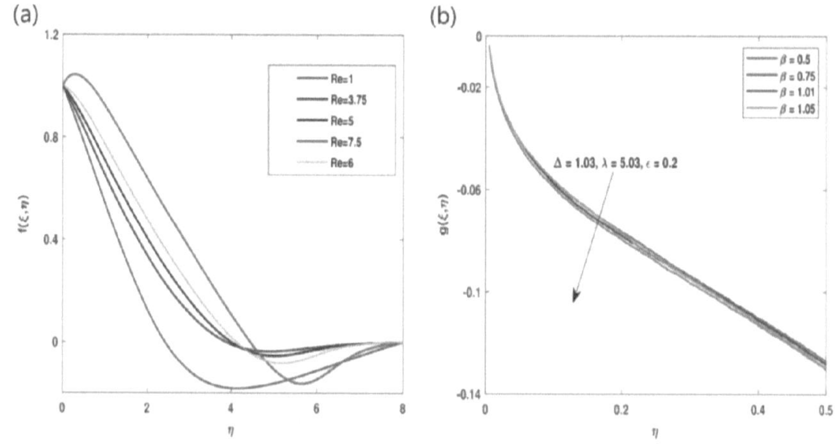

Figure 11.6 Effect of Reynolds number variation on micro-rotational profile for $\beta = 0.5$, $\epsilon = 0.2$, St = 0.2, Rd = 0.5, q = 0.4, $\Delta = 5.5$ and $Pr = 0.7$.

the buoyancy parameter is increased. The Lorentz drag act as an opposing force which help to increase the frictional heating between the fluid layers and release thermal energy and hence the micropolar profile is increased.

Figure 11.5(d) depicts the effect of skin-friction profiles for the non-identical values of Casson parameter β in the range $0.5 \leq \beta \leq 1.05$ by keeping other physical parameters fixed. The skin friction profile is often described by a dimensionless parameter called the Reynolds number (Re), which represents the ratio of inertial forces to viscous forces. In general, for low Reynolds numbers, the flow is dominated by viscosity, leading to a smooth and laminar skin friction profile. It is noted that for increased values of β, the skin friction profiles are increased and a steady condition is reached for $\eta \geq 0.4$. Figure 11.6(a) depicts the effect of heat transfer profiles for the various values of Casson parameter in the range $0.5 \leq \beta \leq 1.05$ by keeping other physical parameters fixed. Figure 11.6(b) justified the variation of Casson parameter β on micropolar profiles. The Casson parameter β strengthened the skin friction profiles and weakened the micro-rotational profile. An asymptotic heat transfer profile is observed for increased values of β. The physical behaviour of β interpreted that for increased values of β, the thermal boundary layer is thickened and due to this reason the heat transfer profile is increased and lowered in a steady state.

11.6 CONCLUDING REMARKS

The present model derived a set of boundary layer equations for heat and mass transfer characteristics of laminar boundary layer micropolar fluid over a linearly stretching surface. The model considered the effects of constant heat flux, viscous dissipation, Joule heating together with variable viscosity and thermal conductivity.

- Numerical simulations reveal that the micro-rotational parameter Δ enhanced the temperature profile.
- The micropolar parameter Δ lowered the skin friction coefficient and the velocity profiles.
- The buoyancy parameter λ enhanced the temperature profile near the wall region.
- The non-Newtonian Casson parameter β lowered the micropolar profile and increased the skin friction profile.
- The thermal boundary layer is thickened for increased values of β and hence the heat transfer profile is increased and decreased in a steady state.

REFERENCES

1. Ahmed Mohammed Alshehri, Hasan Huseyin Coban, Shafiq Ahmad, Umair Khan and Wajdi Mohamad Alghamdi. Buoyancy effect on a micropolar fluid flow past a vertical riga surface comprising water-based SWCNT–MWCNT hybrid nanofluid subject to partially slipped and thermal stratification: Cattaneo–Christov Model. Math. Probl. Eng. 2021. https://doi.org/10.1155/2021/6618395
2. Rama Subba Reddy Gorla, Richard Pender and John Eppichs. Heat transfer in micropolar boundary layer flow over a flat plate. Inl. J. Engg Sci. 1983 Vol. 21, No. 7, pp. 791–798.
3. C.H Chen. Laminar mixed convection adjacent to vertical, continuously stretching sheets. Heat Mass Transf. 1998 Vol. 33, pp. 471–476.
4. Yusuf O. Tijani, Mojeed T. Akolade, Hammed A. Ogunseye, Adeshina T. Adeosun and Oluwaseun Farotimi. On the generalized Fick's and Fourier's Laws for an unsteady Casson-Williamson fluids over a stretching surface: A spectral approach. J. Nanofluids. 2023 Vol. 12, pp. 91–103.
5. Govind R. Rajput, Bipin P. Jadhav and S. N. Salunkhe. Magnetohydrodynamics boundary layer flow and heat transfer in porous medium past an exponentially stretching sheet under the influence of radiation Heat Transf. 2020, pp. 1–15 https://doi.org/10.1002/htj.21752
6. Showkat Ahmad Lone, Maryam Ahmed Alyami, Anwar Saeed, Abdullah Dawar, Poom Kumam and Wiyada Kumam. MHD micropolar hybrid nanofluid flow over a flat surface subject to mixed convection and thermal radiation. Sci. Rep. 2022 Vol. 12. https://doi.org/10.1038/s41598-022-21255-8
7. Farizza Haniem Sohut, Anuar Ishak and Siti Khuzaimah Soid. MHD stagnation point of Blasius flow for micropolar hybrid nanofluid toward a vertical surface with stability analysis. Symmetry 2023 Vol 15. https://doi.org/10.3390/sym15040920
8. Rama Subba Reddy Gorla. Magnetohydrodynamic free convection boundary layer flow of a thermo micropolar fluid over a vertical plate. Int. J. Engg Sci. 1998 Vol. 36, No. 3, pp. 315–327.
9. Hossam A. Nabwey and A. Mahdy. Transient flow of micropolar dusty hybrid nanofluid loaded with Fe_3O_4-Ag nanoparticles through a porous stretching sheet. Results Phys. 2021 Vol. 21. https://doi.org/10.1016/j.rinp.2020.103777
10. Rama Subba Reddy Gorla, Paul P. Lin and An-Jen Yang. Asymptotic boundary layer solutions for mixed convection from a vertical surface in a micropolar fluid. Int J Engg Sci. 1990 Vol. 28, No. 6, pp. 525–533.

11. Mohammed Abd El-Aziz. Viscous dissipation effect on mixed convection flow of a micropolar fluid over an exponentially stretching sheet. Can J. Phys. 2009 Vol. 87, pp. 359–368.
12. Nadeem Abbas S. Nadeem and M.Y. Malik. Theoretical study of micropolar hybrid nanofluid over Riga channel with slip conditions. Phys. A 2020. https://doi.org/10.1016/j.physa.2019.124083.
13. G. P. Vanitha, U. S. Mahabaleshwar, M. Hatami and Xiaohu Yang. Heat and mass transfer of micropolar liquid flow due to porous stretching/shrinking surface with ternary nanoparticles. Sci. Rep. 2023 Vol. 13. https://doi.org/10.1038/s41598-023-29469-0
14. Iskandar Waini, Anuar Ishak, Yian Yian Lok and Ioan Pop. Micropolar nanofluid flow in a stagnation region of a shrinking sheet with Fe_3O_4 nanoparticles. Mathematics. 2022 Vol. 10. https://doi.org/10.3390/math10173184
15. A. Mahdy, E. R. El-Zahar, A. M. Rashad, W. Saad and H. S. Al-Juaydi. The magneto-natural convection flow of a micropolar hybrid nanofluid over a vertical plate saturated in a porous medium. Fluids. 2021 Vol. 6. https://doi.org/10.3390/fluids6060202
16. Maryam Subhani and Sohail Nadeem. Numerical analysis of micropolar hybrid nanofluid. Appl. Nanosci. https://doi.org/10.1007/s13204-018-0926-2
17. James Lau Tze Chen, A. N. Oumer and Azizuddin A. A review on thermophysical properties of bio, non-bio and hybrid nanofluids. J. Mech. Eng. Sci. 2019 Vol. 13, No. 4, pp. 5875–5904.
18. Rama Subba Reddy Gorla. Combined forced and free convection in micropolar boundary layer flow on a vertical flat plate. Int. J. Engg Sci. 1988 Vol. 26, No. 4, pp. 385–391.
19. A. Mohammadein and R.S Gorla. Heat transfer in micropolar fluid with viscous dissipation and internal heat generation. Int. J. Numer. Methods Heat Fluid Flow. 2001 Vol. 11, No. 1, pp. 50–58.
20. Umair Khan, Aurang Zaib, Ioan Pop, Sakhinah Abu Bakar and Anuar Ishak. Unsteady micropolar hybrid nanofluid flow past a permeable stretching/shrinking vertical plate. Alexandria Engg J. 2022 Vol. 61, pp. 11337–11349.

LIST OF NOTATIONS

f	dimensionless stream function
F	dimensionless velocity
Gr	Grashof number due to temperature
Le	Lewis number
H	magnetic parameter
K^*	Rosseland mean absorption coefficient
Nu_x	Nusselt number
Pr	Prandtl number
Re	Reynolds number
St	Magnetic parameter
Ec	Eckert number
Rd	Thermal radiation flux
q_r	radiation heat flux
T	temperature of the nanofluid
T_w	uniform temperature over the surface of the sheet
u	velocity component in the x direction

U_w moving plate velocity
$U\infty$ free stream velocity
U reference velocity
v velocity component in the y direction

Greek symbols

ρ density kg
ψ Streamfunction
σ^* Stefan-Boltzmann constant
λ buoyancy parameter
Δ micropolar parameter
S vortex viscosity
T ratio between heat capacity of nanoparticle and base fluid
μ viscosity
ν kinematic viscosity
ϵ the ratio between free stream velocity and the reference velocity
N Total spin of microstructure and the fluid medium in the flow field
j micro inertia

Chapter 12

Flow features of radiating ternary hybrid nanofluid flow under the influence of induced magnetic field and nanoparticle shapes

Pinaki Ranjan Duari and Kalidas Das

12.1 INTRODUCTION

Non-renewable energy sources like coal, oil, and natural gas have been dominant in global energy consumption for several decades due to their proportionately restricted accessibility and well-established infrastructure. However, they also have significant environmental, social, and economic drawbacks, including greenhouse gas production, habitat devastation, and resource depletion associated with energy production and distribution. A growing concern over deviations in climate, air quality, and energy assurance has led to the increase in renewable energy sources, including solar, wind, hydroelectric, and warmth. These renewable alternatives offer sustainable and environmentally sound options for meeting energy needs while lessening the dependency on finite and polluting fossil fuels and nuclear substances. Solar energy (Acharya, 2020; Das et al., 2014, 2022; Kandasamy et al., 2013) holds great promise as a fresh and sustainable energy source to diminish greenhouse gas emanations and the impacts of global warming. Various technologies are employed in the production of electricity, heat, and light for residential, commercial, and industrial uses. The electromagnetic field is made up of an extensive range of wavelengths and frequencies that the sun emits. The energy source used to generate solar electricity is solar radiation. which has a vital role in absorption using nanoparticles. Hunt (1978) marks a significant milestone in the field of numerous natural processes, such as photosynthesis in plants and the water cycle, and the introduction of radiation technology and its applications in several industries.

Nanofluids (Choi, 1995; El-Khouly et al., 2012; Pordanjani et al., 2019; Wole-osho et al., 2020) are indeed engineered liquids that have nanoparticles suspended in a base liquid. These particles are typically solid and can be made from various materials, including metals, oxides, carbides, carbon-based materials, etc. Nanofluids offer unique opportunities for enhancing heat transfer (HT), energy efficiency, and performance in a variety of applications, which demands the balance of several qualities, which is where hybrid nanofluids are useful. The term "hybrid nanofluid" (Sarkar et al., 2015) refers to a novel type of nanofluid that incorporates additional components or features beyond traditional nanofluids, resulting in enhanced properties or

DOI: 10.1201/9781003534785-12

functionalities. HT is greatly impacted by such mono-nanofluid evaluated the effects of magnetic and hydrodynamic flows on vertical plates using hybridized nanoliquids. An investigation of magnetohydrodynamics nanofluid flow through the rotating discs was carried out by Zangooee et al. (2019). Acharya and Mabood (2012) monitored the flow of a hybrid nanofluid made of Fe_3O_4, graphene, and water on a surface. According to their findings, biot numbers in hybrid nanofluids (Fe_3O_4-graphene-H_2O) were augmented by 74.25% while in conservative nanofluids (Fe_3O_4-H_2O) were augmented by 69.17%. There have been several studies published in recent years (Chakraborty et al., 2023; Das et al., 2022; Khashi'ie et al., 2022) that have revealed interesting results from hybridized nanofluids.

In recent years, ternary hybrid nanofluids have gained significant attention due to their enhanced thermal properties and versatile applications. These nanofluid interactions between multiple types of nanoparticles and the base liquid can lead to aggregation. These have garnered attention due to their potential for enhanced thermal conductivity, stability, and other desirable properties compared to binary nanofluids. The widespread use of ternary hybrid nanofluids across multiple engineering disciplines highlights their versatility and potential for various applications, ranging from controlling heat and energy conservation to biomedical technologies. The impact of $ZnFe_2O_3 + MnZnFe_2O_3 + NiZnFe_2O_3 - H_2O$ on dispersion of heat on a dovetailing fin has been observed by Goud et al. (2022). To determine ternary nanoparticles' thermal properties, Al Oweidi (2022) looked at their heat transmission and fluid movement. Several parameters were investigated using flow-described equations. Numerous studies (Madkhali, 2023; Mahmood et al., 2022a, 2022b, 2023; Rafique et al., 2023; Shamshuddin et al., 2023) have been carried out lately to look into the potential applications of ternary nanofluid flow.

The implications of hybrid nanofluids' shape factor are being emphasized by Ghadikolaei et al. (2018). In the following year, Ghadikolaei and Gholinia (2019) extended the study, taking into account how the structure of nanoparticles affects three-dimensional radiating flow. Khashi'ie et al. (2012) have demonstrated that shape features may influence the heat conductivity of nanoliquids containing radiative Cu-Al_2O_3/H_2O. A study by Rashid et al. (2012) investigated how shape factor influences HT across horizontal cylinders while being stretched and contracted. The outline of nanoparticles may contribute to enhancing HT by considering the flow regime, according to Zahmatkesh et al. (2012). Researchers have recently reported interesting results regarding nanoparticle shape (Imran et al., 2022; Rashid et al., 2020; Hamilton and Crosser, 1962; Yang and Xu, 2017).

The term "multiple slip" typically pertains to the slip conditions experienced by nanoparticles suspended in the fluid. Three categories of slip circumstances exist and velocity slip: In conventional fluid dynamics, slip velocity is typically determined by the interplay of solid barriers and liquid particles. This interaction results in a velocity progression near the boundary, with the fluid velocity being different from the solid surface's velocity. Slip length (both velocity and thermal) plays a vital part in determining the behavior of fluids at small scales, especially in systems involving nanofluids. These phenomena challenge traditional boundary circumstance assumptions and introduce new factors that need to be considered for accurate modeling and optimization of heat transmission and fluid flow processes. Different thermal characteristics of nanoparticles in relation to the base fluid might lead to increased concentration slip and slide effects. Concentration gradients of nanoparticles near a solid surface can indeed induce slip, known as concentration slip. The concentrations of nanoparticles in the bulk fluid are different from those close to the surface, which causes this phenomenon. These concentration gradients may have an impact on the fluid flow behavior, leading to slip velocities at the fluid-solid interface. This is particularly relevant in nanofluids, where the distribution of nanoparticles is not uniform throughout the flow domain. Multi-slip mechanisms play a crucial role in modeling and predicting nanofluid behavior in micro/nanofluidic systems, HT enhancement, and other applications. Investigation and characterization of these slip events in nanofluid flows are frequently accomplished by experimental research and computational fluid dynamics (CFD) simulations. Numerous investigators have examined multiple slips (Gul et al., 2012; Khan et al., 2019; Mabood and Shateyi, 2019; Ramana et al., 2012; Seid et al., 2022).

Magnetic fields serve a significant part in an extensive range of applications, including energy generation and medical therapies. The hydromagnetic boundary layer is governed by induced magnetized fields in numerous accurate methods, such as magneto-plasma dynamics, nuclear engineering thermohydraulic, etc. A magneto-convection flow was demonstrated by (Kumar et al., 2019) using induced magnetic fields. Numerous investigations have been conducted on the behavior of wedges in magnetic fields (Mukhopadhyay et al., 2013; Nadeem et al., 2018; Raju and Sandeepa, 2016). In a subsequent investigation, Basha et al. (2019) demonstrated that nonlinear thermal radiation and magnetic fields can convert sunlight energy into electricity. Dolui et al. (2023) assert that heat radiation has a negative impact on catheterized multiple stenoses in the artery. A study of non-linear radiant HT in porous surfaces with applied magnetic fields was carried out by Jalili et al. (2023).

In light of the previously mentioned literature, this study explores the hydrothermal alterations in radiating ternary nanofluidic transportation across a wedge while being heated by solar radiation and subjected to generated magnetic fields during slip circumstances. The shape factors of nanoparticles were examined in this study.

12.2 MATHEMATICAL MODEL

An incompressible Fe_3O_4-Ag-MgO-H_2O flow toward a wedge is considered. The current model includes non-linear solar radiation as well as heat generation/absorption. The movement of flow and velocity are affected by

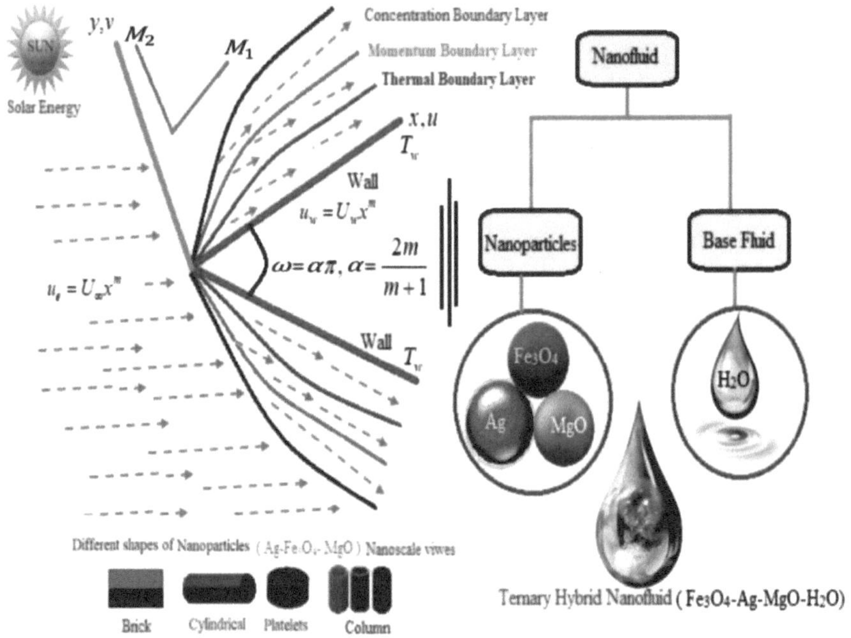

Figure 12.1 Physical illustrations of the problem.

moving wedges with velocity $u_w = U_w x^m$ in the unimpeded stream $u_e = U_\infty x^m$, $0 \le m \le 1$. In Figure 12.1, $u_w > 0$ represents a diminishing surface velocity of the wedge, while $u_w < 0$, represents an expanding surface velocity of the wedge. It is assumed that the wedge angle to be $\omega = \alpha\pi$, where $\alpha = \frac{2m}{m+1}$. Usually, the magnetic field is permitted to apply to mutual restraints, $\bar{M}(M_1, M_2)$ with the free-stream assumption $M_e = M_0 x$. In contrast, M_0 and M_e are both the magnetized field at the boundary layer edge and the homogeneous upstream magnetized field far from the wedge. The components of the induced magnetized field, M_1 and M_2, are portrayed in the flow's parallel and normal directions, respectively. In this model, the Hall current and ohmic dissipation effects are deserted as the magnetic Reynolds number is presumed small.

Under the above circumstances, the governing equations are (Mukhopadhyay et al., 2013; Nadeem et al., 2018; Raju and Sandeepa 2016).

$$\frac{\partial u}{\partial x} + \frac{\partial u}{\partial y} = 0 \tag{12.1}$$

$$\frac{\partial M_1}{\partial x} + \frac{\partial M_2}{\partial y} = 0 \tag{12.2}$$

$$u\frac{\partial u}{\partial x} + v\frac{\partial u}{\partial y} = u_e\frac{\partial u_e}{\partial x} + \frac{\mu_{Thnf}}{\rho_{Thnf}}\frac{\partial^2 u}{\partial y^2} + \frac{\mu_m}{4\pi\rho}\left[M_1\frac{\partial M_1}{\partial x} + M_2\frac{\partial M_1}{\partial y} - M_e\frac{\partial M_e}{\partial x}\right] \tag{12.3}$$

$$u\frac{\partial M_1}{\partial x}+v\frac{\partial M_1}{\partial y}=\mu_e\frac{\partial^2 M_1}{\partial y^2}+M_1\frac{\partial u}{\partial x}+M_2\frac{\partial u}{\partial y} \tag{12.4}$$

$$\begin{aligned}u\frac{\partial T}{\partial x}+v\frac{\partial T}{\partial y}&=\alpha_{Thnf}\frac{\partial^2 T}{\partial y^2}+\tau_{Thnf}\left[D_B\left(\frac{\partial C}{\partial y}\frac{\partial T}{\partial y}\right)+\frac{D_T}{T_\infty}\left(\frac{\partial T}{\partial y}\right)^2\right]\\&-\frac{1}{\left(\rho C_p\right)_{Thnf}}\frac{\partial q_{rad}}{\partial y}+\frac{Q^*}{\left(\rho C_p\right)_{Thnf}}\left(T-T_\infty\right)\end{aligned} \tag{12.5}$$

$$u\frac{\partial C}{\partial x}+v\frac{\partial C}{\partial y}=D_B\frac{\partial^2 C}{\partial y^2}+\left(\frac{D_T}{T_\infty}\right)\frac{\partial^2 T}{\partial y^2} \tag{12.6}$$

It is accomplished here by transferring heat through heat waves as opposed to direct interaction with heat radiation. Equation (12.5) shows the amount of heat transferred per unit area by the term q_{rad}. Rosseland's approximation for nonlinear radiation q_{rad} is (Basha et al., 2019; Das et al., 2022; Ghadikolaei and Gholinia, 2019):

$$q_{rad}=-\frac{4\sigma_1}{3k^*}\frac{\partial T^4}{\partial y}=-\frac{16\sigma_1}{3k^*}T^3\frac{\partial T}{\partial y} \tag{12.7}$$

where σ_1 signifies the Stefan-Boltzmann constant and k^* designates the mean absorption coefficient.

Now, we placed $T=T_\infty[1+(K-1)\theta]$, whereas $K=\frac{T_w}{T_\infty}$ attached the boundary conditions:

1. Static wedge

$$\left.\begin{array}{l}u=0, v=0, \dfrac{\partial M_1}{\partial y}=0, M_2=0, T=T_w+N_1\left(\dfrac{\partial T}{\partial y}\right), C=C_w+N_2\left(\dfrac{\partial C}{\partial y}\right) \text{at } y=0\\ u\to u_e, M_1\to M_e, T\to T_\infty, C=C_\infty \text{ as } y\to\infty\end{array}\right\} \tag{12.8}$$

2. Moving wedge

$$\left.\begin{array}{l}u=u_w, v=0, \dfrac{\partial M_1}{\partial y}=0, M_2=0, T=T_w+N_1\left(\dfrac{\partial T}{\partial y}\right), C=C_w+N_2\left(\dfrac{\partial C}{\partial y}\right) \text{at } y=0\\ u\to u_e, M_1\to M_e, T\to T_\infty, C=C_\infty \text{ as } y\to\infty\end{array}\right\} \tag{12.9}$$

where x and y stand for along and normal to the wedge, respectively, (u, v) denote the velocity parallel to (x, y), T_w and C_w represents temperature

Table 12.1 Thermophysical characteristics of H_2O and Fe_3O_4, Ag, and MgO nanoparticles

Physical properties	H_2O	$Fe_3O_4(\chi_1)$	$Ag(\chi_2)$	$MgO(\chi_3)$
C_p (J/Kg K)	4179	670	235	955
ρ (Kg/m^3)	997.1	5180	10500	3560
κ (W/mK)	0.613	9.7	429	45

and concentration at the wedge wall, respectively. The ternary hybrid nanofluids' thermal diffusivity is represented by $\alpha_{Thnf}\ \frac{\kappa_{Thnf}}{(\rho C_p)_{Thnf}}$. The thermophoretic and Brownian diffusion coefficients are denoted by D_T and D_B, respectively,* heat generation coefficients N_1 and N_2 represent thermal and solutal slip, respectively.

12.2.1 Thermophysical characteristics

The ternary hybrid nanofluid consists of stable nano-dispersions of Fe_3O_4, Ag, and MgO nanoparticles with the basic fluid being water, each with a diameter of 50 nm. Table 12.1 displays related thermophysical characteristics, and the linked thermophysical relationships are displayed in Table 12.2.

Table 12.2 Thermophysical relationships of various types of nanofluid

Properties	**Mono-nanofluid (Fe_3O_4-H_2O)**
Density	$\rho_{nf} = (1-\chi_1)\rho_f + \chi_1\rho_{s1}$
Heat capacity	$(\rho C_p)_{nf} = (1-\chi_1)(\rho C_p)_f + \chi_1(\rho C_p)_{s1}$
Viscosity	$\mu_{nf} = \dfrac{\mu_f}{(1-\chi_1)^{2.5}}$
Thermal conductivity	$\dfrac{\kappa_{nf}}{\kappa_f} = \dfrac{\kappa_{s1} + (p-1)\kappa_f - (p-1)\chi_1(\kappa_f - \kappa_{s1})}{\kappa_{s1} + (p-1)\kappa_f + \chi_1(\kappa_f - \kappa_{s1})}$
Properties	**Hybrid nanofluid (Fe_3O_4-Ag-H_2O)**
Density	$\rho_{hnf} = (1-\chi_2)\{(1-\chi_1)\rho_f + \chi_1\rho_{s1}\} + \chi_2\rho_{s2}$
Heat capacity	$(\rho C_p)_{hnf} = (1-\chi_2)\{(1-\chi_1)(\rho C_p)_f + \chi_1(\rho C_p)_{s1}\} + \chi_2(\rho C_p)_{s2}$
Viscosity	$\mu_{hnf} = \dfrac{\mu_f}{(1-\chi_1)^{2.5}(1-\chi_2)^{2.5}}$
Thermal conductivity	$\kappa_{hnf} = \dfrac{\kappa_{s2} + (p-1)\kappa_{nf} - (p-1)\chi_2(\kappa_{nf} - \kappa_{s2})}{\kappa_{s2} + (p-1)\kappa_{nf} + \chi_2(\kappa_{nf} - \kappa_{s2})} \times \kappa_{nf}$ where $\kappa_{nf} = \dfrac{\kappa_{s1} + (p-1)\kappa_f - (p-1)\chi_1(\kappa_f - \kappa_{s1})}{\kappa_{s1} + (p-1)\kappa_f + \chi_1(\kappa_f - \kappa_{s1})} \times \kappa_f$

(Continued)

Table 12.2 Thermophysical relationships of various types of nanofluid *(Continued)*

Properties	**Ternary hybrid nanofluid (Fe_3O_4-Ag-MgO-H_2O)**
Density	$\rho_{Thnf} = (1-\chi_1)\{(1-\chi_2)[(1-\chi_3)\rho_f + \chi_3\rho_{s3}] + \chi_2\rho_{s2}\} + \chi_1\rho_{s1}$
Heat capacity	$(\rho C_p)_{Thnf} = (1-\chi_1)\{(1-\chi_2)[(1-\chi_3)(\rho C_p)_f + \chi_3(\rho C_p)_{s3}] + (\rho C_p)_f + \chi_2(\rho C_p)_{s2}\} + \chi_1(\rho C_p)_{s1}$
Viscosity	$\mu_{Thnf} = \dfrac{\mu_f}{(1-\chi_1)^{2.5}(1-\chi_2)^{2.5}(1-\chi_3)^{2.5}}$
Thermal conductivity	$\kappa_{Thnf} = \dfrac{\kappa_{s3} + (p-1)\kappa_{hnf} - (p-1)\chi_3(\kappa_{hnf} - \kappa_{s3})}{\kappa_{s3} + (p-1)\kappa_{hnf} + \chi_3(\kappa_{hnf} - \kappa_{s3})} \times \kappa_{hnf}$
	where $\kappa_{hnf} = \dfrac{\kappa_{s2} + (p-1)\kappa_{nf} - (p-1)\chi_2(\kappa_{nf} - \kappa_{s2})}{\kappa_{s2} + (p-1)\kappa_{nf} + \chi_2(\kappa_{nf} - \kappa_{s2})} \times \kappa_{nf}$
	and $\kappa_{hnf} = \dfrac{\kappa_{s1} + (p-1)\kappa_f - (p-1)\chi_1(\kappa_f - \kappa_{s1})}{\kappa_{s1} + (p-1)\kappa_f + \chi_1(\kappa_f - \kappa_{s1})} \times \kappa_f$

Here, the volume fractions of Fe_3O_4, Ag, and MgO are signifying as χ_1, χ_2, and χ_3, respectively. Consequently, the suffixes *Thnf*, *hnf*, *nf*, *f*, *s1*, *s2*, and *s3* are committed to comprehending Fe_3O_4-Ag-MgO-H_2O, Fe_3O_4-Ag-H_2O, Fe_3O_4-H_2O, H_2O, Fe_3O_4, Ag, and MgO solid particles, respectively.

Based on the Hamilton-Crosser model proposed in (1962; Yang and Xu 2017), the shape factor (p) is denoted by $3/\psi$, where ψ denotes the nanoparticles' sphericity. Table 12.3 illustrates the sphericity values for the different shapes of nanoparticles for which Maxwell's model has been reduced to Hamilton-Crosser's model.

Table 12.3 Nanoparticle shape factor

Nanoparticles type	*Shape structure (nanoscale view)*	*Sphericity ψ*	*Shape factor $p = 3/\psi$*
Brick		0.81	3.7
Cylindrical		0.62	4.9
Platelets		0.82	5.7
Column		0.4710	6.4

To tackle the Equations (12.1–12.6) with (12.8–12.9), we introduce the following similarity conversions:

$$\left.\begin{aligned}
&\zeta = y\sqrt{\frac{(m+1)u_e}{2v_f x}}, \psi = f(\zeta)\sqrt{\frac{2v_f x u_e}{(m+1)}}, u = U_\infty x^m \frac{df(\zeta)}{d\zeta}\\
&v = -\sqrt{\frac{(m+1)v_f U_\infty x^{m-1}}{2}}\left\{f(\zeta) + \frac{m-1}{m+1}\zeta\frac{df(\zeta)}{d\zeta}\right\}, M_1 = M_0 x^m \frac{dh(\zeta)}{d\zeta}\\
&M_2 = -M_0\sqrt{\frac{2v_f x^{m-1}}{U_\infty (m+1)}}\left\{mh(\zeta) + \frac{m-1}{2}\zeta\frac{dh(\zeta)}{d\zeta}\right\}, \theta = \frac{T - T_\infty}{T_w - T_\infty}, \phi = \frac{C - C_\infty}{C_w - C_\infty}
\end{aligned}\right\} \tag{12.10}$$

So, using the conversions (12.10) in (12.1–12.6), we get the ODEs as:

$$\left.\begin{aligned}
&\frac{1}{\vartheta_1\vartheta_2}\frac{d^3 f(\zeta)}{d\zeta^3} + f(\zeta)\frac{d^2 f(\zeta)}{d\zeta^2} + \frac{2m}{m+1}\left[1 - \left(\frac{df(\zeta)}{d\zeta}\right)^2\right]\\
&- \beta m\left[h(\zeta)\frac{d^2 h(\zeta)}{d\zeta^2} - \left(\frac{dh(\zeta)}{d\zeta}\right)^2 + 1\right] = 0\\
&\frac{d^3 h(\zeta)}{d\zeta^3} + Pr_m h(\zeta)\frac{d^2 f(\zeta)}{d\zeta^2}\left(1 - \frac{2m}{m+1}\right) = 0\\
&\vartheta_3\frac{d^2\theta(\zeta)}{d\zeta^2} + \vartheta_4 Pr\, f(\zeta)\frac{d\theta(\zeta)}{d\zeta} + \frac{4}{3}Nr\frac{d}{d\zeta}\left(\left[1 + (K-1)\theta(\zeta)\right]^3\frac{d\theta}{d\zeta}\right)\\
&+ \mathrm{Pr}\left[\mathrm{N}_b\frac{d\theta(\zeta)}{d\zeta}\frac{d\phi(\zeta)}{d\zeta} + \mathrm{N}_t\left(\frac{d\theta(\zeta)}{d\zeta}\right)^2 + Q^*\theta(\zeta)\right] = 0\\
&\frac{d^2\phi(\zeta)}{d\zeta^2} + \left(\frac{\mathrm{N}_t}{\mathrm{N}_b}\right)\frac{d^2\theta(\zeta)}{d\zeta^2} + Scf(\zeta)\frac{d\phi(\zeta)}{d\zeta} = 0
\end{aligned}\right\} \tag{12.11}$$

where $\vartheta_1 = (1-\chi_1)^{2.5}(1-\chi_2)^{2.5}(1-\chi_3)^{2.5}$

$$\vartheta_2 = (1-\chi_1)\left\{(1-\chi_2)\left[(1-\chi_3) + \chi_3\frac{\rho_{s3}}{\rho_f}\right] + \chi_2\frac{\rho_{s2}}{\rho_f}\right\} + \chi_1\frac{\rho_{s1}}{\rho_f}$$

$$\vartheta_3 = \frac{\kappa_{s3} + (p-1)\kappa_{hnf} - (p-1)\chi_3\left(\kappa_{hnf} - \kappa_{s3}\right)}{\kappa_{s3} + (p-1)\kappa_{hnf} + \chi_3\left(\kappa_{hnf} - \kappa_{s3}\right)} \times \frac{\kappa_{s2} + (p-1)\kappa_{nf} - (p-1)\chi_2\left(\kappa_{nf} - \kappa_{s2}\right)}{\kappa_{s2} + (p-1)\kappa_{nf} + \chi_2\left(\kappa_{nf} - \kappa_{s2}\right)} \times \frac{\kappa_{s1} + (p-1)\kappa_{f} - (p-1)\chi_1\left(\kappa_{f} - \kappa_{s1}\right)}{\kappa_{s1} + (p-1)\kappa_{f} + \chi_1\left(\kappa_{f} - \kappa_{s1}\right)}$$

$$\vartheta_4 = (1-\chi_1)\left\{(1-\chi_2)\left[(1-\chi_3) + \chi_3\frac{\left(\rho C_p\right)_{s3}}{\left(\rho C_p\right)_f}\right] + \chi_2\frac{\left(\rho C_p\right)_{s2}}{\left(\rho C_p\right)_f}\right\} + \chi_1\frac{\left(\rho C_p\right)_{s1}}{\left(\rho C_p\right)_f}$$

and

1. Static wedge

$$\left.\begin{array}{l} f(0)=0, f'(0)=0, h''(0)=0, h(0)=0, \theta(0)=1+\gamma\theta'(0), \phi(0)=1+\varepsilon\phi'(0) \\ f \to 1, h' \to 1, \theta \to 0, \phi \to \infty \ as \ \zeta \to \infty \end{array}\right\} \tag{12.12}$$

2. Moving wedge

$$\left.\begin{array}{l} f(0)=0, f'(0)=S, h''(0)=0, h(0)=0, \theta(0)=1+\gamma\theta'(0), \phi(0)=1+\varepsilon\phi'(0) \\ f \to 1, h' \to 1, \theta \to 0, \phi \to \infty \ as \ \zeta \to \infty \end{array}\right\} \tag{12.13}$$

Here, $\beta = \frac{M_0{}^2\mu_e}{4\pi U_\infty{}^2\rho}$ induced magnetic parameter, $Pr_m = \frac{\nu_f}{\mu_e}$ magnetic Prandtl number, $Pr = \frac{\nu_f}{\alpha_{Thnf}}$ Prandtl number. $Nr = \frac{4\sigma_1 T_\infty^3(T_w - T_\infty)}{3k^*\kappa_{hbf}}$ is the solar radiation parameter and temperature ratio, $K = \frac{T_w}{T_\infty}$. A wedge (S) travels in the free stream's direction as well as the opposing one (S > 0 and S < 0), respectively; however, a static wedge (S = 0) doesn't move, $N_b = \frac{(\rho C_p)_s D_B (C_w - C_\infty)}{(\rho C_p)_f \nu_f}$ Brownian diffusion parameter, $N_t = \frac{D_T(\rho C_p)_s (T_w - T_\infty)}{(\rho C_p)_f \nu_f T_\infty}$ thermophoresis parameter, $Sc = \frac{\nu_f}{D_T}$ Schmidt number, $\gamma = N_1\sqrt{\frac{a}{\nu_f}}$ is the thermal slip, and $\varepsilon = N_2\sqrt{\frac{a}{\nu_f}}$ is solutal slip.

For engineering aids, we must calculate the following quantities:

Friction drag: $C_f = \frac{\tau_s}{\rho u_e^2}, \tau_s = \mu\left(\frac{\partial u}{\partial y}\right)_{y=0}$, which converts into reduced skin friction $C_{fr} = C_f \mathrm{Re}^{\frac{1}{2}} = qf''(0)$ *where* $q = \sqrt{\frac{m+1}{2}}$

Heat transfer coefficient: $Nu = \frac{xq_s}{\kappa(T_w - T_\infty)}, q_s = -\kappa\left(\frac{\partial T}{\partial y}\right)_{y=0}$, which becomes $Nur = NuRe^{-\frac{1}{2}} = -q[1 + \frac{4}{3}Nr[1 + (\theta_w - 1)\theta(0)]^3]\theta'(0)$ *where* $q = \sqrt{\frac{m+1}{2}}$

Table 12.4 Values of C_{fr} for Hartree pressure gradient, m

Hartree pressure gradient, m	*Mukhopadhyay et al. (2013)*	*Raju and Sandeepa (2016)*	*Present result*
−0.05	0.123802	0.1234791	0.1234612
0.0	0.332206	0.3332312	0.3334231
0.333	0.757586	0.7574542	0.7575163
1	1.232710	1.23258887	1.2322462

12.3 NUMERICAL SOLUTION METHODOLOGY

In this chapter, two successful mathematical techniques have been used in order to address the nonlinear ODEs (Equation 12.11) with boundary conditions as in Equations (12.12 and 12.13) using RK-4 and Newton's strategy. The entire simulation is conducted using MAPLE-17 software. A computation is done mathematically for numerous of pertinent parameters. Table 12.4 presents the numerically determined quantities of C_{fr} for different values of "m", with $\beta = Nr = 0$. The findings are compared with the conclusions made by Mukhopadhyay et al. (2013) and Raju and Sandeepa (2016). We see from Table 12.4, a strong concurrence with the above-discussed points. Figure 12.2 is an illustration of a flow chart.

12.4 RESULT AND DISCUSSION

The system of nonlinear ODEs (Equation 12.11) is solved using numerical techniques along with BCs in Equations (12.12) and (12.13). The graphs and charts represent the results of the calculations and provide an accurate visual

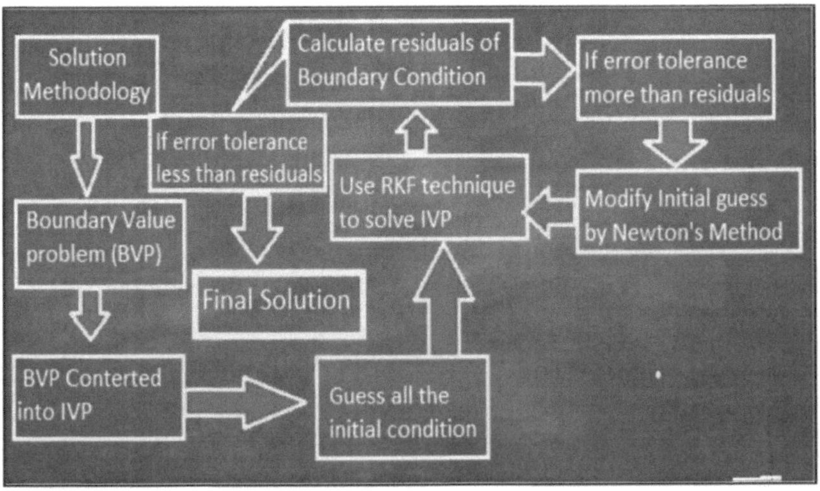

Figure 12.2 RK-4 base flow chart.

representation of the data. The graphs and charts are also useful for analyzing the data and providing insights into potential trends or patterns.

12.4.1 The velocity profile

According to Figure 12.3(i–iii), velocity profile (f') curves were presented for mono-nanofluids (Fe_3O_4-H_2O), hybrid nanofluids (Fe_3O_4-Ag-H_2O), and ternary hybrid nanofluids (Fe_3O_4-Ag-H_2O) against m, h', S. Nanofluid flow can be significantly affected by the induced magnetic parameter. It is possible to introduce a magnetic dipole moment into the suspension of nanoparticles by subjecting a nanofluid to a magnetic field, altering the fluid flow behavior. These formations may hinder the flow of the fluid and make it more resistant to deformation, which raises the viscosity of the fluid. Consequently, as β increases, f' progressively decreases, as illustrated in Figure 12.3(i). Nanofluids exhibit MHD behavior when a magnetic field is present. This implies that electromagnetic forces have an impact on fluid motion. Specifically, the Lorentz

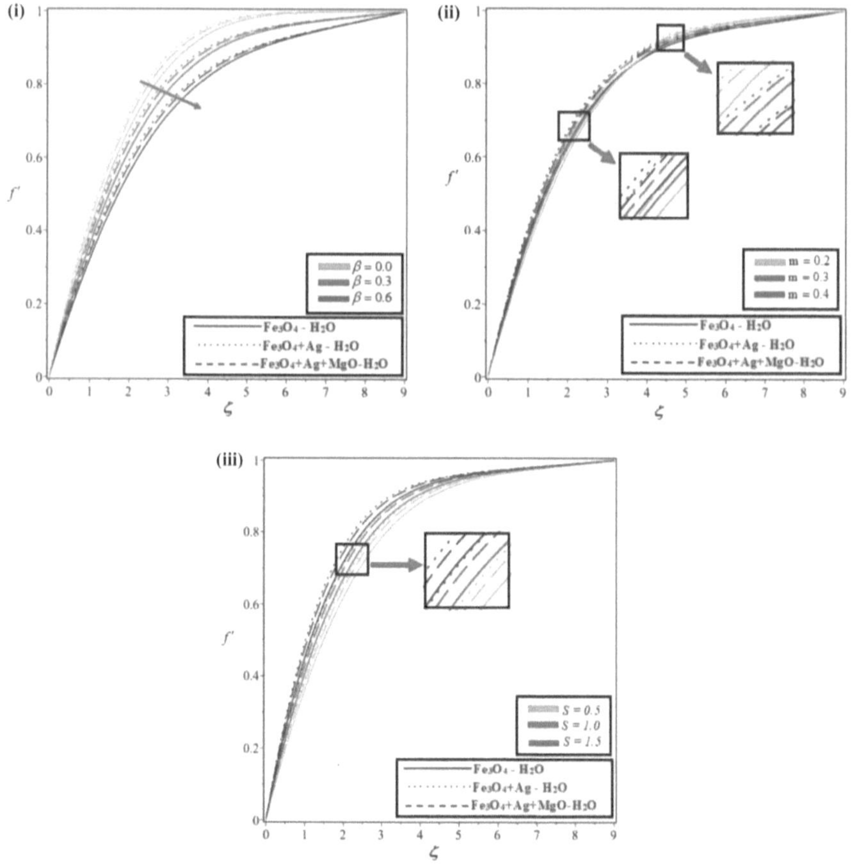

Figure 12.3 Velocity profile for (i) β, (ii) m, and (iii) S.

force is an important factor in the changing f' of nanofluids. A variation in f' of m is shown in Figure 12.3(ii). The fluid velocity seems to escalate for the Hartree pressure gradient in the region $0 \le \zeta \le 4.5$, and a reverse result has been observed in the region $\zeta > 4.5$. Based on a comparison of Fe_3O_4-H_2O, Fe_3O_4-Ag-H_2O, and Fe_3O_4-Ag-MgO-H_2O nanofluids, Figure 12.3(i and ii) demonstrates that hybrid nanofluids perform better than both mono and ternary hybrid nanofluids. Figure 12.3(iii) illustrates how the wedge parameter affects velocity profiles. Fluid velocity appears to raise the wedge parameter. With reference to aerodynamics, the wedge parameter is a geometric parameter that characterizes the shape of an obstacle or body in fluid flow. It is commonly denoted by the symbol S. S is frequently associated with ω, which is the created angle by the flow direction and the leading edge of the wedge. In response to changes in wedge angle, flow patterns can change considerably around the wedge. It could be compared to a wedge flow through a flat plate (ω near to 0). With larger wedge angles, the flow segregation and recirculating zones may become more apparent.

12.4.2 The induced magnetic field profile

Figure 12.4(i–iv) highlighted the demonstration of h' mono-nanofluid (Fe_3O_4-H_2O), hybrid nanofluid (Fe_3O_4-Ag-H_2O), and ternary hybrid nanofluid (Fe_3O_4-Ag-MgO-H_2O) against the variation of Pr_m, h', S, and m, respectively.

Figure 12.4(i) illustrates β variation for h'. A fluctuating magnetic flux across a surface generates an electromotive force (EMF) in any loop or circuit bounded by that surface. For that reason, β seems to have escalated for h' in Figure 12.4(i). A comparison of mono-nanofluid (Fe_3O_4-H_2O), hybrid nanofluid (Fe_3O_4-Ag-H_2O), and ternary hybrid nanofluid (Fe_3O_4-Ag-MgO-H_2O) indicates that in mono-nanofluid, the β profile is higher, while in hybrid nanofluid, it is lower. Variations in the generated magnetic field caused by Pm_r become visible in Figure 12.4(ii). It is witnessed that β decreases as Pm_r grows. Pm_r may exert a substantial influence by means of the properties of magnetic fluids, often known as ferrofluids or magnetorheological fluids. Consequently, by modifying the fluid's effective magnetic diffusivity, induced magnetic field profile curves can modify the Pm_r. The β changes for S and m are shown in Figure 12.4(iii and iv), respectively. Higher values of S and m appear to result in a decrease in the induced magnetic field. When wedges are constructed, the distribution of β may be impacted by the wedge's angle. The results in both cases clearly show that Fe_3O_4-Ag-H_2O has a substantially faster reduction rate than both to Fe_3O_4-H_2O and Fe_3O_4-Ag-MgO-H_2O.

12.4.3 The temperature profile

Figure 12.5(i–vi) illustrates how different parameters affect temperature profiles. There has been discussion of a comparative analysis of three types of nanofluids: Fe_3O_4-H_2O, Fe_3O_4-Ag-H_2O, and Fe_3O_4-Ag-MgO-H_2O in ternary

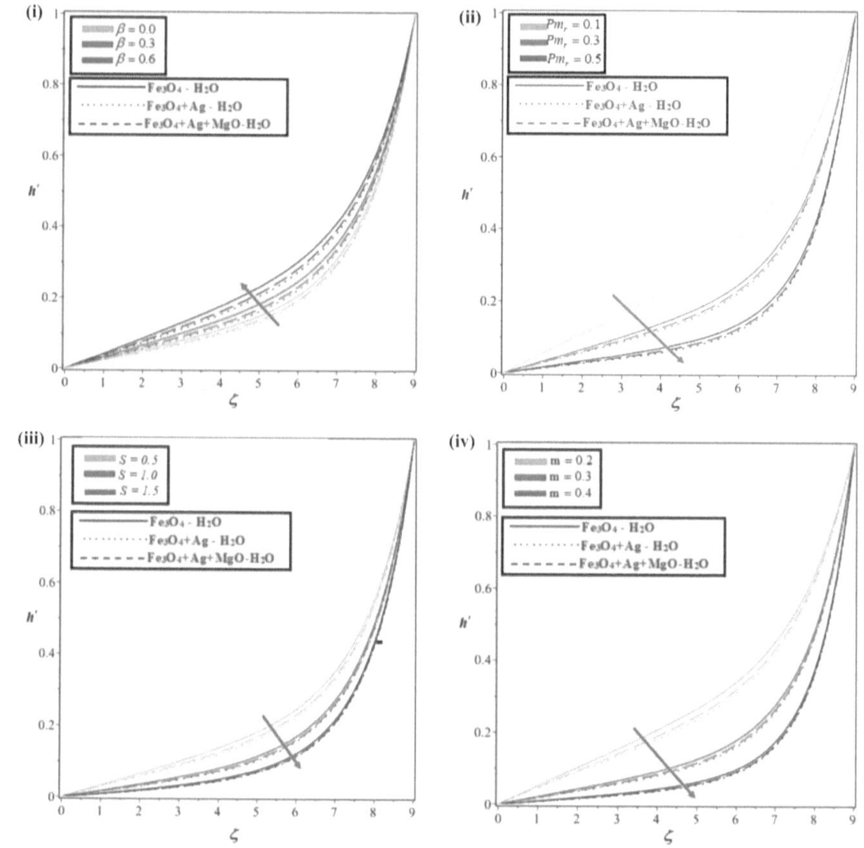

Figure 12.4 Induced magnetic field for (i) β, (ii) Pm_r, (iii) S, and (iv) m.

form. By aligning nanoparticles with a magnetized field, β increases the thermal conductivity of nanofluids, improving HT performance in various applications. This indicates that a greater β raises the flow's temperature, as seen in Figure 12.5(i). The figures clearly show that the increase rate for Fe_3O_4-Ag-MgO-H_2O is significantly faster than that for Fe_3O_4-H_2O and Fe_3O_4-Ag-H_2O. Figure 12.5(ii) illustrates the effects of solar radiation on θ. Radiation temperatures appear to be rising. In comparison to Fe_3O_4-H_2O and Fe_3O_4-Ag-H_2O, Fe_3O_4-Ag-MgO-H_2O investigates a greater thermal profile. Identically, as for the heat source factor, Figure 12.5(iii) displays the thermal pattern swelling. Figure 12.5(iv) shows the reverse trend for the wedge parameter. Fe_3O_4-Ag-MgO-H_2O acquires a high magnitude compared to Fe_3O_4-H_2O and Fe_3O_4-Ag-H_2O.The Hartree pressure gradient temperature variations are shown in Figure 12.5(v). The fluid temperature seems to be a diminishing trend for m.

Figure 12.5(vi) depicts the temperature profile (θ) variations for γ.The fluid temperature seems to be a diminishing trend for γ. Thermal slip generally

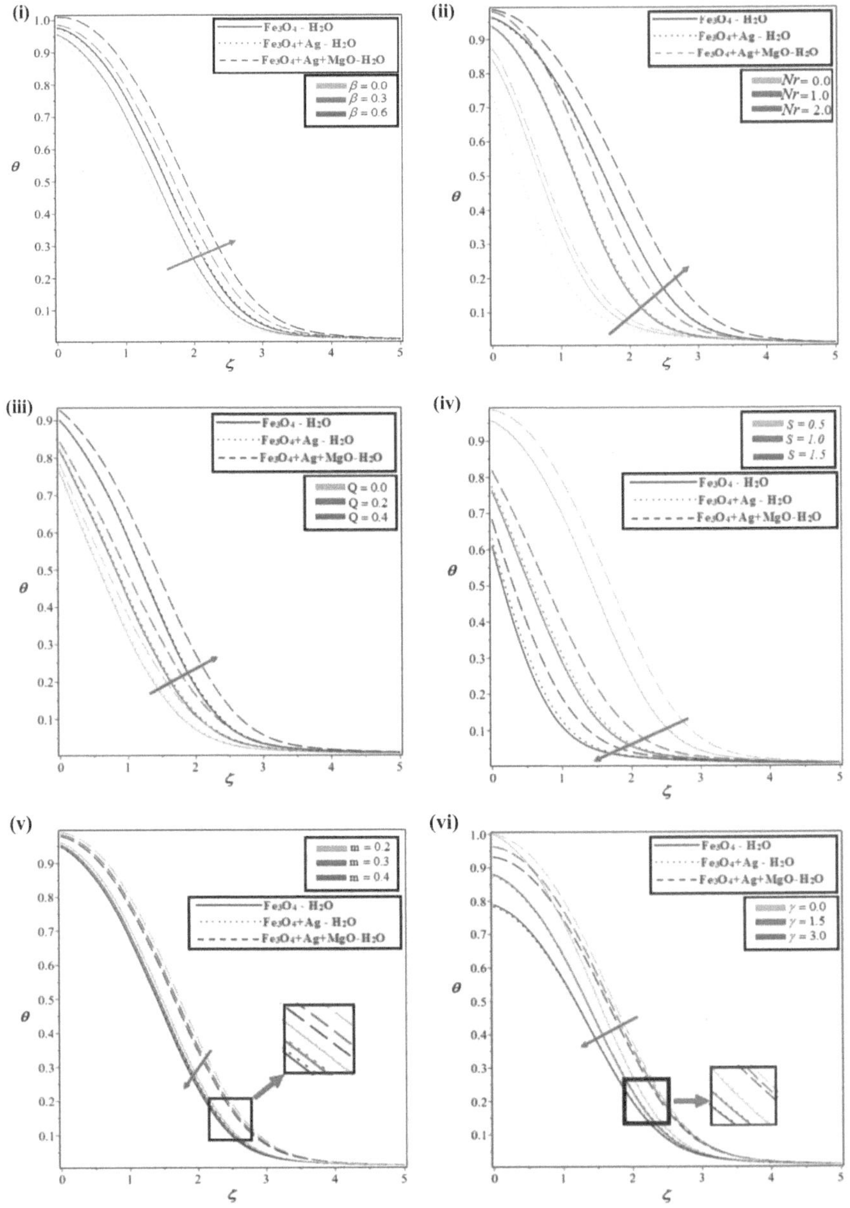

Figure 12.5 Temperature profiles for (i) β, (ii) *Nr*, (iii) Q, (iv) S, (v) *m*, and (vi) γ.

enhances HT at the boundary, leading to a decrease in the temperature gradient close to the interface between solid and liquid. This means that the rate at which temperature changes with distance from the boundary is altered compared to the case without thermal slip.

12.4.4 Analysis of skin friction coefficients (C_{fr})

Figure 12.6(i–iii) shows contour color plots for S and β that show the skin friction, C_{fr} for Fe_3O_4-H_2O, Fe_3O_4-Ag-H_2O, and Fe_3O_4-Ag-MgO-H_2O nanofluids, respectively. Amplification of C_{fr} can be detected with rising values of $S(0.5 \leq S \leq 1.5)$ and declining values of $\beta(0 \leq \beta \leq 0.6)$ for Fe_3O_4-H_2O, Fe_3O_4-Ag-H_2O, and Fe_3O_4-Ag-MgO-H_2O in Figures 12.6(i, ii, and iii), respectively. It is noted from the figure that the maximum value of C_{fr} is 0.55502 for hybrid nanofluids (Fe_3O_4-Ag-H_2O). Furthermore, compared to mono-nanofluids (Fe_3O_4-H_2O) and ternary hybrid nanofluids (Fe_3O_4-Ag-MgO-H_2O), the rate of A has a greater impact on hybrid nanofluid flow.

12.4.5 Analysis of the local Nusselt number (*Nur*)

The visual display for Fe_3O_4-H_2O, Fe_3O_4-Ag-H_2O, and Fe_3O_4-Ag-MgO-H_2O for *Nr* and *Q* on *Nur* is embellished in Figure 12.7(i–iii). Enlargement of

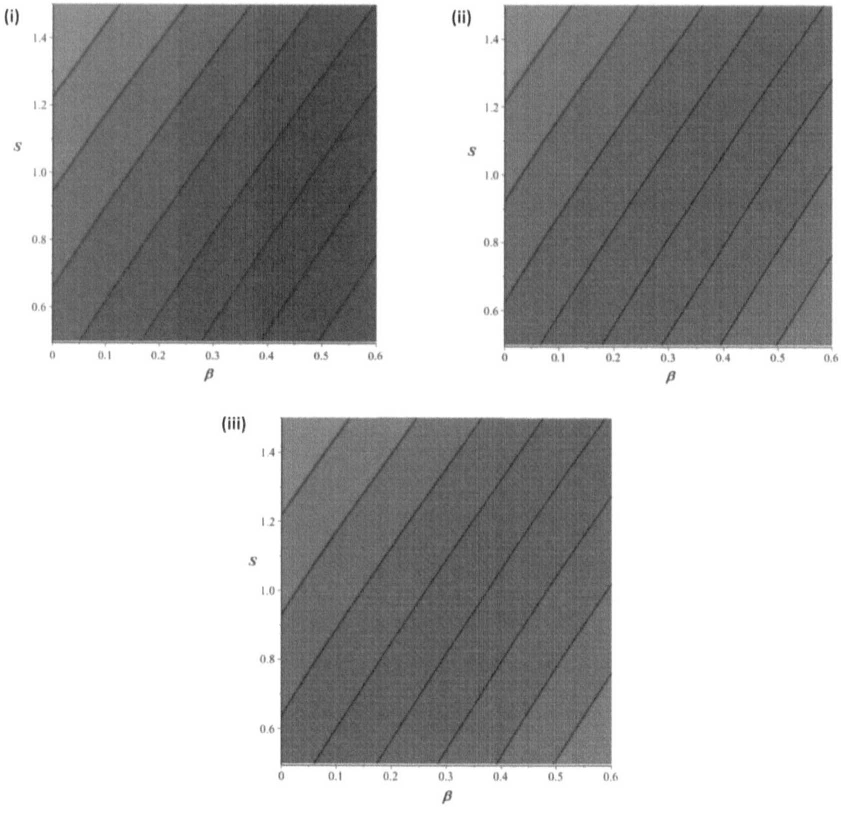

Figure 12.6 Skin friction aspect in contour plots (i) Fe_3O_4-H_2O, (ii) Fe_3O_4-Ag-H_2O, and (iii) Fe_3O_4-Ag-MgO-H_2O.

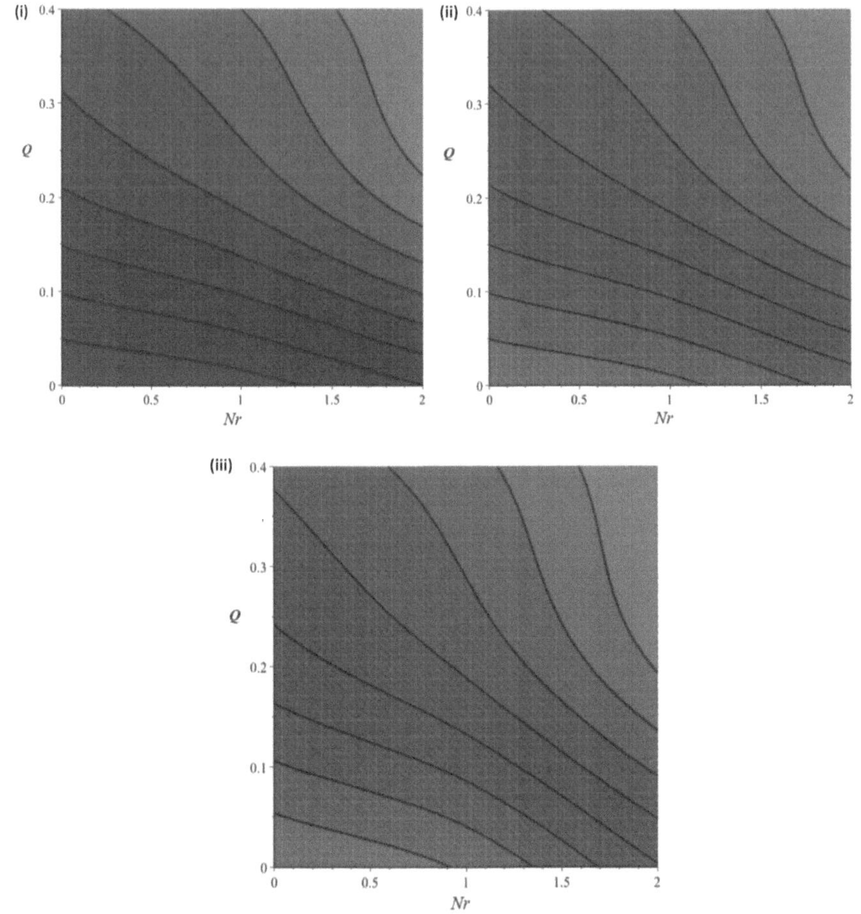

Figure 12.7 Nusselt numbers aspect in contour plots for ((i) Fe_3O_4-H_2O, (ii) Fe_3O_4-Ag-H_2O, and (iii) Fe_3O_4-Ag-MgO-H_2O.

Nur can be observed with both declining values of *Nr* and *Q* for Fe_3O_4-H_2O, Fe_3O_4-Ag-H_2O, and Fe_3O_4-Ag-MgO-H_2O in Figure 12.7(i–iii). The figure illustrates that *Nur* is maximum between *Nr*(=0.0 approximately) and *Q* (=0.0 approximately) and the highest possible rate is 34.98% for Fe3O4-Ag-MgO-H2O, whereas the HT rates of Fe_3O_4-$H_2O$and Fe_3O_4-Ag-H_2O are 16.95% and 20.30%, respectively. *Nur* primarily characterizes convective HT, thermal radiation can influence the transfer of heat processes, especially in situations involving high temperatures or transparent media. It is crucial for the accurate analysis of HT and the design of systems to understand how convection and radiation interact.

A 2D ternary hybrid nanofluidic streamline is shown in Figures 12.8(i–iv) and stream function values are presumably reduced for the more slippery regime.

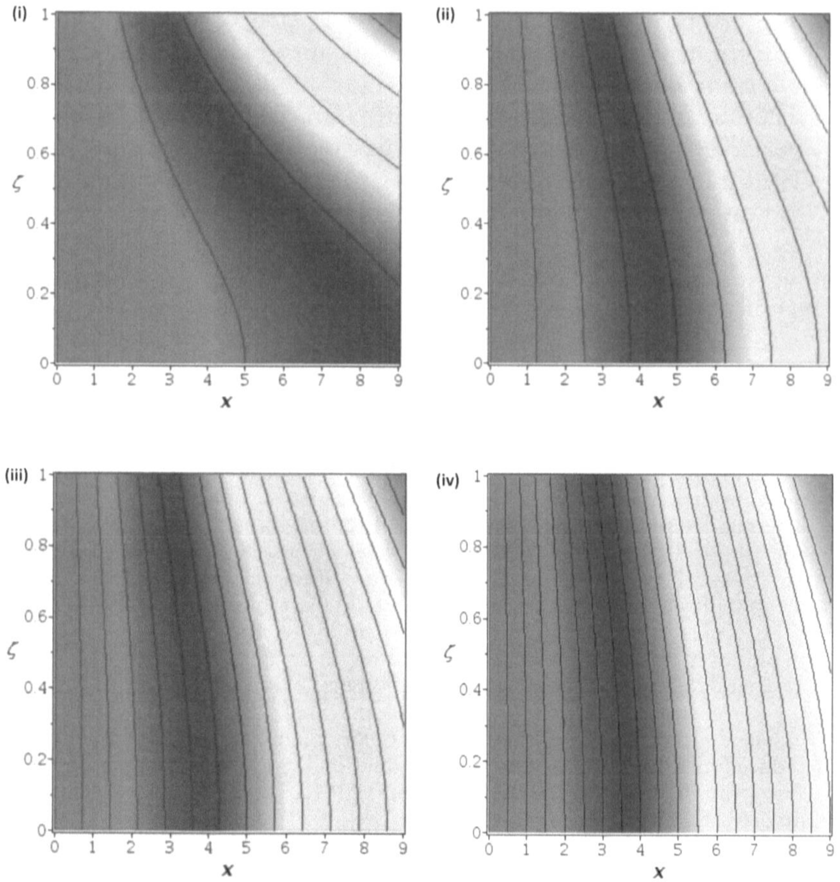

Figures 12.8 (i–iv) Streamlines for $S = 0.1$, $S = 0.4$, $S = 0.7$, and $S = 1.0$.

12.5 CONCLUSIONS

The goal of this study is to better understand the hydrothermal scenario of Fe_3O_4-Ag-MgO-H_2O ternary hybrid nanofluidic movement in water toward a wedge under the influence of induced magnetic fields. In the current model, solar radiation and nanoparticle shape are taken into account. There are several streamline plots, contour plots, and graphs that illustrate the important findings. The study's principal conclusions are as follows:

- The velocity is declining for the induced magnetic parameter and swelling for the wedge parameter, despite the fact that Hartree pressure gradients appear to have a dual character.
- Prandtl number and Hartree pressure gradients result in a smaller magnetic induced field, but the induced magnetic parameter reveals the opposite result.

- Thermal profiles are trending for the parameters of solar radiation and induced magnetic field, and Hartree pressure gradients display the conflicting trend.
- The drag friction boost for the wedge parameter, and the contrary trend is marked for the induced magnetic parameter.
- Heat transference decreases for solar radiation. The highest magnitude in heat transport is shown in the Fe_3O_4-Ag-MgO-H_2O sightsee.

In the future, the proposed study should include statistical analyses such as correlation coefficients, probable error, and entropy generation.

REFERENCES

Acharya, N. (2020). On the flow patterns and thermal behavior of hybrid nanofluid flow inside a microchannel in presence of radiative solar energy. Journal of Thermal Analysis and Calorimetry, 141, 1425–1442.

Acharya, N., and Mabood, F. (2012). On the hydrothermal features of radiative Fe_3O_4–graphene hybrid nanofluid flow over a slippery bended surface with heat source/sink. Journal of Thermal Analysis and Calorimetry, 143, 1273–1289.

Al Oweidi, K. F. (2022). Partial differential equations of entropy analysis on ternary hybridity nanofluid flow model via rotating disk with hall current and electromagnetic radiative influences. Scientific Reports, 12 (1), 20692.

Basha, T., Sivaraj, R., Reddy, A. S., and Chamkha, A. J. (2019). SWCNH/diamond-ethylene glycol nanofluid flow over a wedge, plate and stagnation point with induced magnetic field and nonlinear radiation – Solar energy application. The European Physical Journal Special Topics, 228, 2531–2551.

Chakraborty, T., Duari, P. R., and Acharya, N. (2023). Unfolding flow features of MHD hybrid nanofluid (Ag–Al_2O_3–H_2O) and mono nanofluid (Al_2O_3–H_3O) flow over exponentially expanded sheet soaked in a Darcy–Forchheimer absorbent medium coexisting non-uniform heat generation/absorption. *Waves in Random and Complex Media*, 1–28. https://doi.org/10.1080/17455030.2023.1271153.

Choi, S. U. S. (1995). Enhancing thermal conductivity of fluids with nanoparticles in developments and applications of non-Newtonian flows. FED 231/MD 66, 99–105.

Das, K., Acharya, N., SK, M. T., Duari, P. R., and Chakraborty, T. (2022a). Slip flow of hybrid nanofluid in presence of solar radiation. International Journal of Modern Physics C, 33 (2), 2250017.

Das, K., Duari, P. R., and Kundu, P. K. (2014). Solar radiation effect on Cu–water nanofluid flow over a stretching sheet with surface slip and temperature jump. Arabian Journal for Science and Engineering, 39, 9015–23.

Dolui, S., Bhaumik, B., and De, S. (2023). Combined effect of induced magnetic field and thermal radiation on ternary hybrid nanofluid flow through an inclined catheterized artery with multiple stenosis. Chemical Physics Letters, 63, 140209.

El-Khouly, M. M., El Bouz, M. A., and Sultan, G. I. (2012). Experimental and computational study of using nanofluid for thermal management of electronic chips. Journal of Energy Storage, 39, 102630.

Ghadikolaei, S. S., Hosseinzadeh, K., and Ganji, D. D. (2018). Investigation on three-dimensional squeezing flow of mixture base fluid (ethylene glycol-water) suspended by hybrid nanoparticle (Fe_3O_4-Ag) dependent on shape facto. Journal of Molecular Liquids, 262 (15), 376–388.

Goud, J. S., Srilatha, P., Kumar, R. S. V., Kumar, K. T., Khan, U., Raizah, Z., Gill, H. S., and Galal, A. M. (2022). Role of ternary hybrid nanofluid in the thermal distribution of a dovetail fin with the internal generation of heat. Case Studies in Thermal Engineering, 35, 101213.

Gul, H., Ramzan, M., Chung, J. D., Chu, Y. M., and Kadry, S. (2012). Multiple slips impact in the MHD hybrid nanofluid flow with Cattaneo–Christov heat flux and autocatalytic chemical reaction. Scientific Reports, 11 (1), 1–14.

Hamilton, R., and Crosser, O. (1962). Thermal conductivity of heterogeneoustwo-component systems. Industrial & Engineering Chemistry Fundamentals, 1, 187–191.

Hunt, A. J. (1978). Small particle heat exchangers, Report LBL-78412 for the US Department of Energy. Lawrence Berkeley Laboratory.

Imran, M., Yasmin, S., Waqas, H., Ali Khan, S., Muhammad, T., Alshammari, N., Hamadneh, N. N., and Khan, I. (2022). Computational analysis of nanoparticle shapes on hybrid nanofluid flow due to flat horizontal plate via solar collector. Nanomaterials, 12, 663.

Jalili, P., Azar, A. A., Jalili, B., and Ganji, D. D. (2023). Study of nonlinear radiative heat transfer with magnetic field for non-Newtonian Casson fluid flow in a porous medium. Results in Physics, 48, 106371.

Kandasamy, K., Muhaimin, I., Khamis, A. B., and Roslan, R. B. (2013). Unsteady Heimenz flow of Cu nanofluid over a porous wedge in the presence of thermal stratification due to solar energy radiation: Lie group transformation. International Journal of Thermal Science, 65, 196–205.

Khan, S. H., Nie, Y., and Ali, B. (2019). Multiple slip effects on magnetohydrodynamic axisymmetric buoyant nanofluid flow above a stretching sheet with radiation and chemical reaction. Symmetry, 11, 1171.

Khashi'ie, N. S., Md Arifin, N., Sheremet, M., and Pop, I. (2012). Shape factor effect of radiative $Cu–Al_2O_3/H_2O$ hybrid nanofluid flow towards an EMHD plate. Case Studies in Thermal Engineering, 26, 101199.

Khashi'ie, N. S., Md Arifin, N., and Pop, I. (2022). Magnetohydrodynamics (MHD) boundary layer flow of hybrid nanofluid over a moving plate with Joule heating. Alexandria Engineering Journal, 61 (3), 1938–1945.

Kumar, B., Seth, G. S., Nandkeolyar, R., and Chamkha, A. J. (2019). Outlining the impact of induced magnetic field and thermal radiation on magneto-convection flow of dissipative fluid. International Journal of Thermal Sciences, 146, 106101.

Mabood, F., and Shateyi, S. (2019). Multiple slip effects on MHD unsteady flow heat and mass transfer impinging on permeable stretching sheet with radiation. Modelling and Simulation in Engineering, *2019*(1), 3052790.

Madkhali, H. A. (2023). Investigation on the role of tri-nanoparticles for non-Fourier thermal enhancement by MHD non-Newtonian rheological fluid composed charged particles. Case Studies in Thermal Engineering, 41, 102655.

Mahmood, Z., Ahammad, N. A., Alhazmi, S. E., Khan, U., and Bani-Fwaz, M. Z. (2022a). Ternary hybrid nanofluid near a stretching/shrinking sheet with heat generation/absorption and velocity slip on unsteady stagnation point flow. International Journal of Modern Physics B, 2250209.

Mahmood, Z., Alhazmi, S. E., Khan, U., Bani-Fwaz, M. Z., and Galal, A. M. (2022b). Unsteady MHD stagnation point flow of ternary hybrid nanofluid over a spinning sphere with Joule heating. International Journal of Modern Physics B. https://doi.org/10.1142/S0217979222502307

Mahmood, Z., Khan, U., Saleem, Rafique, K., and Eldin, S. M. (2023). Numerical analysis of ternary hybrid nanofluid flow over a stagnation region of stretching/shrinking curved surface with suction and Lorentz force. Journal of Magnetism and Magnetic Materials. https://doi.org/10.1016/j.jmmm.2023.170654

Mukhopadhyay, S., Mondal, I. C., and Chamkha, A. J. (2013). Casson fluid flow and heat transfer past a symmetricwedge. Heat Transfer Research, 42 (8), 665–675.

Nadeem, S., Ahmad, S., and Muhammad, N. (2018). Computational study of Falkner-Skan problem for a static and moving wedge. Sensors and Actuators B: Chemical, 263, 69–76.

Pordanjani, A. H., Aghakhani, S., Afrand, M., Mahmoudi, B., and Mahian, O. (2019). An updated review on application of nanofluids in heat exchangers for saving energy. Energy, 198, 111886.

Rafique, K., Mahmood, Z., Khan, U., Eldin, S. M., Oreijah, M., Guedri, K., and Abd El-Wahed Khalifa, A. E. (2023). Investigation of thermal stratification with velocity slip and variable viscosity on MHD flow of Al_2O_3–Cu–TiO_2/H_2O nanofluid over disk. Case Studies in Thermal Engineering, 49, 103292.

Raju, C. S. K., and Sandeepa, N. (2016). Falkner-Skan flow of a magnetic-Carreau fluid past a wedge in the presence of cross-diffusion effects. The European Physical Journal – Plus, 131, 267.

Ramana, R. M., Raju, K. V., and Kumar, J. G. (2012). Multiple slips and chemical reaction effects on unsteady MHD heat and mass transfer flow over a permeable stretching sheet with radiation. Turkish Journal of Computer and Mathematics Education, 12 (6), 4489–4498.

Rashid, U., Abdeljawad, T., Liang, H., Iqbal, A., Abbas, M., and Siddiqui, M. J. (2020). The shape effect of gold nanoparticles on squeezing nanofluid flow and heat transfer between parallel plates. Mathematical Problems in Engineering, 2020. https://doi.org/10.1155/2020/9584854.

Rashid, U., Liang, H., Ahmad, H., Abbas, M., Iqbal, A., and Hamed, Y. S. (2012). Study of (Ag and TiO_2)/water nanoparticles shape effect on heat transfer and hybrid nanofluid flow toward stretching shrinking horizontal cylinder. Results in Physics, 12, 03812.

Sarkar, J., Ghosh, P., and Adil, A. (2015). Review on hybrid nanofluids: Recent research, development and applications. Renewable and Sustainable Energy Reviews, 43, 164–177.

Seid, E., Haile, E., and Walelign, T. (2022). Multiple slip, Soret and Dufour effects in fluid flow near a vertical stretching sheet in the presence of magnetic nanoparticles. International Journal of Thermofields, 13, 100136.

Shamshuddin, M. D., Akkurt, N., Saeed, A., and Kumam, P. (2023). Radiation mechanism on dissipative ternary hybrid nanoliquid flow through rotating disk encountered by hall currents: HAM solution. Alexandria Engineering Journal, 65, 543–559.

Wole-osho, I., Okonkwo, E. C., Abbasoglu, S., and Kavaz, D. (2020). Nanofluids in solar thermal collectors: Review and limitations. International Journal of Thermophysics, 41, 157.

Yang, L., and Xu, X. (2017). A renovated Hamilton–Crosser model for the effective thermal conductivity of CNTs nanofluids. International Communications in Heat and Mass Transfer, 81, 42–50.

Zahmatkesh, I., Sheremet, M., Yang, L., Heris, S. Z., Sharifpur, M., Meyer, J. P., Ghalambaz, M., Wongwises, S., Jing, S. D., and Mahian, O. (2012). Effect of nanoparticle shape on the performance of thermal systems utilizing nanofluids: A critical review. Journal of Molecular Liquids, 312, 114430.

Zangooee, M. R., Hosseinzadeh, K., and Ganji, D. D. (2019). Hydrothermal analysis of MHD nanofluid (TiO_2-GO) flow between two radiative stretchable rotating disks using AGM. Case Studies in Thermal Engineering, 14, 100460.

Chapter 13

Enhancing refrigeration system performance through condenser heat recovery and thermal energy storage

Avesahemad S.N. Husainy, Sonali Kale, Atharv Joshi, and Vaibhav Chougule

13.1 INTRODUCTION

In modern energy management, maximizing efficiency and minimizing waste are critical goals, particularly in systems that consume significant amounts of energy, such as refrigeration systems. Refrigeration plays a pivotal role in various sectors, including residential, commercial, and industrial applications, by preserving food, maintaining comfortable indoor climates, and supporting industrial processes. However, the energy demands of refrigeration systems can be substantial, leading to high operational costs and increased environmental impact.

Condenser heat recovery involves capturing the waste heat generated by the refrigeration system's condenser and repurposing it for other uses. Traditionally, this heat is expelled into the atmosphere as a byproduct of the refrigeration process. By recovering and utilizing this heat, it is possible to offset the energy requirements of additional heating systems, thereby reducing overall energy consumption and operational costs.

Thermal energy storage (TES), on the other hand, focuses on storing excess thermal energy produced during off-peak hours for use during peak demand periods. This technology allows for the balancing of energy loads and can significantly enhance the efficiency of refrigeration systems. By leveraging lower energy rates during off-peak times and utilizing stored energy during peak periods, TES systems help manage energy consumption more effectively.

The condenser in refrigeration systems is in charge of releasing the heat that is taken in by the chilled area. By capturing and utilizing this waste heat for different purposes, condenser heat recovery seeks to increase system efficiency as a whole. The fundamental principle behind condenser heat recovery is the concept of heat transfer. In refrigeration systems, the refrigerant absorbs heat from the refrigerated space and transfers it to the condenser. By using a heat exchanger, this heat can be redirected to another medium (e.g., water or air) for reuse. TES is a well-known and essential technology that helps balance energy supply and demand on a daily, weekly, or even seasonal basis by recovering heat from industrial waste and capturing renewable energy sources (Tao and He, 2018). Heat recovery can be implemented by

DOI: 10.1201/9781003534785-13

expanding a conventional installation or by designing it appropriately, which will recover wasted energy. Such a process can result in a significant reduction of the facility's operational costs related to heating and hot domestic water preparation (Zając, 2019).

Sensible heat storage involves heating a medium and using its temperature change to store energy. Latent heat storage involves storing energy in the phase change of a material (e.g., ice melting). During periods of low energy demand or lower energy costs (charging phase), excess energy is used to store heat in the TES system. During high-demand periods (discharging phase), the stored thermal energy is released to supplement the refrigeration system or provide heating. In this innovation, phase change material (PCM) in the form of copper compartments or a rectangular box is positioned to function as a latent heat storage device. When the setup is activated, the cycle runs in a no-load state, improving the vapor compression refrigeration system (VCRS) cycle's performance. The storage is turned on as we continue to increase the loading circumstances. Starting the experiment included replacing the PCM in the copper chambers with water.

13.2 LITERATURE SURVEY

The waste heat recovery system in a domestic refrigerator was tested experimentally at different load levels, including no load, 1 liter, 3 liters, 5 liters, and 10 liters. The performance metrics were computed and compared for each load level. When the experiment duration is extended under all loading scenarios, the waste heat recovery rate rises. A maximum heat recovery of between 31.15 and 33.66 kJ/kg is achieved (Thangavel et al., 2020). TES and waste heat recovery are two of the HVAC industry's biggest problems. Earlier, refrigeration systems' condensers used to waste heat, resulting to a significant energy waste (Stalin et al., 2012). However, new studies have shown that this waste heat may be used for a variety of purposes, including food drying and process warmth maintenance. This approach saves money by preserving energy without adding to costs, in addition to offering a technically viable solution. A study indicates that air conditioners and refrigerators use 15% or so of the world's electricity. Also proposed reusing waste heat from refrigeration condensers to enhance COP and perform tasks like food drying economically, offering a technically feasible and economically viable solution to save energy and balance electricity demand. Vapor compression refrigeration offers a huge range of applications in both household and industrial settings because of its extremely high coefficient of performance (COP). The primary focus of the research is on repurposing condenser waste heat and applying it to specific tasks such as food drying or food warming while concurrently optimizing the COP. Both technically and financially, this approach is doable. This is the most practical way to save energy and balance the supply and demand for electricity without using more energy (Husainy et al., 2017). Romdhane Ben

Slama (2009) presented results showing that their system could produce water at a temperature of 60°C, alongside analyzing the economic importance of waste heat recovery systems from an energy-saving standpoint. Elumalai et al. (2015) emphasized the significance of experimental investigations for optimizing certain design parameters.

A significant body of research has been conducted on the design, development, and testing of refrigerating systems containing PCMs for various applications such as cold storages, ice factories, beverage storages, domestic refrigeration systems, and industrial refrigeration systems. Odeh and Abu-Mulaweh (2012) introduced an experimental setup aimed at aiding mechanical engineering undergraduates in understanding practical heat transfer principles by harnessing waste heat from a window air conditioner to devise a thermosiphon system for heating water efficiently.

PCMs can retain surplus heat and keep temperatures stable during power outages because of their capacity to absorb and release significant quantities of thermal energy during phase transitions (Reddy et al., 2016). Integrating PCMs as latent heat thermal energy storage systems appears to be a potential way to address these issues (Sharma et al., 2022). Because of their superior heat transport properties and huge latent heat storage capacity, nano-enhanced phase change materials (NEPCMs) are a good option for enhancing the thermal efficiency of refrigeration systems. In order to increase NEPCM's latent heat, the research attempts to determine its thermal properties. The PCM material utilized is a graphene (CuO) nanoparticle-infused eutectic combination of potassium chloride (KCL) and sodium chloride (NaCl) that has been integrated into a refrigeration test rig.

13.3 EXPERIMENTAL SET UP

Waste heat recovery from the condenser involves capturing the thermal energy expelled during the condensation process and redirecting it for beneficial uses. Traditionally, this waste heat is released into the environment, but innovative technologies now allow for its recovery and application in various fields, from domestic hot water heating to industrial processes. By harnessing this otherwise wasted energy, systems can improve their overall efficiency, reduce operational costs, and lower their environmental impact. Thermal storage in the evaporator refers to the technique of capturing and storing thermal energy absorbed during the refrigeration cycle. This energy can be stored in various forms, such as PCMs or ice, and used during periods of high demand or to balance energy consumption. Thermal storage not only allows for load shifting, which can lead to significant cost savings, but also enhances the operational efficiency of refrigeration systems by mitigating peak load pressures and reducing system wear and tear. For the primary enhancement, installing a thermal heating box above the simple refrigerator is carried out. This heating box will consist of modified coils from the

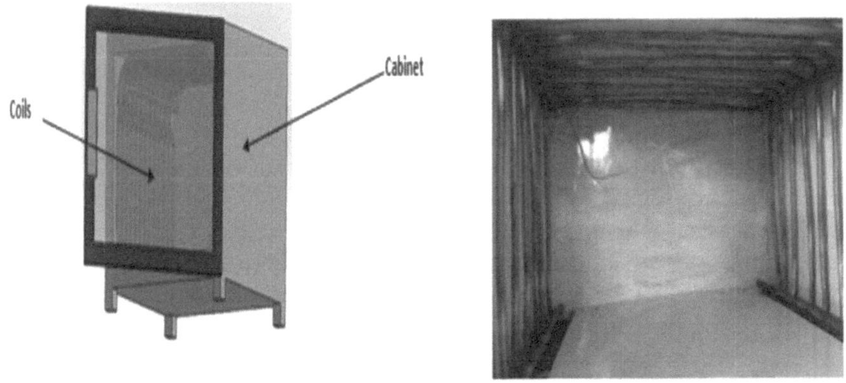

Figure 13.1 Thermal heating box.

refrigerator's condenser, arranged as a heat exchanger. These coils, when brazed in the cabinet, will act as hot coils to keep food warm. Additionally, the refrigerator can still function as a conventional one by keeping the cabinet door open when the heat sink is not needed. This setup allows for increased COP, and since less heat is rejected to the environment compared to conventional refrigerators with air-cooled condensers, it's safer in environmental aspects. The developed separate rectangular hot box as shown in Figure 13.1 is to keep it on the refrigerator surrounding round through thermocol. Thereby making a panel side to so that it could fit indicators such as voltmeter, ammeter, temperature indicator, HP-LP cut out, energy meter, two pressure gauges.

In the subsequent enhancement, experimental setup for testing PCM-enhanced evaporator section as shown in Figure 13.2, refrigeration includes essential equipment such as a hermetically sealed compressor, condenser, capillary or expansion valve, and evaporator. The wall cabinet, made of corrosion-resistant MS material is leak-proof and has a proper finish. CO_2 welding ensures strong joints in the cabinet. Various instruments like ammeters, voltmeters, temperature sensors, pressure gauges, and watt meters are used to

Figure 13.2 PCM-enhanced evaporator section.

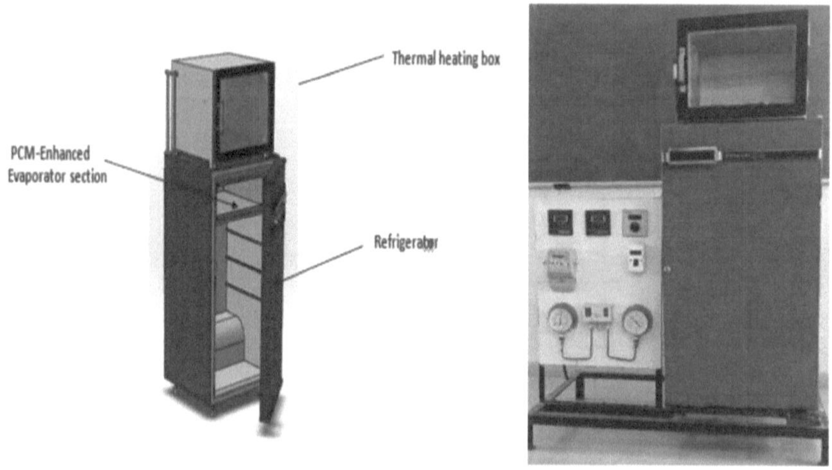

Figure 13.3 Outline of experimental set up.

measure and record data accurately. For experimentation, the use of eutectic/inorganic PCM combinations like $CaCl_2$, NaOH, and KCl mixed with water as the base fluid was done.

The thermal heating box and the PCM-enhanced evaporator section refrigeration arrangement are integrated in the combined experimental setup as shown in (Figure 13.3). Within the thermal heating box, it uses modified refrigerator condenser coils as a heat exchanger. A number of instruments are used in addition to necessary equipment such a compressor, condenser, and evaporator to ensure precise data measurement. Combinations of eutectic and inorganic PCM are used in experiments, offering a thorough assessment of the integrated improvements in refrigeration systems.

13.4 CLASSIFICATION OF TES

The following are a few typical kinds of thermal energy storage systems (Figure 13.4):

1. Sensible heat storage

   Systems with a liquid or solid basis for sensible heat storage (SHS) are both feasible. Liquid systems are less expensive than solid systems, but they are more thermally sensitive and less energy dense. The system and application features are taken into consideration when making the choice. Modern SHS technology includes high-temperature systems. With concrete units that have pipes implanted, rock bed units transfer heat through boreholes and molten salts. Large-scale water systems store excess thermal energy during off-peak hours (Almendros-Ibáñez

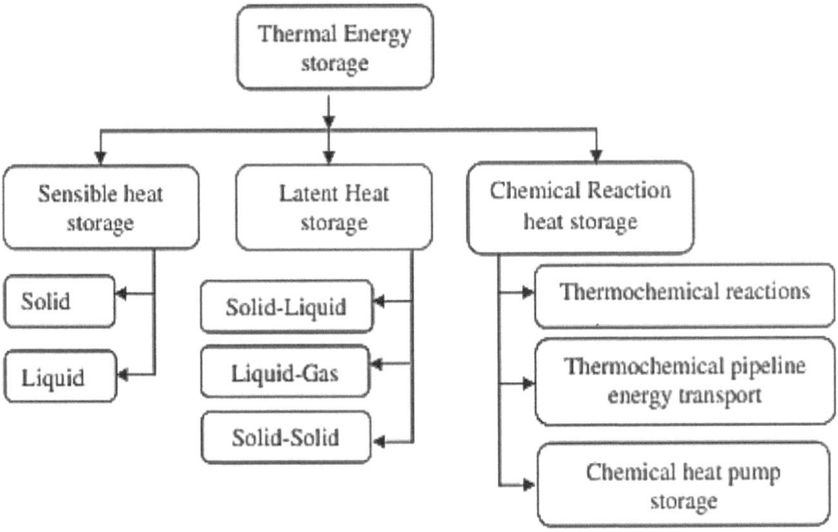

Figure 13.4 Classifications of TES (Dheep and Sreekumar, 2014).

et al., 2019; Koçak et al., 2020; Li et al., 2022; Shafieian et al., 2018; Tao and He, 2018)

$$Q_{sensible} = \int_{T_1}^{T_2} C_p \cdot dT$$

where $Q_{sensible}$ is the sensible heat stored, C_p the specific heat of the material, and dT the temperature change.

Rock beds: Sensible heat storage uses packed stone beds or subterranean rock formations. Rocks are suited for medium- to large-scale applications because of their effective heat-absorbing and heat-releasing properties.

2. Latent heat storage:

Phase transition materials (PCMs): At a constant temperature, PCMs undergo a phase transition from solid to liquid or vice versa, storing thermal energy in the process. Paraffin wax, salt hydrates, and eutectic mixes are examples of common PCMs. They are utilized in many different applications, including solar thermal systems, building envelopes, and electronics cooling, and they provide high-energy storage densities.

$$Q_{latent} = \int_{T_1}^{T_{pc}} C_{p,s} \cdot dT + \Delta H_{pc} + \int_{T_{pc}}^{T_2} C_{p,l} \cdot dT$$

where Q_{latent} is the sensible and latent heat stored and ΔH_{pc} is the heat of fusion at the phase change temperature dT_{pc}. Latent heat storage (LHS) is a technique for storing energy that is utilized in power generation, heating, and cooling by taking advantage of phase shifts in materials like salts and paraffins. It depends on the efficient absorption and release of heat during phase shift, which is influenced by temperature and storage medium, among other things (Muhammad et al., 2023).

13.5 APPLICATION OF TES

TES finds application in solar thermal power plants, defense applications, cold storage, refrigeration and air conditioning, solar water heating systems, solar air heating systems, solar cookers, greenhouse construction, and the materials utilized to store that thermal energy efficiently as shown in Figure 13.5.

- Solar energy systems
  TES systems can store excess thermal energy generated by solar collectors during the day for use during the night or periods of low sunlight. This improves the reliability and dispatch ability of solar energy systems.
- Cold chain and food preservation
  TES systems minimize energy usage and food waste by maintaining the temperature of perishable commodities during storage and transportation in cold chain logistics and food preservation.

13.6 WASTE HEAT RECOVERY

After conducting this experiment, we got some results relating to the objectives. The results show:

- Gradual increase in temperature after some time.
- Cooling effect as well as heating effect in the same application.

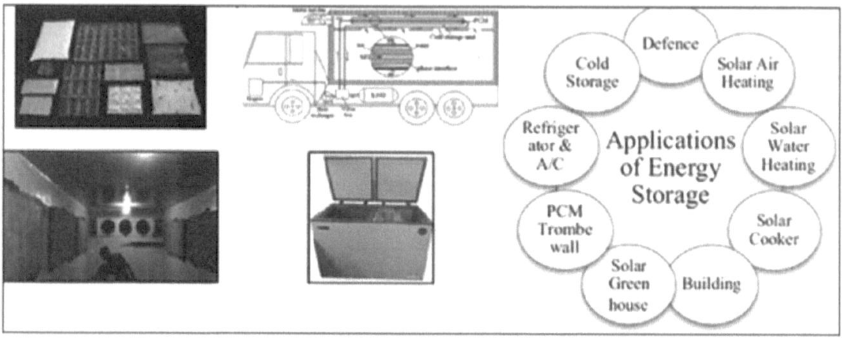

Figure 13.5 Applications of TES.

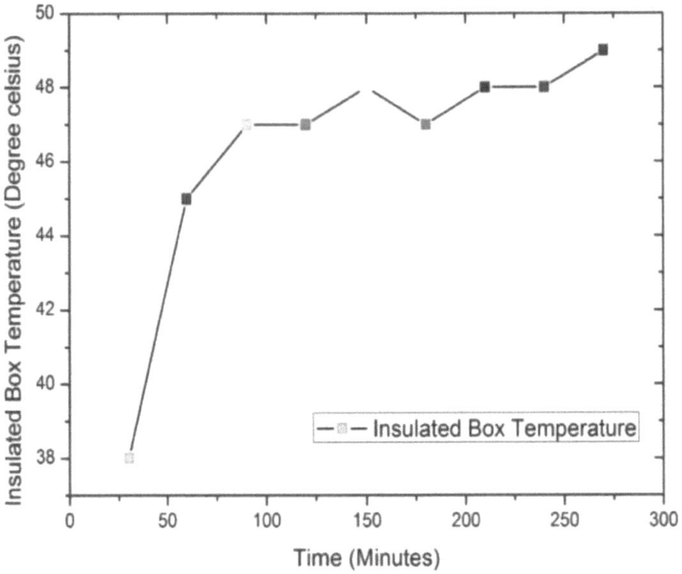

Figure 13.6 Time vs temp.

- System works with greater efficiencies as compared to the conventional refrigeration unit.

Heat recovered in insulated box where temperature increased by 18°C (Figure 13.6). This gives us an opportunity to save and to conserve energy and reuse the same for various purposes. This study shows that we can save considerable amount of energy in the form of electricity. The model described above can work as dual working device that can be used for refrigeration as well as for keeping the food warm or reheating the food. There is also an increase in COP by almost 10% in the new setup where the condenser is mounted on the top of the refrigeration.

13.7 THERMAL ENERGY STORAGE

Table 13.1 Following combinations are used for experimentation

Sr no	*Name of PCM*	*Percentage*
1	Water	NA
2	NaCl	1%, 3%, 5%
3	KCl	1%, 3%, 5%

Table 13.2 Comparison between without PCM and water as a PCM

Time (min)	*COP theoretical water PCM*	*COP theoretical without PCM*	*COP Carnot water PCM*	*COP Carnot without PCM*	*Compressor work with water PCM*	*Compressor work without PCM*	*T4 evaporator outlet without water as a PCM*	*T4 evaporator outlet with water as a PCM*
20	5.7	4.8	4.2	4	5.4	6.1	0	0
40	5.1	4.9	4.1	3.7	5.4	5.9	−1	1
60	5.2	4.4	4.2	3.7	5.4	5.6	2	−1
80	4.3	4.9	3.9	3.8	5.5	5.8	1	−1
100	4.1	4.6	3.9	3.8	5.5	5.7	0	−2
120	4.9	4.6	4	3.7	5.2	5.3	0	0
140	4.9	4.9	4	3.6	5.6	5.6	0	0
160	4.6	4.9	4	3.7	5.4	5.5	0	0
180	4.7	5.3	3.9	3.9	5	5.5	0	0
200	4.9	4.2	4.1	3.7	5.2	5.4	0	0

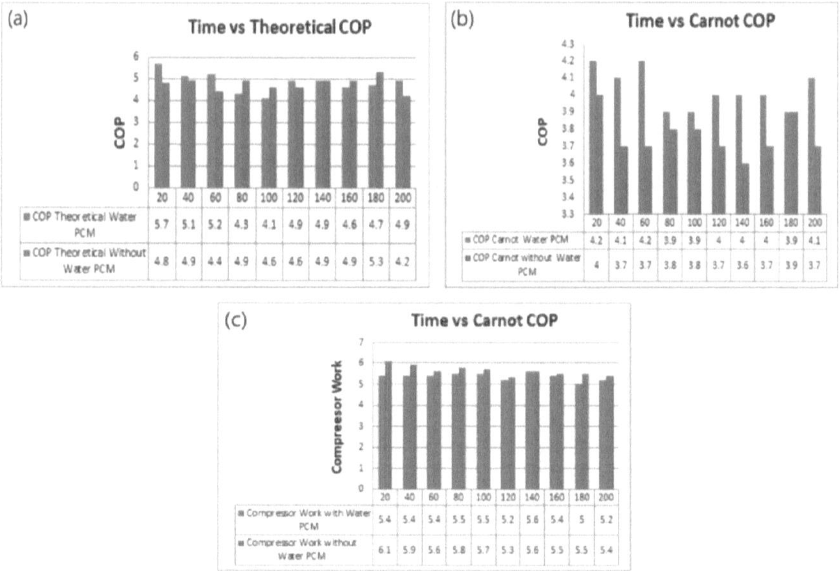

Figure 13.7 (a) Comparison between without PCM and water as a PCM (theoretical COP). (b) Comparison between without PCM and water as a PCM (Carnot COP). (c) Comparison between without PCM and Water as a PCM (compressor work).

Table 13.3 Comparison between evaporator outlet temperature with and without water as PCM

Time	*T4 Evaporator outlet temp without PCM*	*T4 Evaporator outlet temp with water as a PCM*	*Difference In Temperature*
10	27	23	4
20	32	25	7
30	30	28	2
40	30	28	2
50	31	28	3
60	32	28	4
70	31	28	3
80	32	28	4
90	32	29	3
100	32	28	4
110	31	29	2
120	31	29	2
130	32	28	4
140	32	29	3
150	33	29	4

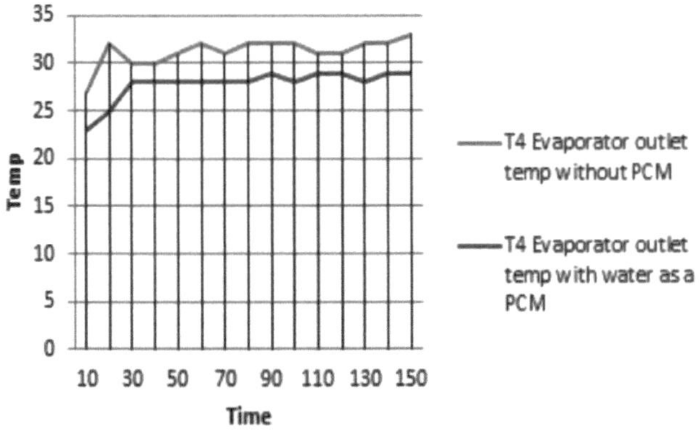

Figure 13.8 Comparison between evaporator outlet temperature with and without water as PCM.

Table 13.4 Experimentation with NaCl as a phase change material 1%

NaCl 1% (When system ON)														
Sr. No	*Voltage*	*Current*	*Cond. pressure*	*Evap. pressure*	*Time period (10 blink)*	*Temperature*					*COP theoretical*	*COP Carnot*	*Compressor work*
						T1	*T2*	*T3*	*T4*	*T5*			
1	257	1.25	2	19	55.94	71	32	1	4	25	4.7	3.9	4.9
2	255	1.23	2	19	60	70	33	1	3	23	4.9	3.9	5.3
3	254	1.2	2	19	65.16	68	35	−1	2	22	4.9	3.9	5.7
4	252	1.18	2	19	62.96	70	35	0	2	20	4.5	3.9	5.5
5	253	1.17	1.8	18	63.38	69	35	−3	1	18	4.4	3.7	5.6

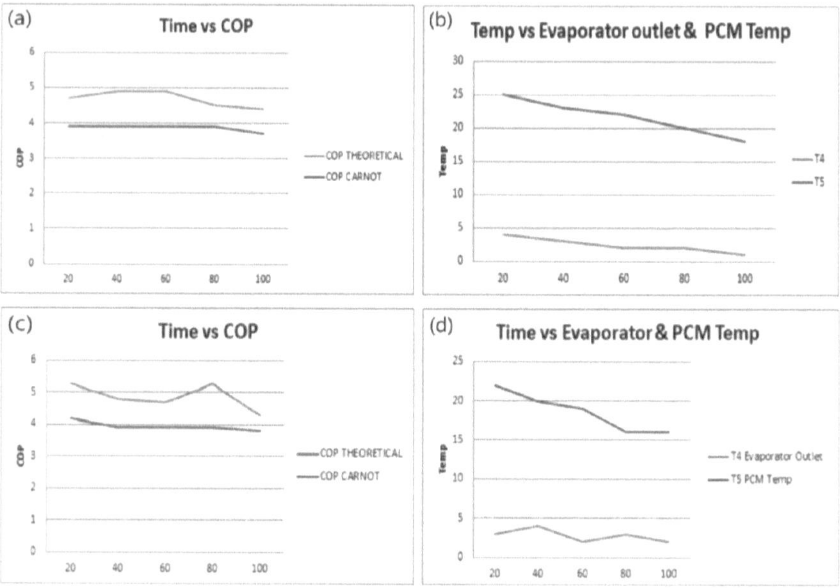

Figure 13.9 (a) Time vs COP with considering NaCl PCM 1%. (b) Time vs evaporator and PCM temp NaCl PCM 1%. (c) Time vs COP with considering NaCl PCM 3%. (d) Time vs evaporator and PCM temp NaCl PCM 3%.

Table 13.5 Experimentation with NaCl as a phase change material 3%

NaCl 3% (When system ON)													
Sr No	Voltage	Current	Cond. pressure	Evap. pressure	Time period (10 Blink)	Temperature					COP theoretical	COP Carnot	Compressor work
						T1	T2	T3	T4	T5			
1	252	1.23	2.2	19.2	56.87	67	35	2	3	22	5.3	4.2	5
2	248	1.2	2.4	19	58.36	71	36	1	4	20	4.8	3.9	5.1
3	250	1.21	2.4	19.4	56.88	70	33	1	2	19	4.7	3.9	5
4	253	1.23	2.4	20	57.06	69	34	0	3	16	5.3	3.9	5
5	254	1.23	2.4	20	56.9	71	33	0	2	16	4.3	3.8	5

13.8 COMPARISON BETWEEN WITHOUT PCM AND NaCl 3% AS A PCM

Time	COP Carnot without PCM	COP Carnot with 3% NaCl	Compressor work without PCM	Compressor work with 3% NaCl	COP theoretical without PCM	COP theoretical with 3% NaCl
20	4	4.2	6.1	5	4.8	5.3
40	3.7	3.9	5.9	5.1	4.9	4.8
60	3.7	3.9	5.6	5	4.4	4.7
80	3.8	3.9	5.8	5	4.9	5.3
100	3.8	3.8	5.7	5	4.6	4.3

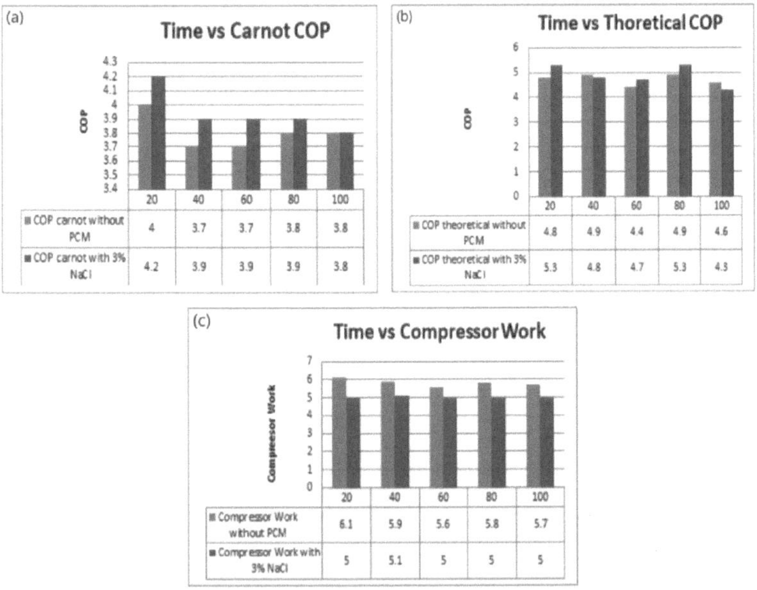

Figure 13.10 (a) Comparison between without PCM and NaCl 3% as a PCM (Carnot COP). (b) Comparison between without PCM and NaCl 3% as a PCM (theoretical COP). (c) Comparison between without PCM and NaCl 3% as a PCM (compressor work).

Table 13.6 Experimentation with NaCl as a phase change material 5%

NaCl 5%(when system ON)													
Sr. No	Voltage	Current	Cond. pressure	Evap. pressure	Time period (10 Blink)	Temperature T1	T2	T3	T4	T5	COP theoretical	COP Carnot	Compressor work
1	253	1.2	1.8	18.2	61.82	57	24	−1	0	17	5.2	4.6	5.4
2	254	1.22	2	19	57.46	66	26	−1	0	22	5.5	4	5.1
3	253	1.18	2	19	64.57	65	29	−1	0	22	6.1	4.1	5.7
4	252	1.17	1.9	18.2	64.45	68	30	−1	−1	22	4.8	3.9	5.7
5	249	1.16	1.8	19	62.08	73	30	0	0	23	4.1	3.7	5.5

Table 13.7 Experimentation with NaCl as a Phase change material 5%

NaCl 5%(when system off)					
Sr. no	T1	T2	T3	T4	T5
1	41	34	3	18	23
2	31	34	13	28	24
3	30	34	15	26	19
4	30	34	15	26	19
5	30	33	14	26	19
6	30	34	15	26	19
7	30	34	15	27	20
8	30	33	15	27	19
9	31	35	16	28	20
10	33	37	17	28	20

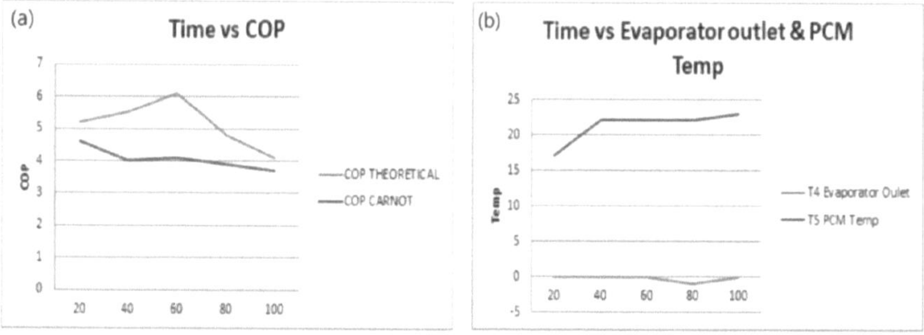

Figure 13.11 (a) Time vs COP with considering NaCl PCM 5%. (b) Time vs evaporator and PCM temp NaCl PCM 5%.

Table 13.8 Comparison of evaporator temperature with and without NaCl 5%

Sr. no.	*Time (min)*	*T4 evaporator outlet temp without PCM*	*T4 evaporator outlet temp with NaCl as PCM 5%*	*Temperature difference*
1	10	27	18	9
2	20	32	28	4
3	30	30	26	4
4	40	30	26	4
5	50	31	26	5
6	60	32	26	6
7	70	31	27	4
8	80	32	27	5
9	90	32	28	4
10	100	32	28	4

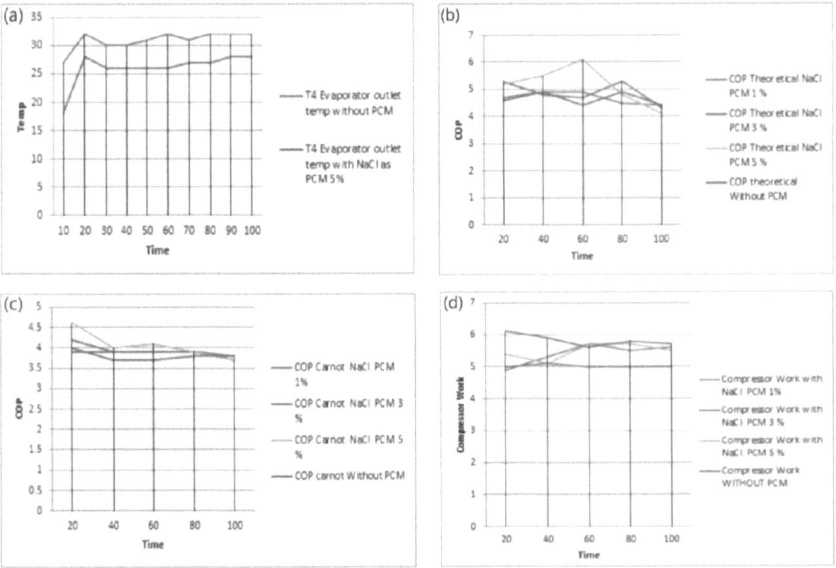

Figure 13.12 (a) Comparison of evaporator temperature with and without NaCl 5%. (b) Comparison between NaCl (PCM) with without PCM (Theoretical COP). (c) Comparison between NaCl (PCM) with without PCM (Carnot COP). (d) Comparison between NaCl (PCM) with without PCM (compressor Work).

13.9 COMPARISON BETWEEN NaCl (PCM) WITH WITHOUT PCM

Time (min)	COP theoretical NaCl PCM 1%	COP theoretical NaCl PCM 3%	COP theoretical NaCl PCM 5%	COP theoretical without PCM	COP Carnot NaCl PCM 1%	COP Carnot NaCl PCM 3%	COP Carnot NaCl PCM 5%	COP Carnot without PCM	Compressor work with NaCl PCM 1%	Compressor work with NaCl PCM 3%	Compressor work with NaCl PCM 5%	Compressor work without PCM
20	4.7	5	5.2	4.6	4	4	5	4	5	5	5.4	6.1
40	4.9	5	5.5	4.9	4	4	4	4	5	5	5.1	5.9
60	4.9	5	6.1	4.4	4	4	4	4	6	5	5.7	5.6
80	4.5	5	4.8	4.9	4	4	4	4	6	5	5.7	5.8
100	4.4	4	4.1	4.4	4	4	4	4	6	5	5.5	5.7

13.10 COMPARISON OF EVAPORATOR OUTLET TEMPERATURE (T4) BETWEEN DIFFERENT VOLUMES OF KCl AND WITHOUT PCM

Time (Min)	*T4 evaporator outlet temp without PCM*	*T4 evaporator outlet temp with KCl as PCM 1%*	*T4 evaporator outlet temp with KCl as PCM 3%*	*T4 evaporator outlet temp with KCl as PCM 5%*
10	27	21	23	23
20	32	23	23	21
30	30	23	23	23
40	30	23	23	24
50	31	24	23	24
60	32	23	23	25
70	31	23	23	24
80	32	23	22	20
90	32	23	22	21
100	32	24	21	22

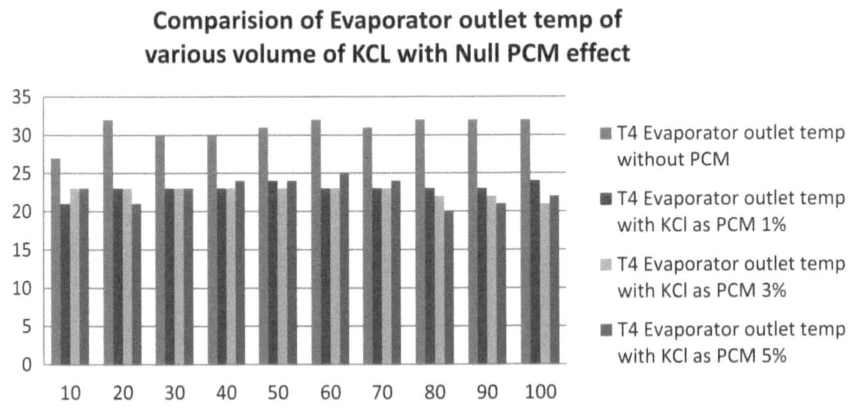

Figure 13.13 Comparison of evaporator outlet temp of various volume of KCL with null PCM effect.

13.11 CONCLUSION

By integrating condenser heat recovery systems, we not only improve the overall efficiency of refrigeration systems but also capitalize on previously wasted heat energy, transforming it into a valuable resource for various applications. Additionally, the incorporation of thermal energy storage systems further augments system performance by providing greater operational flexibility and reducing peak energy demands. Experiments were carried out under different mass fraction of NaCl (1%, 3%, and 5%), KCl (1%, 3%, and 5%) and H_2O

PCM. The number of compressor on off-cycle within a certain periodic time for different PCMs and without PCM can be pointed up. Use 3% KCl as PCM imposes a great impact on COP improvement at certain thermal loads. Using water as PCM and certain thermal load, it is found that the COP improvement has been achieved by the PCM in respect without PCM in conventional refrigerator. Condenser heat recovery captures and repurposes wasted heat, while thermal energy storage improves operational flexibility and reduces peak energy use. Together, these advancements promote sustainability and cost savings, making them crucial for modern refrigeration systems.

REFERENCES

Almendros-Ibáñez, J. A., Fernández-Torrijos, M., Díaz-Heras, M., Belmonte, J. F., & Sobrino, C. (2019). A review of solar thermal energy storage in beds of particles: Packed and fluidized beds. *Solar Energy*, *192*, 193–237.

Dheep, G. R., & Sreekumar, A. J. E. C. (2014). Influence of nanomaterials on properties of latent heat solar thermal energy storage materials–A review. *Energy Conversion and Management*, *83*, 133–148.

Elumalai, P., Vijayan, R., Ramasamy, K. K., & Premkumar, M. (2015). Experimental study on energy recovery from condenser unit of small capacity domestic refrigerator. *Middle-East Journal of Scientific Research*, *23*(3), 417–420.

Husainy, A. S., Kale, S. K., Sawant, S. M., Sur, A., & Patil, D. (2024). Heat transfer phenomenon of nano-enhanced phase change material (NEPCM) incorporated in refrigeration test rig. *ES Energy & Environment*, *23*, 1099.

Husainy, M. A. S., Patil, M. A., Suganawar, M. S., Sannake, M. P., & Surywanshi, M. S. (2017). Review of waste heat recovery from household refrigerator. *Journal for Research*, *3*(08).

Koçak, B., Fernandez, A. I., & Paksoy, H. (2020). Review on sensible thermal energy storage for industrial solar applications and sustainability aspects. *Solar Energy*, *209*, 135–169.

Li, Y. C., Zhang, L., & Feng, B. (2022). Optimization of thermal performance of high temperature sensible heat thermal energy storage system for direct steam generation: A simulation work. *Applied Thermal Engineering*, *217*, 119225.

Muhammad, Y., Saini, P., Knobloch, K., Frandsen, H. L., & Engelbrecht, K. (2023). Rock bed thermal energy storage coupled with solar thermal collectors in an industrial application: Simulation, experimental and parametric analysis. *Journal of Energy Storage*, *67*, 107349.

Odeh, S., & Abu-Mulaweh, H. (2012). Design and development of experimental setup of hybrid PV/thermal collector. *Global Journal of Engineering Education*, *14*(2), 170–176.

Reddy, M. S., Venkatesh, G., Reddy, K. J., & Suresh, B. (2016). Performance improvement of a domestic refrigerator by using phase change material. *International Journal of Advance Research In Science and Engineering*, *5*, 166–173.

Shafieian, A., Khiadani, M., & Nosrati, A. (2018). A review of latest developments, progress, and applications of heat pipe solar collectors. *Renewable and Sustainable Energy Reviews*, *95*, 273–304.

Sharma, S. K., Mallick, A., Singh, D. B., & Tiwari, G. N. (2022). Experimental study of solar energy–based water purifier of single-slope type by incorporating a number of similar evacuated tubular collectors. *Environmental Science and Pollution Research*, *29*(5), 6837–6856. http://dx.doi.org/10.1007/s11356-021-16123-3

Slama, R. B. (2009, November). Water-heater coupled with the refrigerator to develop the heat of the condenser. In *International Renewable Energy Congress November* (pp. 5–7).
Stalin, M. J., Krishnan, S. M., & Kumar, G. V. (2012). Efficient usage of waste heat from air conditioner. *International Journal of Advances in Engineering & Technology*, *4*(1), 414.
Tao, Y. B., & He, Y. L. (2018). A review of phase change material and performance enhancement method for latent heat storage system. *Renewable and Sustainable Energy Reviews*, *93*, 245–259.
Thangavel, P., Mokesh, S., Reavanth, S. A., & Pragatheeswaran, K. (2020, November). Performance Evaluation of Domestic Refrigerator with Waste Heat Recovery System. In *IOP Conference Series: Materials Science and Engineering* (Vol. 995, No. 1, p. 012029). IOP Publishing.
Zając, A. (2019). Use of waste heat recovery from refrigeration system in a commercial facility—A case study. In *E3S web of conferences* (Vol. 116, p. 00103). EDP Sciences.

Index